Electron–Atom and Electron–Molecule Collisions

PHYSICS OF ATOMS AND MOLECULES

Series Editors:

P. G. Burke, *The Queen's University of Belfast, Northern Ireland and Daresbury Laboratory, Science Research Council, Warrington, England*
H. Kleinpoppen, *Institute of Atomic Physics, University of Stirling, Scotland*

Editorial Advisory Board:

R. B. Bernstein *(New York, U.S.A.)*
J. C. Cohen-Tannoudji *(Paris, France)*
R. W. Crompton *(Canberra, Australia)*
J. N. Dodd *(Dunedin, New Zealand)*
G. F. Drukarev *(Leningrad, U.S.S.R.)*
W. Hanle *(Giessen, Germany)*
C. J. Joachain *(Brussels, Belgium)*
W. E. Lamb, Jr. *(Tucson, U.S.A.)*
P.-O. Löwdin *(Gainesville, U.S.A.)*
H. O. Lutz *(Bielefeld, Germany)*
M. R. C. McDowell *(London, U.K.)*
K. Takayanagi *(Tokyo, Japan)*

1976:
ELECTRON AND PHOTON INTERACTIONS WITH ATOMS
Edited by H. Kleinpoppen and M. R. C. McDowell

1978:
PROGRESS IN ATOMIC SPECTROSCOPY, Parts A and B
Edited by W. Hanle and H. Kleinpoppen

1979:
ATOM–MOLECULE COLLISION THEORY: A Guide for the Experimentalist
Edited by Richard B. Bernstein

1980:
COHERENCE AND CORRELATION IN ATOMIC COLLISIONS
Edited by H. Kleinpoppen and J. F. Williams

VARIATIONAL METHODS IN ELECTRON–ATOM SCATTERING THEORY
R. K. Nesbet

1981:
DENSITY MATRIX THEORY AND APPLICATIONS
Karl Blum

INNER-SHELL AND X-RAY PHYSICS OF ATOMS AND SOLIDS
Edited by Derek J. Fabian, Hans Kleinpoppen, and Lewis M. Watson

1982:
INTRODUCTION TO THE THEORY OF LASER–ATOM INTERACTIONS
Marvin H. Mittleman

1982:
ATOMS IN ASTROPHYSICS
Edited by P. G. Burke, W. B. Eissner, D. G. Humer, and I. C. Percival

1983:
ELECTRON–ATOM AND ELECTRON–MOLECULE COLLISIONS
Edited by Juergen Hinze

A Continuation Order Plan is available for this series. A continuation order will bring delivery of each new volume immediately upon publication. Volumes are billed only upon actual shipment. For further information please contact the publisher.

Electron–Atom and Electron–Molecule Collisions

Edited by

Juergen Hinze

University of Bielefeld
Bielefeld, Federal Republic of Germany

Plenum Press • New York and London

Library of Congress Cataloging in Publication Data
Main entry under title:

Electron-atom and electron-molecule collisions.

Proceedings of a workshop held May 5-14, 1980, at the Centre for Interdisciplinary Studies, University of Bielefeld, Bielefeld, Germany.
Includes bibliographical references and index.
1. Electron-atom collisions—Congresses. 2. Electron-molecule collisions—Congresses. I. Hinze, Jürgen, 1937–
QC794.6.C6E37 1983 539.7′2112 82-18927
ISBN 0-306-41188-1

Proceedings of a workshop on Electron-Atom and Molecule Collisions, held May 5-14, 1980, at the Centre for Interdisciplinary Studies of the University of Bielefeld, in Bielefeld, Federal Republic of Germany

Printed in the United States of America

PREFACE

The papers collected in this volume have been presented during a workshop on "Electron-Atom and Molecule Collisions" held at the Centre for Interdisciplinary Studies of the University of Bielefeld in May 1980. This workshop, part of a larger program concerned with the "Properties and Reactions of Isolated Molecules and Atoms," focused on the theory and computational techniques for the quantitative description of electron scattering phenomena.

With the advances which have been made in the accurate quantum mechanical characterisation of bound states of atoms and molecules, the more complicated description of the unbound systems and resonances important in electron collision processes has matured too. As explicated in detail in the articles of this volume, the theory for the quantitative explanation of elastic and inelastic electron molecule collisions, of photo- and multiple photon ionization and even for electron impact ionization is well developed in a form which lends itself to a complete quantitative ab initio interpretation and prediction of the observable effects. Many of the experiences gained and the techniques which have evolved over the years in the computational characterization of bound states have become an essential basis for this development.

To be sure, much needs to be done before we have a complete and detailed theoretical understanding of the known collisional processes and of the phenomena and effects, which may still be uncovered with the continuing refinement of the experimental techniques. The first article, the only one in this volume presenting experimental work, is a fine example of this. The following seven papers give an overview of the theory as it has been developed to date for the description of electron atom and molecule collisions with examples of the application of these theories to specific systems presented in the next six articles. Progress made towards the understanding of electron impact ionization is presented in the paper by Jacubowicz and Moores, which is followed by papers dealing with the important problems of photo- and multiphoton ionization. The volume concludes with a contribution by Jung and Taylor with a suggestion of an experimental technique which should permit the suppression of the background in the observation of normally hidden resonances.

I do hope the volume reflects the active development in this field and provides the reader with the stimulation as it existed in the lively discussions during the workshop due to the engaged contributions of all participants. I wish to acknowledge gratefully the assistance of the directors and the staff of the Centre for Interdisciplinary Studies contributing greatly to the success of the workshop, and my special thanks go to K. Mehandru and E. Ronchi for their untiring help in the preparation of this volume.

J. Hinze
Dept. of Chemistry
University of Bielefeld

CONTENTS

ELECTRON-PHOTON ANGULAR CORRELATION AND SPIN EFFECTS IN ELECTRON-ATOM COLLISIONS

K. Blum
Institut für Theoretische Physik I
Universitat Münster, Münster, West Germany

H. Kleinpoppen*
Institute of Atomic Physics
University of Stirling
Stirling FK9 4LA, Scotland

ABSTRACT

In this paper we draw attention to the importance of the following effects in electron-atom collisions:

a) spin-orbit effects extracted from the electron-photon angular correlations in electron impact excitation of heavy atoms;

b) interference effects in electron impact ionization of atoms extracted from spin effects in collisions with polarised electrons and polarised atoms.

A theoretical parametrisation of excited atomic states is outlined which provides five parameters (σ, λ, $\overline{\chi}$, ε and Δ) for P state excitation of heavy atoms as the basis for the analysis of electron-photon angular correlations. An experimental polarisation analysis of electron-photon angular correlations from the excitation/de-excitation of the mercury intercombination line λ = 2537Å ($6^3P_1 \rightarrow 6^1S_o$ transition) is described which can be used to extract $\lambda = \sigma_o/\sigma$, $\overline{\chi}$ and the spin-orbit phase angles ε and Δ.

Spin effects extracted from ionization processes with polarised electrons and polarised atoms can be used to determine "interference" ionization cross sections. Most recent results of such effects in hydrogen and alkali atoms are presented.

*Temporary address: Zentrum für Interdisziplinäre Forschung, Universität Bielefeld, Postfach 8640, D-4800 Bielefeld 1, West Germany.

The sophistication of experimental technology in electron-atom collisions has reached a state where we now can extract scattering or excitation amplitudes and their phases, target parameters such as orientation, alignment and state multipoles, and also coherence parameters, spin effects and spin-orbit phase angles from the observables. This information obtained on details of atomic collisions goes far beyond that from usual cross section measurements.

In this article we report on recent information that is extracted from experiments in which electron-photon angular correlations and spin effects are observed in electron-atom collisions. Our report is not an exhaustive review since most of the previous results on such data were already summarized in different journals.[1)2)] We therefore mainly concentrate on very recent new types of progress and divide our article into the two sections, namely on angular correlations and on spin effects in electron-atom collisions.

I. Electron-Photon Angular Correlations

The first motivation for studying angular correlations from electron impact excitations of atoms[3)] was primarily concerned with the question of how the momentum transfer direction is related to the symmetry axis of the photon angular distribution in the scattering plane. For the case of the $^1S_0 \rightarrow {}^1P_1 \rightarrow {}^1S_0$ excitation/de-excitation of light atoms the Born approximation predicts that the electron-photon angular correlation is characterised by a selection rule $\Delta m_\ell = 0$ with regard to the momentum transfer as axis of quantisation. This means that the photon emission of the $^1P_1 \rightarrow {}^1S_0$ transition from a single atom is equivalent to the emission of a classical oscillator oriented along the momentum transfer. Superposing two classical oscillators in phase with each other parallel and perpendicular to the direction of the incoming beam would then determine the partial cross sections $\sigma_{m\ell=0}$ and $\sigma_{m\ell=\pm1}$ for exciting the magnetic sublevel cross sections of the $^1P_1$ state. However, when Macek and Jaecks[4)] developed an exact theory for particle-photon angular correlations coherent excitation of magnetic sublevels was required with a finite phase difference χ between the amplitudes $a_{m\ell=0}$ and $a_{m\ell\pm1}$ describing the excitation of the magnetic sublevels (contrary to $\chi = 0$ for Born approximation). This model of coherent excitation in analysing electron-photon coincidence experiments was first applied by Eminyan et al.[5)] and confirmed experimentally by Standage and Kleinpoppen[6)]. In the latter experiment the coherency in the excitation of the 3^1P_1 state was measured in the decay process, in the emission of the photons from the $3^1P_1 \rightarrow 2^1S_0$ transition of helium. The photons turned out to be completely coherent (degree of coherence 100%) which was the consequence of the initially completely coherent excitation of the magnetic sublevels of the 3^1P_1 state. Following this confirmation of the basic assumption of coherent sublevel excitation the technique of electron-photon angular correlation now represents a most powerful tool for studying impact excitation. Since the first round of experiments the remaining problems now with light atoms are related to the extension of the range of energies and scattering angles of the electrons. Recent progress in applications of angular correlations from impact excitation of light atoms (atomic hydrogen and helium) has been summarised by McDowell[7]. A new aspect, however, is connected with the study of spin-orbit interactions and the way in which they may be extracted from angular correlations in excitations of heavy atoms.

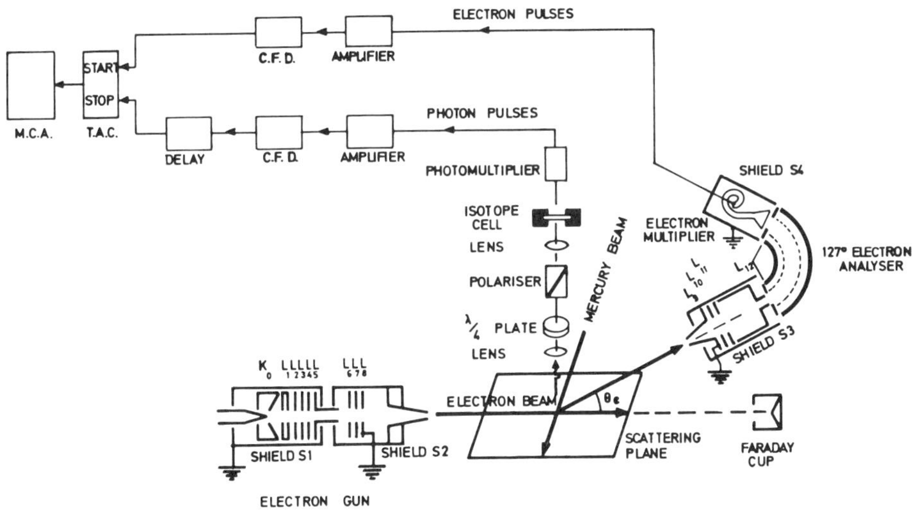

Figure 1. Schematic diagram of an apparatus for measuring electron-photon angular correlations[20]: the incoming electron beam crosses an atomic beam of mercury atoms. The scattered electrons having excited the atoms pass an 127° monochromator and they are detected by an electron channeltron. The photons observed perpendicular to the scattering plane, pass a $\lambda/4$ wave plate, a linear polariser and a mercury isotope cell. MCA-TAC coincidence electronics is applied (MCA = multichannel analyser; TAC = time-to-amplitude converter). The isotope cell contains mercury isotopes ^{199}Hg(nuclear spin I=3/2), ^{201}Hg(I=3/2) in the proportion of 48.7%, 48.1% and a small amount, 3.2% of even isotopes (I=0). At a temperature of about $40^{o}C$ the isotope cell is opaque for the radiation of the intercombination line (2537A) from the odd isotopes (I≠0) whereas the radiation from the even isotopes of the atomic beam is only slightly absorbed.

Recent improvements in experimental technology were mainly concerned with the following aspects (see Figure 1 as an example). Instead of an electron monochromater as initially used by Eminyan et al.[5] multistage electron guns have recently been applied which provide higher initial electron beam currents. In order to avoid complications from the various hyperfine structures of the different isotopes of a given atomic system (such as mercury in Fig.1) an innovation is the application of an isotope absorption filter. This filter contains a vapour of odd mercury isotopes with a pressure so that the resonance line radiation from the odd isotopes excited by electron bombardment in the interaction region is completely absorbed (details of this filter have been described by Zaidi et al.[8]) Angular correlations were also applied to measure differential inelastic cross sections for transitions which cannot be separated from each other by the common energy loss spectrometry. For example the separation between the 3^3D and the 3^1D states of helium is only about 0.6 meV. A successful experiment in separating the inelastically scattered electrons with this small energy difference consists in detecting the unresolved electrons in coincidence with their photons from the transitions $3^3D \rightarrow 2^3P$ (λ = 5876Å) and $3\,{}^1D \rightarrow 2^1P$ (λ = 6678Å). This experiment so far has provided cross sections for the excitation of the 3^3D state;[9] the technique applied is similar to that of Pochet et al.[10]

As mentioned above, a new aspect with regard to applications of angular correlation measurements is related to spin-orbit effects in excitation of heavy atoms. It is expected that spin-orbit effects would require a new type of parametrisation for the analysis of angular correlations from heavy-atom excitations. We therefore first discuss a new parametrisation suitable for the angular correlation experiments with heavy atoms.

1. Parametrisation of the atomic states (including spin-orbit effects)

The number of parameters which are necessary for a complete characterisation of excited states in atomic excitation processes has been discussed by various authors.[1] [11-16] In this section a systematic discussion of the parameters will be given which are necessary for a complete characterisation of the excited atomic states. In general, when no spins are detected in an angular correlation experiment the state of the atoms is derived by a set of parameters which are bilinear combinations of the scattering amplitudes summed over all final spin components M_{S_1} and m_1 of atoms and electrons and averaged over the initial spin S_0 and s_0 and their components M_{S_0} and m_0. In LS-coupling we write

$$< f_{M'} f_M^* > = \frac{1}{2(2S+1)} \sum_{\substack{M_{S_1} m_1 \\ M_{S_0} m_0}} f(M' M_{S_1} M_{S_0} m_1 m_0)\, f(M\, M_{S_1} M_{S_0} m_1 m_0)^* \quad (1)$$

where M' and M are the magnetic quantum numbers. In eq.(1) it has been assumed that initial and detected electrons have sharp momentum and that the atoms, excited by these detected electrons, have sharp values of orbital and spin angular momentum L and S_1, respectively.

We will normalise according to the condition:

$$< |f_M|^2> = \frac{1}{2(2S_o + 1)} \sum_{M_{S_1} m_1 M_{S_o} m_o} |f(M\, M_{S_1} m_1 m_o)|^2 = \sigma(M) \qquad (2)$$

where $\sigma(M)$ is the differential cross section for excitation of the substate with quantum number M.

Recently, coincidence experiments have been performed where heavy atoms (e.g. xenon, krypton and mercury) are excited by electron impact. Here, spin orbit effects will be significant during the scattering process. Of particular interest is the case where states with total electronic angular momentum $J=1$ have been selected in the electronic channel. Assuming $S_0=J_0=0$ we define the parameters:

$$< f_{M'} f_M^* > = \tfrac{1}{2} \sum_{m_1\, m_o} f(M\, m_1 m_o)\, f(M m_1 m_o)^* \qquad (3)$$

where now M denotes the z-component of J.

The array of parameters (1) or (3) is often called the atomic density matrix.

The parameters $< f_{M'} f_M'^*>$ are not all independent. From the definition (1) or (3) follows the "hermiticity" condition:

$$< f_{M'} f_M^* > = < f_M f_{M'}^* >^* \qquad (4)$$

Reflection invariance in the scattering plane gives the condition (see for example Rodberg, Thaler[17]):

$$< f_{M'} f_M^* > = (-1)^{M'+M} < f_{-M'} f_{-M}^* > \qquad (5)$$

Hence, for $L=1$ ($J=1$) these conditions reduce the number of independent parameters to five. A convenient set is the following[12]

$$\lambda = \frac{\sigma(0)}{\sigma} \quad , \quad \cos\chi = \frac{\mathrm{Re} < f_1 f_o^* >}{\sqrt{\sigma(0)\,\sigma(1)}}$$

$$\sin\Phi = \frac{\mathrm{Im} < f_1 f_o^* >}{\sqrt{\sigma(0)\,\sigma(1)}} \quad , \quad \cos\varepsilon = -\frac{< f_1 f_{-1}^* >}{\sigma_1} \qquad (6a)$$

(note that $\langle f_1 f_{-1}^* \rangle$ is real). Here the quantities $\langle f_{M'} f_M^* \rangle$ are either given by eq.(1) or (3) depending on whether spin orbit effects can be neglected or not. Together with the differential cross section

$$\sigma = 2\sigma(1) + \sigma(0) \qquad (6b)$$

this constitutes a set of five parameters characterising the state of the excited atoms. From σ and λ the population of the states with $M = \pm 1, 0$ can be obtained. χ and ϕ characterise the interference between the states with $M = \pm 1$ and $M = 0$ and ε the interference between the states $M = +1$ and $M = -1$.

It has been shown (Hermann and Hertel[12], Blum[13]) that the number of independent parameters is further reduced in the following cases:

i) when LS-coupling holds during the scattering process then standard formulae of scattering theory give:

$$\langle f_{M'} f_M^* \rangle = \frac{1}{2(2S_o + 1)} \sum_S (2S + 1)\, f_{M'}^{(s)} f_M^{(s)*} \qquad (7)$$

where S is the total spin: $S = S_0 + 1/2$ and $f_M^{(S)}$ is the amplitude for excitation of the substate with magnetic quantum number M in the channel with total spin S. In this case the amplitudes are independent of all spin components and reflection invariance gives a relation between the amplitudes

$$f_M^{(S)} = (-1)^M f_{-M}^{(S)} \qquad (8a)$$

resulting in

$$\cos \varepsilon = 1 \qquad (8b)$$

and reducing the number of independent parameters to four.

ii) when, in addition, $S_o = 0$ there is only one allowed spin channel $S=1/2$. In this case it follows from eqs.(6):

$$\cos \chi^2 + \sin^2 \phi = 1 \qquad (9)$$

and we are left with three independent parameters σ, λ, χ which completely characterise the state of the excited atoms. In the case of singlet-singlet transitions the state of the atoms can be represented by a single wave function (pure state)

$$|\psi\rangle = f_1 \{ |+1\rangle - |-1\rangle \} + f_o |0\rangle \qquad (10)$$

In singlet-triplet transitions it has been shown that the orbital part of the atomic state vectors is pure (Blum and Kleinpoppen[15]).

The relations (8) and (9) do not apply when spin-orbit interact is important during the collision. Here the full set (6) of parameters is required for a complete description of the atoms. In this case we have in particular $\cos \varepsilon < 1$ and the derivation from 1 is a measure of the strength of the spin-orbit coupling.

Recently, da Paixao et al.[16] have introduced the parameters

$$\cos \Delta = \frac{|\langle f(1) f(o)^* \rangle|}{\sqrt{\sigma(1)\sigma(o)}} , \quad \langle f(1)\ f(0)^* \rangle = |\langle f_1 f_o \rangle| e^{i\bar{\chi}} \tag{11}$$

Together with σ, λ, $\cos \varepsilon$ these form another set of five independent parameters which are particularly suitable for the study of spin-orbit effects. Δ and ε show an interesting behaviour close to the electron scattering angles $\theta_e = 0^o$ and $\theta_e = 180^o$ as was first pointed out by Gy. Csanak and his group[16]. In these cases the excitation process is axially symmetric with respect to the direction of motion of the incident electrons. As a consequence we have $\langle f_{M'} f_M^* \rangle = 0$ for $M' \neq M$. When spin-orbit effects are significant $\sigma_1 \neq 0$ (whereas, in the case of LS-coupling, only the substate M = 0 can be excited by forward or backward scattered electrons and $\sigma_1 = 0$). Hence, for $\theta_e \to 0^o$ and $\theta_e \to 180^o$,

$$\cos \varepsilon \to 0, \ \cos \Delta \to 0 \tag{12a}$$

or

$$\varepsilon \to \frac{\pi}{2} , \quad \Delta \to \frac{\pi}{2} \tag{12b}$$

For a further discussion of these parameters see the paper by da Paixao et al.[16] where also numerical values are given.

2. Experimental determination of the (λ, χ, ε, Δ) parameters

The parameters introduced above can be determined from a determination of the angular distribution and polarisation of the emitted photons. Let us first consider the case where the emitted radiation is observed in the scattering plane. The angular distribution (which is proportional to the coincidence rate) is given by the expression:

$$I = \frac{W\sigma}{3\gamma} \{ \frac{1-\lambda}{2} (2 - \sin^2\theta) - \frac{1-\lambda}{2} \cos \varepsilon \sin^2\theta \quad \lambda \sin^2\theta + \sqrt{\lambda(1-\lambda)} \cos \chi \sin 2\theta \} \tag{13}$$

where γ is the decay constant, W a spectroscopic factor, and θ is the angle between the direction of observation and the Z-axis in the collision system. λ, ε, χ can be determined by measuring I at various angles.

In order to see the physical significance of cos ε in particular we will consider some special cases of eq.(13). First of all, consider an experiment where the electrons are observed in forward direction. Here, the interference terms $\sim$ cos ε and cos χ vanish and eq.(13) becomes:

$$I = \frac{W\sigma}{3\gamma}\left\{\frac{1-\lambda}{2}(2-\sin^2\theta) + \lambda\sin^2\theta\right\} \tag{14}$$

This expression can readily be interpreted. The spatial distribution of radiation emitted in a transition from the state with M = 0 to the ground state with $J_0 = 0$ is

$$I \sim \sin^2\theta \tag{15a}$$

and for a transition from one of the states with M = ±1 to the ground state

$$I \sim \tfrac{1}{2}(2 - \sin^2\theta) \tag{15b}$$

Eq. (15a) can be understood as the spatial intensity distribution due to a classical electric dipole oscillating in Z-direction. Eq.(15b) can be thought of as the radiation field due to a classical electric rotor, for example an orbiting electron in a circular uniform motion in the z-y-plane with a positively charged particle at the centre of the circle. The first term in eq.(14) can therefore be interpreted as the radiation emitted by an ensemble of rotors, half of them rotating clockwise and the other half anticlockwise in the x-y-plane corresponding to the states with M = ± 1 where the radiation from these two groups of rotors overlaps incoherently. The second term is the radiation distribution emitted by an ensemble of dipoles oriented in Z-direction.

We will now discuss the other extreme case of a single-singlet transition in the LS-coupling scheme. Here, cos ε = 1 and eq.(13) becomes

$$I = \frac{W\sigma}{3\gamma}\{(1-\lambda)(1-\sin^2\theta) + \lambda\sin^2\theta + \sqrt{\lambda(1-\lambda)}\cos\chi\sin 2\theta\}$$

$$= \frac{W}{3\gamma}\{2\sigma_1\cos^2\theta + \sigma_0\sin^2\theta + 2\sqrt{\lambda(1-\lambda)}\sigma\cos\chi\sin\theta\cos\theta\} \tag{16}$$

The first two terms of eq.(13) can be combined to the first term in eq.(16) which can be interpreted as the spatial intensity distribution of radiation emitted by a classical dipole oscillating in x-direction. This is due to the coherence between the states with M = ±1. Both states superpose with a definite phase of 180° (see eq.(10)) and, in the classical picture, this results in a vibration along the x-axis.

Eq.(16) can then be written in the form:

$$I = \frac{W}{3\gamma} \mid \sqrt{2}\, f_1 \cos\theta + f_o \sin\theta \mid^2 \tag{17}$$

with $\sqrt{\sigma_1} = |f_1|$, $\sqrt{\sigma_o} = |f_0|$, and $f_1 = |f_1|e^{i\chi}$. Eq.(17) corresponds to the angular distribution of radiation emitted by a coherent superposition of two ensembles of dipoles, one oscillating in x-and the other in z-direction.

In general, for $\cos\varepsilon < 1$ it cannot be expected that the angular distribution I can be represented by a coherent superposition similar to eq.(17). When spin-orbit interaction is significant during the collision the scattering amplitudes depend explicitly on the spin components m_o and m_1. Since the initial electrons are unpolarised and the scattered electrons are not spin selected these spin states overlap incoherently and this incoherence is transferred to the atoms to a certain degree. The incomplete determination of initial and final spin states results in a loss of coherence between the excited atomic states which, in turn, gives rise to a loss of polarisation of the emitted radiation.

Finally, we will give the general expression for the Stokes parameters[18] where the emitted radiation is observed in a direction with polar angle θ and azimuth ϕ. Using the parameters of eq.(11) together with λ and ε we obtain:

$$I = \frac{W\sigma}{3\gamma} \{ \frac{1-\lambda}{2} (2-\sin^2\theta) - \frac{1-\lambda}{2} \cos\varepsilon \sin^2\theta \cos 2\phi + \lambda \sin^2\theta$$
$$+ \sqrt{\lambda(1-\lambda)} \cos\Delta \cos\bar{\chi} \sin 2\theta \cos\phi\} \tag{18a}$$

$$IP_1 = \frac{W\sigma}{3\gamma} \{\frac{1-\lambda}{2} (-\sin^2\theta + \cos\varepsilon(1 + \cos^2\theta) \cos 2\phi) + \lambda \sin^2\theta$$
$$+ \sqrt{\lambda(1-\lambda)} \cos\Delta \cos\bar{\chi} \sin 2\theta \cos\phi\} \tag{18b}$$

$$IP_2 = -\frac{W\sigma}{3\gamma}\{(1-\lambda) \cos\varepsilon \cos\theta \sin 2\phi + 2\sqrt{\lambda(1-\lambda)} \cos\Delta \cos\bar{\chi} \sin\theta \sin\phi\} \tag{18c}$$

$$IP_3 = -\frac{W\sigma}{3\gamma}\, 2\sqrt{\lambda(1-\lambda)} \cos\Delta \sin\bar{\chi} \sin\theta \sin\phi \tag{18d}$$

In principle, when σ is known, the parameters can be obtained by measuring I at three different set of angles θ, ϕ and the degree of circular polarisation IP_3 at one angle. Alternatively, all four Stokes parameters may be measured at the same angle.

First experimental approaches for measuring the parameters λ, $\bar{\chi}$, ε and Δ as introduced in this sub-chapter have been reported by Zaidi et al.[19]. Their experimental arrangement was similar to that of Standage and Kleinpoppen[6] in which coincident photons are observed perpendicular to the scattering plane (see Fig.1). The photons of the mercury intercombination line (λ = 2537Å, transition $6^1S_o \rightarrow 6^3P_1$) are observed in coincidence with the electrons and analysed in terms of the normalised polarisation Stokes parameters P_1, P_2 and P_3; the above parameters are related to the linear polarised intensity components $I(\alpha)$ (with α as polarisation angle of the photons referred to the direction of the incoming electrons) in the Stokes parameters P_1 and P_2 and also to the circular polarisation in the Stokes parameter P_3. For the geometry used in the experiments of Zaidi et al.[19] the normalised Stokes parameters {linear polarisations $P_1 = (I(0^o) - I(90^o))/I$, and $P_2 = (I(45^o) - I(135^o))/I$ and circular polarisation $P_3 = (I(RHC) - I(LHD))I^{-1}$} are related to the parameters λ, $\bar{\chi}$, ε and Δ as follows:

$$P_1(\phi = 90^o) = \frac{(3\lambda-1) - (1-\lambda)\cos\varepsilon}{(1+\cos\varepsilon) + (1-\cos\varepsilon)}$$

$$P_2(\phi = 90^o) = \frac{-4|\lambda(1-\lambda)|^{\frac{1}{2}}\cos\Delta\cos\bar{\chi}}{(1-\cos\varepsilon) + (1-\cos\varepsilon)\lambda} \tag{19}$$

$$P_3(\phi = 90^o) = \frac{-4|\lambda(1-\lambda)|^{\frac{1}{2}}\cos\Delta\sin\bar{\chi}}{(1+\cos\varepsilon) + (1-\cos\varepsilon)}$$

Including the coherence correlation factor

$$\mu(\phi = 90^o) = |\mu|e^{i\beta} = (P_2 + iP_3) / (1-P_1^2)^{\frac{1}{2}}$$

and the total vector polarisation $P = (P_1^2 + P_2^2 + P_3^2)^{\frac{1}{2}}$

we obtain for the above case

$$|\mu(\phi = 90^o)| = \frac{\sqrt{2}\cos\Delta}{(1 + \cos\varepsilon)^{\frac{1}{2}}} \tag{20}$$

$$P(\phi = 90^o) = (1 + \frac{8(1-\lambda)(2\cos^2\Delta - \cos\varepsilon - 1)}{|(1+\lambda) + (1-\lambda)\cos\varepsilon|^2}) \tag{21}$$

In the first attempt by Zaidi et al.[19] to determine the parameters λ, χ, ε and Δ the three normalised Stokes parameters P_1,P_2 and P_3 for the excitation of the mercury intercombination line (2537Å) were measured. While it is not possible to obtain all the above four parameters from the three independent measured quantities P_1, P_2 and P_3 limiting values can be obtained for λ and Δ and

Table 1: Coincident photon parameters for the $6^1S_o \to 6^3P_1 \to 6^1S_o$ excitation/de-excitation process in mercury. Wuoted experimental uncertainties represent one standard deviation[19)19a)].

INCIDENT ELECTRON ENERGY	SCATTER-ING ANGLE	NORMALISED STOKES PARAMETERS				$\|P(\phi=90^o)\|$	$\|\mu\ \phi=90^o)\|$	β (rad)
		$P_1(\phi=90^o)$	$P_2(\phi=90^o)$	$P_3(\phi=90^o)$	$P_1(\phi=135^o)$			
5.5 eV	50°	-0.42 ± 0.08	0.34 ± 0.08	0.37 ± 0.07	-0.17 ± 0.09	0.65 ± 0.08	0.55 ± 0.08	0.83 ± 0.15
	70°	-0.13 ± 0.08	0.52 ± 0.08	0.27 ± 0.04	-0.39 ± 0.10	0.60 ± 0.07	0.59 ± 0.08	0.48 ± 0.08
6.5 eV	50°	-0.26 ± 0.09	0.39 ± 0.09	0.44 ± 0.04	-0.24 ± 0.09	0.64 ± 0.07	0.61 ± 0.07	0.85 ± 0.12
	70°	-0.13 ± 0.06	0.45 ± 0.09	0.42 ± 0.07	-0.45 ± 0.10	0.63 ± 0.08	0.62 ± 0.09	0.75 ± 0.13

accurate values for $\bar{\chi}$. The latter follows from the fact that the polarisation ratio P_2/P_3 directly determines $\bar{\chi}$ while the limits for Δ follows from the degree of coherence $|\mu|$ (eq.20) with the possible values within the intervals $0 \leqq \varepsilon \leqq 180^o$ and $0 \leqq \Delta \leqq 90^o$. In a further experiment Zaidi et al.[19a] measured a fourth quantity, namely the linear polarisation $P_1(\phi=135^o)$ of the photons from the $6^3P_1 - 6^1S_o$ transition observed at an azimuthal angle of $\phi=135^o$. $P_1(\phi=135^o)$ follows from the eqs.(18) to

$$P_1(\phi = 135^o) = \frac{3\lambda - 1}{(\lambda + 1)} \tag{21}$$

Table 2: Values of $\bar{\chi}$, Δ, ε and λ for the $6^3S_o - 6^1P_o$ excitation/de-excitation of mercury. The hyperfine effect has been eliminated[19) 19a]. Since ε and Δ appear in cos ε and cos Δ, the data yields only the absolute values $|\varepsilon|$ and $|\Delta|$ of these parameters.

Incident Energy	Scattering Angle	λ (rad)	$\bar{\chi}(=\beta)$ (rad)	$\lvert\varepsilon\rvert$ (rad)	$\lvert\Delta\rvert$ (rad)
5.5 eV	50°	0.26±0.03	0.83±0.15	0.77±0.62	1.04±0.11
	70°	0.18±0.03	0.48±0.08	2.02±0.16	1.25±0.06
6.5 eV	50°	0.23±0.03	0.85±0.12	1.55±0.26	1.12±0.08
	70°	0.16±0.03	0.75±0.13	2.10±0.15	1.26±0.06

From the data of the four Stokes parameters (Table 1) and the four equations (19) and (21) a full set of the λ, $\bar{\chi}$, ε, Δ parameters could be extracted for the first time (Table 2). Since Δ and ε are unequal to zero spin-orbit effects are unambiguously detected in the 6^3P_1 excitation of mercury.

II. Spin-Effects in Electron-Atom Collision

The application of spin-polarised electrons and spin-polarised target atoms has been introduced into electron-atom collision physics only about a decade ago. Experiments with spin-polarised collision partners lead to an analysis from which both scattering amplitudes describing various collisional interactions (Coulomb or direct, exchange interactions, etc.) and interference effects can be extracted[21,22]. In the first stage of such experiments either the electrons or the target atoms (one-electron atoms) were polarised. Electron scattering experiments with polarised alkali atoms, for example, lead to measurements of "exchange"[23,24] and "direct" scattering cross sections whereas forward scattering of polarised electrons on mercury atoms provided depolarisation ratios (i.e., ratio of electron polarisation before and after the collision which depend on direct and exchange interactions[26]. Since summaries on these investigations have already been given[27-30], in this article we mainly concentrate on new electron scattering experiments in which both the projectile and the target atom are polarised.

1. Collisions in which Both Electrons and Atoms are Polarised

Applications of both partially polarised electrons and partially polarised atoms (one-electron atoms) have been reported from research groups at Yale[31] and at Stirling[32] Universities in 1977 for the first time. The basic idea behind these experiments is the fact that an interference effect in the ionisation of one-electron atoms can be extracted from the measurements of ionisation rates with polarised collision particles. The analysis of spin effects in the ionisation can be related to the following reactions:

$$e(\uparrow) + A(\uparrow) \rightarrow A^+ + e(\uparrow) + e(\uparrow)$$

$$e(\uparrow) + A(\downarrow) \rightarrow A^+ + e(\uparrow) + e(\downarrow)$$

$$A^+ + e(\downarrow) + e(\uparrow)$$

The arrows indicate the two possible quantisized spin directions in the symbols for the electron (e) and the atom (A). Two amplitudes, a Coulomb or direct amplitude f(k',k) and an exchange amplitude g(k',k) describe the ionisation process[33,34] (with k' and k as the wave vectors for the outgoing electrons).

The first of the two ionisation reactions (spins parallel to each other) has an intensity which is proportional to the interference term $|f-g|^2=\sigma^{\uparrow\uparrow}$ since the direct and exchange interactions are indistinguishable in the collision process where both spins are parallel to each other. The second of the two reactions is determined by a cross section which is the sum of the square of the two amplitudes describing the direct and exchange process: $\sigma^{\downarrow\uparrow}=|f|^2 + |g|^2$. The total differential cross section is given by*

$$\sigma = \tfrac{1}{2}|f|^2 + \tfrac{1}{2}|g|^2 + \tfrac{1}{2}|f-g|^2$$

$$= |f|^2 + |g|^2 - \mathrm{Re}(f^*g)$$

The factor $\frac{1}{2}$ results from the fact that σ refers to the case of the ionisation with initially unpolarised electrons and unpolarised atoms (the above spin reactions plus the ones with opposite spin are those together for this case). Taking the above equations we have

$$\sigma^{\uparrow\downarrow} = \sigma + \mathrm{Re}(f^*g), \quad \sigma^{\uparrow\uparrow} = \sigma - \mathrm{Re}(f^*g) \tag{22}$$

$$= \sigma + \sigma_{int}, \quad = \sigma - \sigma_{int} \tag{23}$$

It can be shown that the differential interference cross section σ_{int} integrated over all directions of both the electrons of the ionisation process may remain finite (the integral may be called Q_{int} = integrated total interference cross of ionisation). Equivalent equations as those of (22) and (23) can then be written for the total ionisation cross sections Q_{int}, Q, $Q^{\uparrow\downarrow}$ and $Q^{\uparrow\uparrow}$. The finite polarisation P_e and P_A of the electrons and atoms, respectively, can be taken into account as a factor P_eP_A in front of σ_{int} (or Q_{int}) in the above equations; it then follows a simple relationship for the ratio of the total interference cross section to the total integrated cross section of ionisation:

$$\frac{Q_{int}}{Q} = \frac{Q^{\uparrow\downarrow}-Q^{\uparrow\uparrow}}{2P_eP_AQ}$$

In other words the interference cross section relative to the total cross section can be determined by measuring the difference between the ionisation rates for the parallel and anti-parallel spin configurations (knowing, of course, the degree of polarisations P_e and P_A).

Figure 2 shows the scheme of an apparatus as used at Stirling[32] for experiments with polarised electrons and polarised alkali atoms. A beam of thermal alkali atoms is passing through a hexapole magnet. The hexapole magnet has the property to focus atoms with one spin component whereas the atoms with the other spin components are defocussed. In the strong field at the end of the hexapole magnet the atoms have a measurable polarisation of 72%. However, when the atoms leave the magnet and move into the region of a guiding magnetic field of low intensity the high polarisation will be reduced by a factor $(2I + 1)^{-1}$ (I = nuclear spin of the atom) due to the hyperfine structure of the alkali atoms. The net polarisation for sodium

*The minus sign between f and g results from the anti-symmetrisation of the relevant wave function for the process.

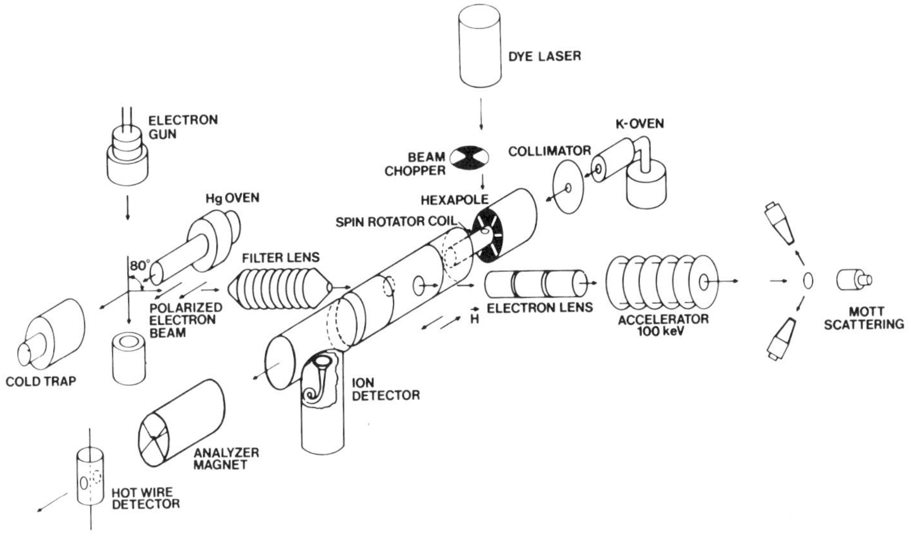

Fig.2: Scheme of the Stirling experiment[32] for the measurement of the relative interference cross section to the total ionisation cross section of alkali atoms.

and potassium (I=3/2) is 18%. The intensity of the polarised atomic beam is measured by a Langmuir-Taylor detector. A laser beam may induce resonance transitions in the atoms which may either destroy (by means of unpolarised laser radiation the population of the spin states can be equalised) or increase (by means of circular polarised laser light[37]) of the atomic spin polarisation

Spin-polarised electrons are produced by elastic electron scattering[35,36] on mercury atoms (e.g., at a typical energy of 80eV and a scattering angle of 80°). An electron filter lens selects the elastically scattered electrons and decelerates them to the wanted energy for the ionisation of the alkali atoms. The polarisation of the electrons is measured by a Mott detector, that of the atoms by a Stern-Gerlach magnet.

Figures 3 and 4 show results of ratios for the interference ionisation cross section to the total cross section of the ionisation for alkali and hydrogen atoms (Yale experiment[30]; in this latter experiment the hydrogen atoms were, too, polarised by a hexapole magnet; the electrons, however, were polarised by the Fano effect[38]). The experimental data are compared with various theoretical predictions. As can be seen from this comparison the relative measurements of the interference cross section represent a very sensitive test on theoretical approximations.

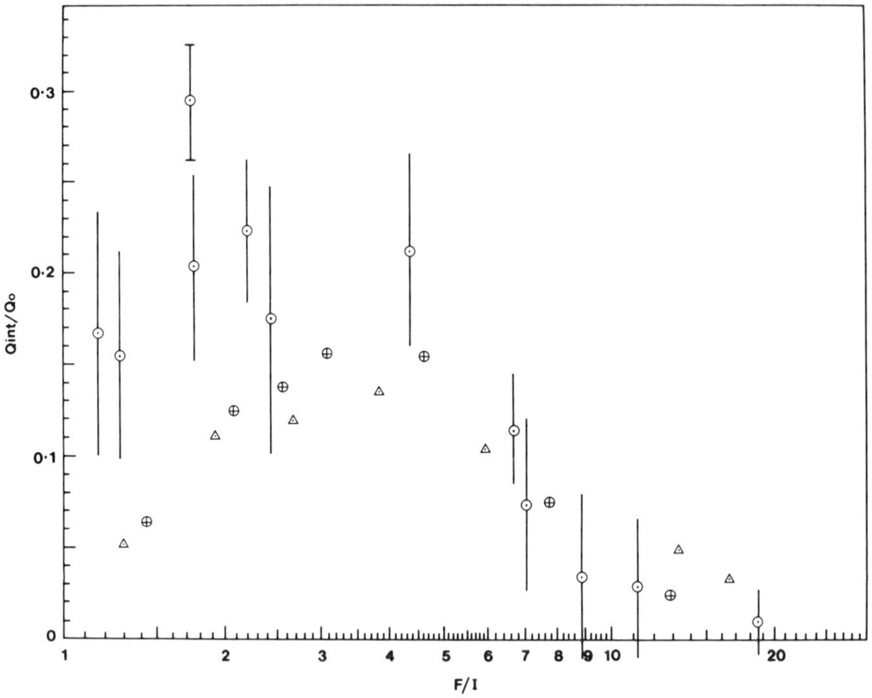

Fig. 3: Ratio of interference cross section to total ionisation cross section by electron impact: experimental data[32)41)] for potassium ⊙ and sodium ⊙ . ⊕ and △ Born approximation[42)] for lithium and Na, respectively.

Concluding, we note that recently a further interference experiment on the ionisation of lithium has been reported by a group from Bielefeld[39)]. The data of all the three interference experiments from Bielefeld, Stirling and Yale Universities are probably too far away from the ionisation threshold (E_{th_r}) in order to test the theoretical prediction for the threshold ratio $Q_{int}(E_{th_r})/Q(E_{th_r})=1$ by Klar and Schlecht[40)].

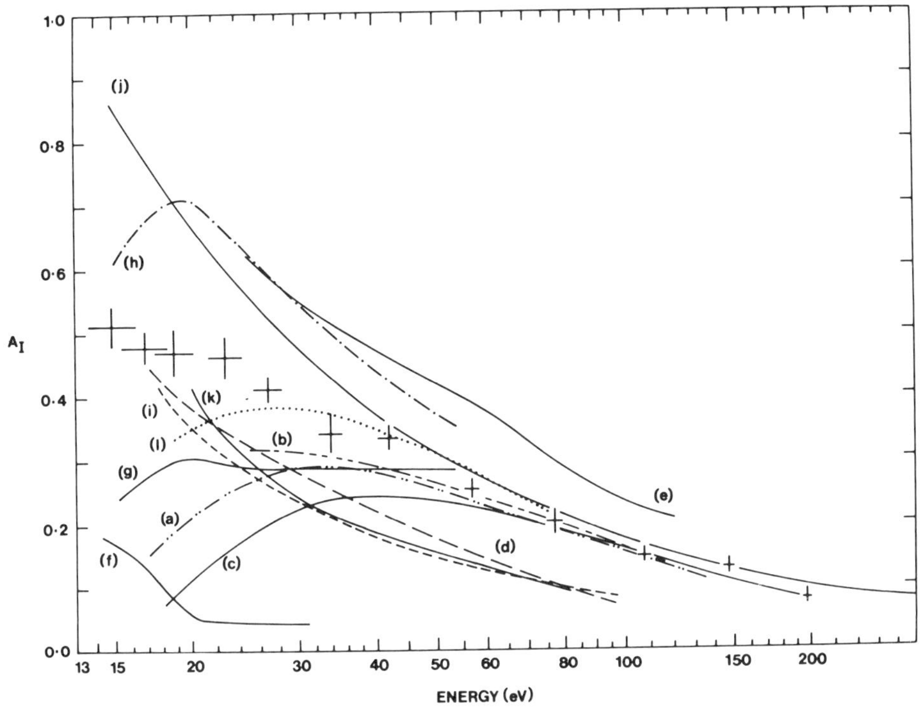

Fig. 4: Experimental and theoretical data of Q^{int}/Q_o for electron impact ionisation of atomic hydrogen: the crosses represent experimental results of Alguard et al[31]. (a-d) Born exchange (BE) calculations from Rudge and Seaton[43], Peterkop[44),45)], Geltman et al.[46], and Golden and McGuire[47], respectively; (e) BE calculation with maximum interference taken from Peterkop[44),45)]; (f) BE calculation with angle-dependent potential from Rudge and Schwartz[48]; (g,h) special average exchange calculations allowing for maximum interference according to Rudge and Seaton[47]; (i-k) Glauber exchange (Goldin and McGuire[47]), modified Born-Oppenheimer (Ochkur[49]) and close-coupling calculations (Gallaher[50]), respectively; (l) BE calculations taken from Rudge[51].

References

1) K Blum, H Kleinpoppen, Physics Reports 52, 203 (1979
2) H Kleinpoppen, Advances in Atomic and Molecular Physics 16, 423 (1979)
3) H Kleinpoppen, Columbia University, unpublished report, 1967
4) M Eminyan, K MacAdam, J Slevin and H Kleinpoppen, Phys. Rev. Lett. 31, 576 (1973); and J.Phys. B7, 1519 (1974)
5) J Macek and D H Jaecks, Phys.Rev. A4, 1288 (1971)

6) M C Standage and H Kleinpoppen, Phys.Rev.Lett. 36, 577 (1975)
7) M R C McDowell, these Proceedings
8) A A Zaidi, I McGregor and H Kleinpoppen, J.Phys. B11, L151 (1978)
9) A Chutjian, R Hippler, I McGregor and H Kleinpoppen, Book of Abstracts, p.172, XIth ICPEAC, Kyoto 1979
10) A Pochet, D Roznel, J Perese, J.Physique 34, 701 (1973)
11) L A Morgan and M R C McDowell, J.Phys. B8, 1073 (1975)
12) H W Hermann and I V Hertel in Coherence and Correlation in Atomic Collisions (Plenum, 1980; H Kleinpoppen and J F Williams, etc.)
13) K Blum, Book of Invited Papers, XIth ICPEAC, Kyoto, 1979
14) K Blum, F.J. da Paixao and Gy. Csanak, J.Phys. B13, L257 (1980)
15) K Blum and H Kleinpoppen in: Electron and Photon Interactions with Atoms, eds. H Kleinpoppen and M R C McDowell (Plenum Press, New York, 1976), p.501.
16) F J da Paixao, N.T. Padial, Gy. Csanak and K Blum, Phys. Rev. Lett. 45, 1164 (1980)
17) J.S Rodberg and R M Thaler, Introduction to the Quantum Theory of Scattering (New York, Academic Press)
18) M Born and E Wolf, Principles of Optics (Pergammon Press, 1970)
19) A A Zaidi, I McGregor and H Kleinpoppen, Phys. Rev.Lett. 45, 1168 and 2078 (1980)
19a) A A Zaidi, S M Khalid, I McGregor and H Kleinpoppen, to be published.
20) A A Zaidi, Ph.D Thesis, Stirling University (1980)
21) B Bederson, Comments on Atomic and Molecular Physics, I 65 (1969) and II, 160 (1970-71)
22) H Kleinpoppen, Phys. Rev. A3, 2013 (1971)
23) D M Campbell, H M Brash and P S Farago, Phys. Rev.A. A36, 449 (1971)
24) R E Collins, B Bederson and M Goldstein, Phys.Rev. A3, 1976
25) D Hils, V McCusker, H Kleinpoppen and S J Smith, Phys.Rev. Lett. 29, 398 (1972)
26) G.F. Hanne, J. Kessler, J.Phys. B9, 791 (1976b)
27) B Bederson, In Atomic Physics 3, 401 (1973) Plenum Press
28) H Kleinpoppen, Adv. Quant. Chemistry, 10, 77 Academic Press (1977)
29) P Farago, Rep. Prog. Phys. 34, 1055 (1971)
30) J Kessler, "Polarised Electrons", Springer-Verlag 1976
31) M J Alguard, V W Hughes, M S Lubell and P E Wainwright, Phys. Rev.Lett. 39, 334 (1977)
32) D Hils and H Kleinpoppen, Satellite Meeting of Xth ICPEAC Conference, Paris, 1977, u. J.Phys. B11 L283 (1978)
33) R Peterkop, Proc. Phys. Soc. London 77, 1220 (1961)
34) M R M Rudge, Rev. Mod. Phys. 40, 564 (1968)
35) H Steidl, E Reichert and H Deichsel, Phys.Lett. 17, 31 (1965)

36) K Jost and J Kessler, Phys.Rev.Lett. 15, 575 (1963)
37) D Hils, W Jitschin and H Kleinpoppen, to be published
38) U Fano, Phys. Rev. 178, 131, 1969
39) G Baum, D Caldwell, E Kisker and Schröder, Verhandl. d. deutschen Physik, Gesellschaft 5, 539 (1980)
40) H Klar and W Schlecht, J.Phys. B9, 1699 (1976)
41) D Hils, K Rubin and H Kleinpoppen in "Coherence and Correlation in Atomic Collisions" (eds. H Kleinpoppen and J F Williams), Plenum Press, N.Y. p.689 (1980)
42) G Peach, Proc. Phys. Soc. 85, 709 (1965) and 87, 381 (1966)
43) M R M Rudge and M J Seaton (1965), Proc. R. Soc. London Ser. A283, 262
44) R Peterkop (1961a), Zh. Eksp. Teor. Fiz. 41, 1938
45) R Peterkop (1962) Sov. Phys. - JETP (Engl.Transl.) 14, 1377
46) S Geltman, M R M Rudge and M J Seaton (1963), Proc.Phys. Soc. London 81, 315
47) J E Goldin and J H McGuire, Phys. Rev. Lett. 32, 1218 (1974)
48) M R M Rudge and S B Schwartz, Proc. Phys.Soc. London 88, 563 (1966)
49) V I Ochkur, Zh. Eksp. Teor. Fiz. 47, 1746 (1964)
50) D F Gallaher, J. Phys. B7, 362 (1974)
51) M R M Rudge, J.Phys. B11, L149 (1978)

VARIATIONAL METHODS IN ELECTRON-MOLECULE COLLISIONS

R. K. Nesbet*

IBM Research Laboratory
5600 Cottle Road
San Jose, California 95193

INTRODUCTION

Variational methods have been used successfully for accurate calculations of cross sections for electron scattering by light atoms. This work has been reviewed in several recent publications (Callaway, 1978; Nesbet, 1977, 1980). It is not necessary to repeat details here. However, in the present context, it is appropriate to examine the applicability of these methods to electron-molecule scattering.

Electron-atom scattering is analogous to electron scattering by a molecule with fixed nuclei. Except for the necessary reduction of symmetry, similar computational methods can be used. Variational methods based on bound-state methods seem most appropriate. This methodology is reviewed in Section II.

Nuclear motion is an essential complicating aspect of electron-molecule scattering. The internal energy levels are so numerous that a direct close-coupling expansion in the rovibrational states becomes impractical. Moreover, even at very low impact energies rotational excitation channels are energetically accessible, so the regime of purely elastic scattering is severely restricted. The one simplifying aspect is that the electron mass is small compared with the effective mass of nuclear motion. Although the usual Born-Oppenheimer approximation breaks down in the electron scattering continuum, the small electron-nuclei mass ratio can be exploited to deduce inelastic

* Work supported in part by the Office of Naval Research

rovibrational transition cross sections from fixed-nuclei electron scattering calculations. Recent theoretical work in this direction will be discussed in Section III.

Recent progress in methods using bound state variational techniques to compute complex energy eigenvalues appropriate to electron scattering resonances has been reviewed by Nuttall (1979) and by Taylor and Yaris (1980). This work will not be considered here, where the discussion will be limited to fully general methods applicable to multichannel scattering including background effects as well as resonances.

VARIATIONAL METHODS FOR FIXED NUCLEI SCATTERING

Two principal methods have been used for accurate electron-atom scattering calculations: the R-matrix method (Burke and Robb, 1975) and the matrix variational method (Nesbet, 1977, 1980), based on the Hulthén-Kohn variational principle. Methods based on the Lippmann-Schwinger equation (Rescigno et. al. 1974) or more specifically on the Schwinger variational principle (Watson and McKoy, 1979; Lucchese et al., 1979) have more recently been shown to be viable. All of these methods can be applied to electron-molecule scattering. The standard close-coupling method for electron-atom scattering is generally unsuitable for molecules, because it requires solution of coupled partial differential equations, or, alternatively, inclusion of coupling between different spherical harmonic terms in a one-center expansion.

At large radii, electron scattering potential functions reduce to simple multipole terms, reciprocal powers of r with coefficients determined by static multipole moments and polarizabilities of the target atom or molecule. Beyond the range of the target wavefunction, exchange terms can be neglected, and only coupled ordinary differential equations occur. Despite the coupling of spherical harmonic terms for a molecular target, it should be feasible to solve such coupled equations for many systems of interest.

In the region of small r, where exchange terms and nonlocal effects due to electronic correlation become important, another approach is needed. It is clearly desirable to exploit techniques of bound-state quantum chemistry in this region, representing the N+1-electron wavefunction as a superposition of configurations, with coefficients to be determined variationally.

The R-matrix method exploits the applicability of different computational techniques for large and small r by explicitly introducing a boundary redius r_0 Basis functions of the inner region are truncated at r_0 and asymptotic close-coupling equations are solved outside r_0 This method has been used successfully for electron-molecule scattering calculations (Schneider et al., 1979b).

In the matrix variational method a basis set of quadratically integrable N+1-electron configurational functions is augmented by auxiliary functions containing oscillatory asymptotic radial components. Bound-state computational techniques are modified only by using the Hulthen-Kohn variational principle to determine the coefficients of these oscillatory functions. Free wave oscillatory functions (spherical Bessel functions) have been used in calculations of electron scattering from neutral atoms. A hybrid version of this method, in which the socillatory basis functions are obtained by explicit integration of the asymptotic close-coupling equations, has been proposed under the name of numerical asymptotic function (NAF) method (Oberoi and Nesbet, 1973). A multichannel version of this method has recently been applied successfully to accurate calculations of electron-atom and electron-ion cross sections (Morgan, 1979, 1980). Details of the multichannel formalism, including explicit asymptotic closed channels, are given by Nesbet (1980).

If the asymptotic close-coupling equations are solved exactly, in the NAF method, only the residual Hamiltonian, a short-range operator, occurs in the integrals required by the Hultken-Kohn method. This should make evaluation of these integrals feasible even for molecules. Similar integrals, defined within a boundary radius or inside a spheroidal surface, must be evaluated in the R-matrix method.

There are two essential difficulties with the R-matrix method, in the usual close-coupling formalism. The first is that the number of coupled channels required, especially for molecules, may be beyond the practical capabilities of the method. The second problem is that the R-matrix method cannot deal with nonlocal potentials that extend outside the boundary radius r_0 This problem becomes especially serious in considering excitation of Rydberg states, since these states have effective radii that increase as the square of the principal quantum number. Lowlying electronic states of some common molecules are known to be partly of Rydberg character. Nonlocal exchange potentials cannot be neglected unless r_0 is outside the target state electronic distribution. The multichannel NAF method may provide a viable alternative approach to these problems, because it uses standard large-scale configuration interaction techniques and avoids a complete change of representation of the wave function at an arbitrary boundary surface.

An alternative hybrid method can be based on the Schwinger variational principle, making use of the multichannel Green's function obtained from numerical solution of the asymptotic close-coupling equations (Nesbet, 1980). All required integrals involve only the residual Hamiltonian, as in the NAF method.

THE INCLUSION OF ROVIBRATIONAL MOTION

The usual Born-Oppenheimer approximation for electronic bound states is valid if electronic energy levels are widely separated compared with coupling matrix elements. This approximation clearly breaks down for a molecular negative ion state in the electron scattering continuum. The implied inherent coupling between electronic and nuclear motion shows up in the form of rovibrational excitation due to electron impact, and in the processes of dissociative attachment and collisional detachment.

These processes are strongly enhanced by an electron scattering resonance. If resonance theory is applied to the rovibrational close-coupling equations (Herzenberg and Mandl, 1962; Bardsley et al., 1966; Chen, 1964a), it is possible to derive a nonlocal optical potential that modifies the Schrödinger equation for nuclear vibrational motion. For practical applications this is simplified to a local complex-valued potential function. This complex energy characterizes a negative ion resonance state that can decay spontaneously by autodetachment, emitting an electron. The standard theory of dissociative attachment, reviewed recently by Bardsley (1979), makes use of such a complex potential function. A version of this theory, the "boomerang" model (Herzenberg, 1968) has been used successfully to describe the irregular vibrational excitation structure observed in e^--N_2 scattering (Birtwistle and Herzenberg, 1971; Dubé and Herzenberg, 1979).

It was shown some time ago by Chen (1964b) that direct application of resonance theory, using the Born-Oppenheimer approximation for nonresonant effects, leads to a Breit-Wigner formula for the rovibrational excitation transition matrix, summed over intermediate states of the negative ion or resonance. With a particular choice of a complex potential function for the $^2\Pi_g$ resonance state of N_2^-, Chen (1964b) derived excitation cross sections for several vibrational states, in good qualitative agreement with expriment. More precise agreement was achieved later with the boomerang model (Dube and Herzenberg, 1979) by adjusting the complex potential function to achieve a close fit to experimental vibrational excitation data.

Examination of the rovibrational close-coupling equations indicates that they are formally identical to the fixed-nuclei electron scattering equation if the residual electronic energy ϵ, which occurs as a numerical parameter in the latter equation, is replaced by the operator $E-H_n$, where E is the total energy and H_n is the Hamiltonian operator for nuclear motion. The fixed-nuclei electronic wave function or scattering matrix becomes an operator function of H_n. This observation provides the heuristic basis of the energy-modified adiabatic (EMA) approximation (Nesbet, 1979). The fixed-nuclei electron scattering matrix $S(\epsilon;R)$ at internuclear coordinate R is converted to the matrix $(\mu|S(E-H_n;R)|\nu)$, which describes transitions between rovibrational states $\chi_{\mu'}$ and χ_μ. This abstract operator formula is approximated by matrix expressions that are unitary, obey correct threshold laws, and have the correct (Breit-Wigner) form for resonances. Both resonant and background contributions to $S(\epsilon;R)$ can be included in the formalism. The purely resonant term, parametrized from the work of Birtwistle and Herzenberg (1971), gives vibrational excitation cross sections in good agreement with experimental data. The EMA approximation and the boomerang model give substantially the same results, as should be expected from the earlier results of Chen (1964b), who used a resonant Breit-Wigner formula.

A purely analytic approach has been considered by Domcke and Cederbaum (1977), using a model Hamiltonian that describes electron-molecule vibrational coupling, for which the wave equation can be solved exactly. This method has been applied successfully to $e^- -N_2$ vibrational excitation.

An alternative theoretical approach (Schneider et al., 1979a) derives a vibrational R-matrix from the fixed-nuclei electronic R-matrix. Vibrational eigenfunctions are computed for each of several R-matrix electronic energy levels, whose energies are functions of the internuclear coordinates. This makes it possible to compute a vibrational R-matrix and the implied scattering matrix for vibrational excitation. Calculations on $e^- -N_2$ vibrational excitation, using electronic orbital wave functions appropriate to the negative ion resonance rather than to the neutral molecule, gave results in good agreement with experiment (Schneider et al., 1979b).

The relationship between these R-matrix results and earlier work has been clarified by recent work of Hazi et al. (1980), who carried out calculations in the boomerang model, using a complex potential function derived from the *ab initio* calculations of Schneider et al. (1979b). Substantially the same results were obtained as with the vibrational R-matrix formalism. As a result of this work, it appears that nonadiabatic rovibrational effects in electron scattering can be deduced from fixed-nuclei electron scattering calculations by use of one of the formal model methods discussed here.

For pure rotational excitation, or in the absence of a resonance, the operator modification of $S(\epsilon;R)$ in the EMA approximation may not be necessary, except in the immediate vicinity of an excitation threshold. The simpler adiabatic-nuclei approximation (Chase, 1956) is known to give good results in such cases (Golden, et al., 1971; Chandra and Temkin, 1976). The latter authors carried out vibrational close-coupling calculations for the resonant $^2\Pi_g$ component of the e^--N_2 wavefunction, using the adiabatic nuclei approximation for other components. The results showed irregular resonance structure similar to that observed experimentally, but the calculations could not be carried to convergence.

REFERENCES

Bardsley, J. N., 1979, The theory of dissociative attachment and recombination, in "Symposium on Electron-Molecule Collisions," I. Shimamura and M. Matsuzawa, eds., University of Tokyo:121.

Bardsley, J. N., Herzenberg, A., and Mandl, F., 1966, Vibrational excitation and dissociative attachment in the scattering of electrons by hydrogen molecules, Proc. Phys. Soc. (London), 89:321.

Birtwistle, D. T. and Herzenberg, A., 1971, Vibrational excitation of N_2 by resonance scattering of electrons, J. Phys. B, 4:53.

Burke, P. G. and Robb, W. D., 1975, The R-matrix theory of atomic processes, Advan. At. Mol. Phys., 11:143.

Callaway, J., 1978, The variational method in atomic scattering, Phys. Repts., 45:89.

Chandra, N. and Temkin, A., 1976, Hybrid theory and calculation of e-N_2 scattering, Phys. Rev. A, 13:188.

Chase, D. M., 1956, Adiabatic approximation for scattering processes, Phys. Rev., 104:838.

Chen, J. C. Y., 1964a, theory of subexcitation electron scattering by molecules. I. Formalism and the compound negative-ion states, J. Chem. Phys., 40:3507.

Chen, J. C. Y., 1964b, Theory of subexcitation electron scattering by molecules. II. Excitation and de-excitation of molecular vibration, J. Chem. Phys., 40:3513.

Domcke, W. and Cederbaum, L. S., 1977, Theory of the vibrational structure of resonances in electron-molecule scattering, Phys. Rev. A, 16:1465.

Dube, L. and Herzenberg, A., 1979, Absolute cross sections from the "boomerang model" for resonant electron-molecule scattering, Phys. Rev. A, 20:194.

Golden, D. E., Lane, N. F., Temkin, A., and Gerjuoy, E., 1971, Low energy electron-molecule scattering experiments and the theory of rotational excitation, Rev. Mod. Phys., 43:642.

Hazi, A. U., Rescigno, T. N., and Kurila, M., 1980, Cross sections for resonant vibrational excitation of N_2 by electron impact, Phys. Rev. A, submitted.

Herzenberg, A., 1968, Oscillatory energy dependence of resonant electron-molecule scattering, J. Phys. B, 1:548.

Herzenberg, A. and Mandl, F., 1962, Vibrational excitation of molecules by resonance scattering of electrons, Proc. Roy. Soc. (London), A270:48.

Lucchese, R. R., Watson, D. K., and McKoy, V., 1980, Iterative approach to the Schwinger variational principle for electron-molecule collisions, Phys. Rev. A, 22:421.

Morgan, L. A., 1979, Electron impact excitation of the n=2 states of He^+, J. Phys. B, 12:L735.

Morgan, L. A., 1980, A hybrid multichannel algebraic variational method. Application to electron scattering by hydrogenic ions, J. Phys. B, 13: in the press.

Nesbet, R. K., 1977, Low-energy electron scattering by complex atoms: theory and calculations, Advan. At. Mol. Phys., 13:315.

Nesbet, R. K., 1979, Energy-modified adiabatic approximation for scattering theory, Phys. Rev. A, 19:551.

Nesbet, R. K., 1980, "Variational Methods in Electron-Atom Scattering Theory," Plenum Press, New York.

Nuttall, J., 1979, The calculation of atomic resonances by the coordinate rotation method, Comments At. Mol. Phys., 9:15.

Oberoi, R. S. and Nesbet, R. K., 1973, Numerical asymptotic functions in variational scattering theory, J. Comput. Phys., 12:526.

Rescigno, T. M., McCurdy, C. W., Jr., and McKoy, V., 1974, Discrete basis set approach to nonspherical scattering II., Phys. Rev. A, 10:2240.

Schneider, B. I., LeDourneuf, M., and Burke, P. G., 1979a, Theory of vibrational excitation and dissociative attachment: an R-matrix approach, J. Phys. B, 12:L365.

Schneider, B. I., LeDourneuf, M., and Vo Ky Lan, 1979b, Resonant vibrational excitation of N_2 by low-energy electrons: an *ab initio* R-matrix calculation, Phys. Rev. Lett., 43:1926.

Taylor, H. S. and Yaris, R., 1980, The rotated-coordinate method and its relation to the stabilization method, Comments At. Mol. Phys., 9:73.

Watson, D. K. and McKoy, V., 1979, Discrete basis function approach to electron-molecule scattering, Phys. Rev. A, 20:1474.

THE SCHWINGER VARIATIONAL PRINCIPLE: AN APPROACH TO ELECTRON-MOLECULE COLLISIONS

R. R. Lucchese, K. Takatsuka, D. K. Watson, and V. McKoy

Arthur Amos Noyes Laboratory of Chemical Physics
California Institute of Technology
Pasadena, California 91125

I. INTRODUCTION

In this contribution we want to discuss several features and applications of the Schwinger variational principle to the study of collisions of low energy electrons with molecules and molecular ions. The Schwinger variational principle has long been known to be a potentially useful formulation of the collision problem but until recently there have been very few applications of this variational principle to electron collision problems.[1] The main drawback to the application of the Schwinger variational principle to more realistic systems has generally been regarded as the occurence of the term $< \Psi_{\underset{\sim}{k}}^{(-)} |V\, G_0\, V|\, \Psi_{\underset{\sim}{k}'}^{(+)} >$ in the variational functional. Historically this drawback seems to have outweighed the possible distinct advantages which the Schwinger variational principle is known to have over other standard variational methods such as the Kohn principle.[2] One of these advantages results from the feature that the trial scattering wave function need not satisfy any specific asymptotic boundary conditions. This feature implies both that the trial wave function could be expanded exclusively in terms of discrete basis functions, if such expansions were advantageous in the particular problem, and that irregular functions, which must be regularized near the origin, are not required in the trial function. Moreover, the Schwinger method is not troubled by the spurious singularities that can arise in the Kohn variational method. Although various ways for avoiding the effects of these singularities in such methods[3] have been developed it is a desirable feature of a method to be free of such singularities.

Here we will discuss the development and several applications of both the Schwinger variational principle as it is usually formulated and an iterative approach which uses the Schwinger variational principle to solve the Lippmann-Schwinger integral equation for electron-molecule scattering. The iterative method uses trial scattering wave functions which contain both discrete basis functions and numerical wave functions which explicitly satisfy the scattering boundary conditions. The discrete basis functions effectively describe the scattering wave function in the molecular region where exchange effects and partial wave coupling can be strong. The procedure usually begins with a trial scattering wave function which is expanded exclusively in a set of discrete basis functions large enough to provide a reasonably accurate estimate of the K matrix. The iterative procedure,which converges rapidly if started with this estimate of the K matrix, can then be applied to obtain results converged to the accuracy required.

In the following section we give a brief discussion of the Schwinger variational principle. In Section III we summarize the results of several applications of this method to the scattering of low energy electrons by simple molecular systems. In Section IV we discuss some formal aspects of the relationship between the Schwinger variational method and Kohn-type methods. We also present some specific numerical comparison between these different variational methods for a simple model problem in order to clarify these relationships. Finally, in Section V we present some results of the application of the Schwinger variational principle to an exactly soluble two-channel model problem in order to assess the effectiveness of the Schwinger method for multichannel cases.

II. THEORY

In the fixed-nuclei approximation the Schrödinger equation for the scattered electron is of the form (in atomic units)

$$[-\frac{1}{2}\nabla^2 + V(R,\vec{r}) - \frac{1}{2}k^2]\,\Psi^{(+)}_{\vec{k}}(R,\vec{r}) = 0 \tag{1}$$

where $V(R,\vec{r})$ is the effective interaction between the target and the scattered electron which depends parametrically on the relative coordinates of the target nuclei. The vector subscript $\vec{k}$ indicates the dependence of the wave function on the direction and magnitude of the incident momentum for a nonspherical target. The partial wave expansion of the scattering wave function $\Psi^{(+)}_{\vec{k}}$ can be written as

$$\Psi^{(+)}_{\vec{k}} = \left(\frac{2}{\pi}\right)^{\frac{1}{2}} \sum_{\ell m} i^{\ell}\, \psi^{(+)}_{k\ell m}(\vec{r})\, Y^{*}_{\ell m}(\hat{k}) \,. \tag{2}$$

The Schwinger variational expression for the partial wave elastic T matrix elements can be written as

$$\tilde{T}_{\ell\ell'm} = \frac{\langle \phi_{k\ell m}|U|\tilde{\Psi}^{(+)}_{k\ell'm}\rangle\langle\tilde{\Psi}^{(-)}_{k\ell m}|U|\phi_{k\ell'm}\rangle}{\langle\tilde{\Psi}^{(-)}_{k\ell m}|U - U\,G_0^{(+)}U|\tilde{\Psi}^{(+)}_{k\ell'm}\rangle} \tag{3}$$

where $U = 2V$ and we have assumed the molecule to be linear. In Eq.(3) the $\tilde{\Psi}^{(\pm)}_{k\ell m}$ are partial wave trial functions and $\phi_{k\ell m}$ are the regular free particle solutions. The expression in Eq.(3) is variationally stable with respect to variations of the partial wave scattering functions about the exact function $\Psi^{(\pm)}_{k\ell m}$. We now expand the trial function $\tilde{\Psi}^{(\pm)}_{k\ell m}$ in a set of basis functions, $\alpha_i(\vec{r})$

$$\tilde{\Psi}^{(\pm)}_{k\ell m}(\vec{r}) = \sum_{i=1}^{N} C^{(\pm)}_{k\ell m,i}\;\alpha_i(\vec{r}) \tag{4}$$

and insert this expansion into Eq.(3). Variation of Eq.(3) with respect to the coefficients, $C^{(\pm)}_{k\ell m,i}$, leads to the stationary result

$$\tilde{T}_{\ell\ell'm} = \sum_{i,j}^{N} \langle \phi_{k\ell m}|U|\alpha_i\rangle\,[\,(D^{(+)})^{-1}]_{ij}\langle\alpha_j|U|\phi_{k\ell'm}\rangle \tag{5}$$

where

$$D^{(+)}_{ij} = \langle\alpha_i|U - U\,G_0^{(+)}U|\alpha_j\rangle\;. \tag{6}$$

The Schwinger variational functional, Eq.(3) has some important properties. First, the functional is homogeneous in $\tilde{\Psi}_{k\ell m}$ and hence multiplication of the trial function by a constant factor does not change the variational expression. Hence we do not need to normalize the trial function. Moreover, $\tilde{\Psi}_{k\ell m}$ occurs in the functional only in the form $U\tilde{\Psi}_{k\ell m}$, which vanishes at infinity because of the properties of U. This implies that there is no constraint on the asymptotic form of the trial function. This feature is particularly attractive for applications to electron-molecule scattering where discrete basis functions can be used to advantage in the solution of the collision problem. Next we note that only regular functions occur in the formulation. Hence the trial scattering wave functions need not contain irregular solutions such as are required in the Kohn variational method. These irregular solutions must be regularized in some suitable way. Some recent calculations with the Schwinger principle which included such regularized irregular solutions suggest that the irregular solutions can interfere with the role of the short-range basis functions and adversely affect the convergence characteristics of the method. Finally, the

Schwinger variational principle is free of the spurious singularities which can arise in the Kohn-like methods.

The initial step in our Schwinger variational procedure is to evaluate Eq.(5) with a set of discrete basis functions which are usually chosen to be Cartesian Gaussian functions.[4] Applications of the Schwinger variational principle in this form to the scattering of low-energy electrons by homonuclear and heteronuclear diatomic systems have shown that the method can provide accurate results with small discrete basis sets.[5,6] In the next section we will discuss some specific results for the elastic scattering of electrons in the static-exchange approximation for some molecules along with some details of the numerical procedures for the evaluation of the required matrix elements.

Generally one tries to obtain reasonably accurate approximations to the partial wave T-matrix elements in Eq.(3) by expanding the trial scattering wave function, $\tilde{\Psi}^{(\pm)}_{k\ell m}$, in an adequate set of discrete basis functions. By increasing the size of the discrete basis set in which the trial scattering wave function is expanded one can hope to judge how well converged the corresponding variational T-matrix elements, $\tilde{T}_{\ell\ell' m}$, are. At this level of approximation the calculations can be carried out very economically and hence such studies of the dependence of the T-matrix elements on the size and composition of the basis set are a useful first step. However, the results of these studies are not necessarily conclusive and, moreover, it would be very desirable to have a procedure which provides systematic convergence to the exact solution for the potential chosen to describe the interaction. We now outline an iterative approach which uses the Schwinger variational principle to solve the Lippmann-Schwinger equation and shows systematic convergence to the scattering solutions for a given potential.[7]

The iterative Schwinger method starts by noting that the variational expression in Eq.(5) is equivalent to solving the Lippmann-Schwinger equation for the T-matrix

$$T = U + U G_0^{(+)} T \tag{7}$$

using the following separable approximation to the potential

$$U^{S_0} = \sum_{\alpha_i, \alpha_j} U|\alpha_i > [U^{-1}]_{ij} < \alpha_j|U . \tag{8}$$

With this separable potential the solution of the Lippmann-Schwinger equation, Eq.(7) is given by

$$T^{s_0} = \sum_{\alpha_i,\alpha_j} U|\alpha_i> [(D^{(+)})^{-1}]_{ij} < \alpha_j|U \tag{9}$$

with $D_{ij}^{(+)s_0}$ defined by Eq.(6). Clearly Eqs.(5) and (9) are equivalent. The errors that exist in this Schwinger variational T-matrix are due to the difference between the exact scattering potential U and the approximate separable potential U^{s_0} given in Eq.(8). It is possible to eliminate these errors due to the difference potential, $U - U^{s_0}$, by an iterative procedure.

First we note that at this stage the trial scattering wave function has been expanded in a set of discrete basis functions only and hence does not satisfy the asymptotic form defined by the variational T-matrix elements, $\tilde{T}_{\ell\ell'm}$, of Eq.(3). However, scattering wave functions, which have the asymptotic form specified by these $\tilde{T}_{\ell\ell'm}$, can be obtained very simply. The procedure does not require the solution of any coupled equations. These scattering solutions are computed using the partial wave expansion of the wave function in Eq.(2). For a linear molecule, the $\Psi_{k\ell m}^{(+)}$ may in turn be expanded in a partial wave series by

$$\Psi_{k\ell m}^{(+)}(\vec{r}) = \sum_{\ell'} \psi_{\ell\ell' m}^{(+)}(k,r)\, Y_{\ell' m}(\hat{r}) \,. \tag{10}$$

The Lippmann-Schwinger equation for the scattering wave function for the potential U^{s_0} is

$$\Psi_{k\ell m}^{(+)s_0}(\vec{r}) = \phi_{k\ell m}(\vec{r}) + < \vec{r}|G_0^{(+)} U^{s_0}| \Psi_{k\ell m}^{(+)s_0} > \tag{11}$$

where $\phi_{k\ell m}(\vec{r})$ are the free particle solutions

$$\phi_{k\ell m}(\vec{r}) = j_\ell(kr)\, Y_{\ell m}(\hat{r}). \tag{12}$$

By using the identity

$$< \vec{r}|U^{s_0}| \Psi_{k\ell m}^{(+)s_0} > = < \vec{r}|T^{s_0}|\phi_{k\ell m} > \tag{13}$$

we obtain the usual equation for the wave function in terms of the T-matrix

$$\Psi_{k\ell m}^{(+)s_0}(\vec{r}) = \phi_{k\ell m}(\vec{r}) + < \vec{r}|G_0^{(+)} T^{s_0}|\phi_{k\ell m} > . \tag{14}$$

The partial-wave functions, $\psi_{\ell\ell' m}^{(+)}$, are given by

$$\psi^{(+)s_0}_{\ell\ell' m}(k,r) = j_\ell(kr)\delta_{\ell\ell'} - \sum_{\alpha_i,\alpha_j} k < j_{\ell'}(kr_<)\, h^{(+)}_{\ell'}(kr_>)$$

$$\times\, Y_{\ell' m}(\hat{r}')|U|\alpha_i >_{\vec{r}'} [(D^{(+)})^{-1}]_{ij}$$

$$\times < \alpha_j|U|j_\ell(kr'')\, Y_{\ell m}(\hat{r}'') >_{\vec{r}''} . \tag{15}$$

The asymptotic form of the partial wave solutions are then

$$\psi^{(+)s_0}_{\ell\ell' m}(k,r) \sim j_\ell(kr)\delta_{\ell\ell'} - k < \phi_{k\ell' m}|T^{s_0}|\phi_{k\ell m} > h^{(+)}_{\ell'}(kr). \tag{16}$$

The radial functions are readily obtained from Eq.(15) by numerical integration. We note that the procedure does not involve the solution of any coupled equations.

The iterative procedure proceeds by augmenting the expansion set of Eq.(4) by the set of functions $\{\Psi^{s_0}_{k\ell_1 m}, \Psi^{s_0}_{k\ell_2 m}, \ldots, \Psi^{s_0}_{k\ell_p m}\}$ which consists of the scattering solutions corresponding to the T-matrix, T^{s_0}. Using this augmented set of functions, the first iteration is completed by calculating a new T-matrix given by

$$T^{s_1} = \sum_{\chi_i,\chi_j} U|\chi_i > [(D^{(+)})^{-1}]_{ij} < \chi_j|U . \tag{17}$$

It is important to note that the expansion basis in Eq.(17) includes both the original set of discrete functions from Eq.(4) and the continuum solutions given by Eq.(15).

A second iteration is begun by constructing the set of solutions $\{\Psi^{s_1}_{k\ell_1 m}, \ldots, \Psi^{s_1}_{k\ell_p m}\}$ which are associated with the T-matrix, T^{s_1}, given by Eq.(17). This set of functions combined with the initial set of discrete basis functions yields a new T-matrix T^{s_2}. This iterative scheme is repeated until the wave functions converge.

If the wave functions do converge such that

$$\Psi^{(+)s_{n+1}}_{k\ell m} = \Psi^{(+)s_n}_{k\ell m} \tag{18}$$

and if we have

$$< \Psi^{(-)s_n}_{k\ell m}|U - U\, G^{(+)}_0 U|\Psi^{(+)s_n}_{k\ell_j m} > = < \phi_{k\ell_i m}|U|\Psi^{(+)s_n}_{k\ell_j m} > \tag{19}$$

for $1 \leq i \leq p$ and $1 \leq j \leq p$, and

$$< \alpha_j | U - U G_0^{(+)} U | \Psi^{(+)s_n}_{k\ell_i m} > = < \alpha_j | U | \phi_{k\ell_i m} > \tag{20}$$

for $1 \leq i \leq p$ and α_j is any one of the original set of discrete basis functions, then it follows that the functions $\Psi^{(+)s_n}_{k\ell m}$ satisfy the Lippmann-Schwinger equation for the exact potential U.

It is of interest to note that Eqs.(19) and (20) are identically satisfied if $\Psi^{s_n}_{k\ell m}$ is the exact solution. The degree of convergence of an approximate wave function can hence be judged by how well the relations given in Eq.(19) and Eq.(20) are satisfied. Moreover each side of Eq.(19) is a nonvariational approximation to the partial wave T-matrix element, $T^{s_n}_{\ell\ell' m}$, where $\Psi^{s_n}_{k\ell m}$ is an approximate trial function. Thus the convergence of the wave function can also be judged by how well the two sides of Eq.(19) compare with each other and with the variationally stable elements $T^{s_n}_{\ell\ell' m}$.

III. SOME APPLICATIONS

In this section we will discuss the results of some applications of the Schwinger variational principle to the elastic scattering of electrons by H_2, N_2, and H_2^+. We have chosen these simple systems primarily to illustrate the important characteristics of the method. The results of applications to other systems are being published elsewhere.[6,8,9,10]

We will first look at some results for the scattering of electrons by H_2 in the static-exchange approximation. The target SCF wave function is constructed from a (5s2z) Cartesian Gaussian basis set as given by Watson et al.[5] The Hartree-Fock energy for H_2 in this basis set is -1.1330 a.u. and the quadrupole moment is 0.452 a.u. Table I shows the variational K-matrix for the $^2\Sigma_g$ channel at $k = 0.5$ a.u. The K-matrix elements, obtained from Eq.(5) using a basis set containing a single s Cartesian Gaussian function with exponent 0.5 centered on each nucleus, are shown in the column with n = 0. The K-matrices resulting from two successive steps in the iterative Schwinger procedure are shown in the columns with n = 1 and 2. For this simple example, these results show that the iterative procedure converges well. Our results agree well with the values of -1.55, 0.015, and 0.018 of Collins et al.[11] for K_{00}, K_{02}, and K_{22} respectively which were obtained by the numerical integration of the single-center expansion of the scattering equations. The variational K-matrices, obtained using only a single s Gaussian basis function on each nucleus and shown in the n = 0 column of Table I, suggest that the Schwinger method may provide accurate solutions of the scattering problem with small discrete basis sets.

Table I. Convergence of the Schwinger variational K matrix starting from one discrete scattering function for $^2\Sigma_g$ symmetry in H_2 with k = 0.5 au.

	$\tilde{K}_{\ell\ell'}$		
(ℓ,ℓ')	n=0	1	2
(0,0)	-2.045	-1.552	-1.548
(0,2)	-0.276(-1)	0.133(-1)	0.134(-1)
(2,2)	-0.372(-3)	0.163(-1)	0.163(-1)

To illustrate some additional convergence characteristics of the iterative Schwinger procedure, Table II shows the behavior of the nonvariational approximation to the K-matrix of Eq.(19). As a practical consideration the degree of convergence of the approximate wave function at any step of the procedure can be judged by how well these nonvariational approximations agree with each other and with the variationally stable estimates of Table I.

In Table III we present K-matrix elements at several energies for the $^2\Sigma_g$ and $^2\Sigma_u$ symmetries for the e-H_2 system. These K-matrix elements were obtained from the iterative Schwinger procedure starting from a discrete basis containing four Gaussian functions centered at the nuclei. These Cartesian Gaussian functions are of the s and z types with exponents of 0.3 and 1.0. All the K-matrices were converged to three significant figures by the first iteration. These K-matrix elements are in good agreement with those of Collins et al.[11] The small discrepancies which exist are probably due to differences in the potentials used in these calculations.

Table IV shows the eigenphases for e-N_2 scattering at three energies. In these calculations we used the SCF wave function of Nesbet[12] and started the iterative procedure from a discrete basis containing s Cartesian Gaussian functions with exponents of 8.0, 4.0, 2.0, 1.0, 0.5, and 0.25 and z functions with exponents of 4.0, 2.0, 1.0, 0.5, and 0.25 centered on each nucleus. These eigenphases are converged in one iteration and eigenphase sums agree well with those calculated by Collins et al.[11] who used the SCF wave function of Cade et al.[13]

Finally we present some results of the application of the Schwinger method to e-H_2^+ scattering. Such applications to electron-molecular ion collisions are important since they relate directly to the calculation of photoionization cross sections of molecules

Table II. Convergence of nonvariational approximations to K matrix starting from one discrete scattering function for $^2\Sigma_g$ symmetry in H_2 with k = 0.5 au.

		$\langle \phi_{k\ell 0} \| U \| \psi^{s_n}_{k\ell' 0} \rangle$	
(ℓ,ℓ')	n=0	1	2
(0,0)	-1.602	-1.567	-1.549
(0,2)	0.179(-1)	0.131(-1)	0.134(-1)
(2,0)	-0.107(-1)	0.136(-1)	0.135(-1)
(2,2)	0.161(-1)	0.163(-1)	0.163(-1)

		$\langle \psi^{s_n}_{k\ell 0} \| U - U G_0^{(P)} U \| \psi^{s_n}_{k\ell' 0} \rangle$	
(ℓ,ℓ')	n=0	1	2
(0,0)	-1.642	-1.586	-1.550
(0,2)	0.667(-2)	0.132(-1)	0.135(-1)
(2,2)	0.155(-1)	0.162(-1)	0.163(-1)

Table III. Iterated Schwinger variational K matrix elements for the $^2\Sigma_g$ and $^2\Sigma_u$ channels for $e\text{-}H_2$.

	$^2\Sigma_g^+$		$^2\Sigma_u^+$	
k	$\tilde{K}_{000}$[a]	K_{000}[b]	$\tilde{K}_{110}$[a]	K_{110}[b]
0.1	-0.217	-0.217	0.123(-1)[c]	0.127(-1)
0.3	-0.722	-0.722	0.113	0.119
1.0	8.04	8.05	1.34	1.34
	$\tilde{K}_{020}$	K_{020}	$\tilde{K}_{130}$	K_{130}
0.1	0.406(-2)	0.39(-2)	0.105(-2)[c]	0.15(-2)
0.3	0.978(-2)	0.11(-1)	0.335(-2)	0.34(-2)
1.0	0.122	0.11	0.304(-1)	0.29(-1)
	$\tilde{K}_{22}$	K_{22}	$\tilde{K}_{330}$	K_{330}
0.1	0.165(-2)	0.21(-2)	0.971(-3)[c]	0.73(-3)
0.3	0.687(-2)	0.74(-2)	0.290(-2)	0.31(-2)
1.0	0.914(-1)	0.93(-1)	0.190(-1)	0.20(-1)

[a] present results.

[b] L. Collins, D. Robb, and M. Morrison (private communication). See reference 11.

[c] A grid extending to 125 au is used to obtain this K matrix element.

Table IV. $^2\Sigma_g$ eigenphases for e-N_2 scattering at energies of 0.1, 0.3, and 1.0 Rydbergs.

ℓ[a] \ E(Rydbergs)	0.1	0.3	1.0
0	2.376	1.901	1.297
2	-0.011	-0.011	-0.120
4	-0.003	-0.001	-0.001
6	-0.001	-0.003	-0.009
Sum[b]	2.361	1.888	1.169
Sum[c]	2.311	1.808	1.043

[a] ℓ corresponds to the principal component of the given eigenphase.

[b] Sum of the eigenphases above.

[c] See reference 11.

including the angular distribution of ejected photoelectrons. To apply the Schwinger method to electron-ion collisions one simply formulates the equations of the previous section using the Coulomb Green's function and V becomes the residual short-range potential obtained by subtracting the Coulomb component from the total ion potential. We use the static-exchange approximation for the scattering potential with the H_2^+ orbital fixed as the $1\sigma_g$ SCF orbital of H_2 given in reference 14.

Table V shows the eigenphases for the $^1\Sigma_u$ symmetry for e-H_2^+ scattering at k = 0.5 au. These are the converged results of the iterative procedure starting from a basis containing 2s and 2z Cartesian Gaussian functions with the same exponents of 1.0 and 0.3 on each nucleus and a single z function with exponent of 1.0 at midpoint of H_2^+. The results were quite well converged even after just one iteration. Further details are discussed in reference 14.

All the matrix elements required in these studies are evaluated with the use of single-center expansions and Simpson's rule quadrature for the radial integrals. Our experience has shown that all the matrix elements, including the $\langle \alpha_i | U\ G\ U | \alpha_j \rangle$ terms, can be evaluated quite accurately and efficiently in this way for linear molecules.

Table V. $^1\Sigma_u$ eigenphases for $e\text{-}H_2^+$ scattering at k = 0.5 au.

ℓ[a]	Variational[b]	Robb and Collins[c]
1	0.349	0.360
3	0.037	0.037
5	0.011	0.011
7	0.004	0.004
Sum	0.401	0.412

[a] ℓ corresponds to the principal component of the given eigenphase.
[b] Present results.
[c] Private communication from D. Robb and L. Collins. See reference 15 for details of their calculation.

IV. RELATIONSHIP BETWEEN THE SCHWINGER AND KOHN VARIATIONAL PRINCIPLES

The Kohn-type variational principles have been applied far more extensively to collision problems than the Schwinger principle. The mathematical relationship between the Schwinger and Kohn principles has not yet been well established. Kato[16] connected the Schwinger principle with the Rubinow method, which to our knowledge is the only direct relationship established between these two groups of variational principles. However, Kato[16] drew no conclusion as to the relative convergence characteristics of these methods. Delves[17] commented briefly on the relationship between the Schwinger and Kohn methods but unfortunately we will show that the implied relationship between these two principles, which was assumed by Delves, is mathematically incorrect. In view of these circumstances and of some recent articles,[18,19] which compared their convergence characteristics solely on the basis of numerical examples for some simple model potentials, we will establish the correct mathematical relationship between the Schwinger and Kohn variational functionals. We will also present some simple numerical results relevant to this comparison.

The usual Kohn variational functional for the tangent of the phase shift λ,[20]

$$[\lambda]_K = \lambda + 2 < \Psi|\hat{H}|\Psi > \tag{21}$$

can be written in the belinear form[21]

$$\frac{1}{2}[\lambda]_k = < \tilde{C}|\hat{H}|\tilde{C} > - < \tilde{C}|V|S > - < S|V|\tilde{C} > - < S|V|S > \quad (22)$$

where $\hat{H} = E - H$, S is the regular solution of the unperturbed Hamiltonian $H_0 = H - V$, and Ψ is the trial wave function which can be written as

$$\Psi = \tilde{C} + S \quad (23)$$

with

$$\tilde{C} = \lambda C + \sum_i a_i \eta_i \quad (24)$$

In Eq.(24) for s-wave scattering

$$C \backsim \begin{cases} k^{-\frac{1}{2}} \cos(kr) & \text{as} \quad r \to \infty \\ 0 & \text{as} \quad r \to 0 \end{cases} \quad (25)$$

and η_i is a discrete basis function. One can generalize the functional in Eq.(22) and write

$$I(\phi,\Psi) = < \phi|\hat{H}|\Psi > - < \phi|V|S > - < S|V|\Psi > - < S|V|S > . \quad (26)$$

A systematic way to select the trial wave functions ϕ and Ψ is as follows. The exact function $\tilde{C}$ of Eq.(24) satisfies a Lippmann-Schwinger equation of the form[21]

$$\tilde{C} = G_0 V S + G_0 V \tilde{C} . \quad (27)$$

This integral equation can be solved by the iterative procedure[22,23]

$$\tilde{C}_{n+1} = G_0 V S + G_0 V \tilde{C}_n \quad (28)$$

We select $\tilde{C}_1$ to be given by the expansion in Eq.(24) and insert this function into the variational functional of Eq.(26). Then we have

$$[\lambda]_k = 2 I(\tilde{C}_1,\tilde{C}_1) . \quad (29)$$

Therefore the functional $2 I(\tilde{C}_1,\tilde{C}_1)$ is just the tangent of the phase shift as given by the Kohn variational principle. Next we consider the higher-rank variational functional $I(\tilde{C}_1,\tilde{C}_2)$. Some manipulation shows that

$$I(\tilde{C}_1,\tilde{C}_2) = \langle \Psi_1|(V - V\, G_0\, V)|\Psi_1\rangle - \langle \Psi_1|V|S\rangle - \langle S|V|\Psi_1\rangle \tag{30}$$

where $\Psi_1 = S + \tilde{C}_1$. The right-hand side of Eq.(30) is just the bilinear form of the Schwinger variational functional. Therefore we have

$$[\lambda]_s = 2\, I(\tilde{C}_1,\tilde{C}_2). \tag{31}$$

If the trial function $\tilde{C}_1$ is good enough so that the iterative provedure converges monotonically, a higher rank functional $I(\tilde{C}_m,\tilde{C}_n)$ should give a more accurate result than any lower rank function. Since the functionals $I(\tilde{C}_1,\tilde{C}_2)$ and $I(\tilde{C}_1,\tilde{C}_1)$ correspond to the Schwinger and Kohn variational principles respectively, this shows that for a given trial function the Schwinger variational principle yields a more accurate result than does the Kohn principle.

Delves[17] has stated without proof that the output from the Schwinger principle $[\lambda]_s$ with the trial function $\Psi^{(1)}$ is identical to the output from the Kohn principle $[\lambda]_k$ with the trial function $\Psi^{(2)}$, where

$$\Psi^{(2)} = S + G_0\, V\, \Psi^{(1)} . \tag{32}$$

Although the realization of the relationship of Eq.(32) is important, the statement itself is incorrect. The output from the Kohn principle with the trial function $\Psi^{(2)}$ corresponds to the functional $I(\tilde{C}_2,\tilde{C}_2)$, and not $I(\tilde{C}_1,\tilde{C}_2)$, in our proof. In fact, $I(\tilde{C}_2,\tilde{C}_2)$ is a higher order functional equivalent to the functional F_0

$$F_0 = \frac{\langle \Psi|V\, G_0\, V|S\rangle \langle S|V\, G_0\, V|\Psi\rangle}{\langle \Psi|V\, G_0\, V - V\, G_0\, V\, G_0\, V|\Psi\rangle} \tag{33}$$

which we have previously discussed[21] and has also been stated by Newton.[24] Finally we note that some of the functionals $I(\tilde{C}_m,\tilde{C}_n)$ $m,n \geqslant 2$ correspond to different stages in the iterative Schwinger method.[7]

To compare the convergence of the Kohn and Schwinger variational principles we have carried out calculations on the same model system as was used by Thirumalai and Truhlar.[18] However, in contrast to their studies[18] we use the same trial scattering wave function in the two variational principles.

The scattering potential is the attractive exponential potential

$$V(r) = -e^{-r} \tag{34}$$

and we consider only s-wave scattering. All comparisons are made in terms of the K matrix element, i.e., tan δ_0. The trial scattering function used in both the Schwinger and Kohn variational principles is

$$\Psi^t(r) = X^{(n)}(r)/r \tag{35}$$

with the function $X^n(r)$ of the form

$$X^n(r) = \alpha_0 \sin kr + \alpha_1(1 - e^{-\beta r}) \cos kr + \sum_{a=1}^{n} C_a r^a \exp(-\alpha r) \tag{36}$$

where if n = 0 no discrete basis functions are included in the trial function. In all results presented here we choose $\alpha = 2.5$ and $\beta = 1.0$.

In Tables VI - VIII we compare the results obtained with the trial function of Eq.(36) in the Schwinger variational principle with the results of several other variational methods.[18] These include the anomaly-free (AF)[25] and optimized anomaly-free (OAF) adaptations of the Kohn methods[26] and the minimum-norm-Kohn (MNK),[27] minimum-norm-inverse-Kohn (MNR),[20a] and optimized-minimum-norm (OMN)[26] versions of the Harris-Michels-type methods. From the results at these three energies the Schwinger variational principle clearly yields superior results to those of the Kohn and Harris-Michels methods. These results are consistent with the mathematical relationship between these variational principles which we have shown above. Further details of these studies will be published elsewhere.[28]

V. A MULTICHANNEL APPLICATION OF THE SCHWINGER VARIATIONAL PRINCIPLE

The applications of the Schwinger variational principle which we have discussed so far in this article have been to single-channel electron collision problems. We believe that these results do demonstrate the considerable potential of the method. We are currently extending the Schwinger method to multichannel problems on the basis of a new formulation.[29] In order to make some preliminary assessment of the effectiveness and accuracy of the Schwinger method for multichannel cases, we have solved the model two-channel problem proposed by Huck.[30] In this application we have found that the Schwinger method is very effective yielding results which show far better convergence than those of the standard variational principles.[25-27] Details can be found in reference 31.

Table VI. Ratio of variational K matrix elements to the accurate value for k = 0.55 au.[a]

n[b]	AF[c]	OAF	MNK	MNR	OMR	Schwinger[d]
0	0.9735	0.9733	0.9733	0.9735	0.9733	0.9972
2	0.9968	0.9940	0.9902	0.9969	0.9941	0.9999
4	0.9999	0.9970	0.9910	0.9999	0.9970	1.0000

[a]Accurate value is K_0 = 2.2003827.

[b]The number of discrete basis functions in the trial function. See Eq.(36). For n = 0 no discrete basis functions are included in the trial function.

[c]The results in the AF, OAF, MNK, MNR, and OMN columns are from Ref. 18 except those for n = 0 which are from Ref. 20a.

[d]Results with the Schwinger variational principle.

Table VII. Ratio of variational K matrix elements to the accurate value for k = 0.35 au.[a]

n	AF	OAF	MNK	MNR	OMN	Schwinger
0	--	--	--	--	--	0.9765
2	0.9879	0.9858	0.9622	0.9902	0.9861	0.9999
4	0.9980	0.9978	0.9878	0.9980	0.9978	1.0000

[a]Accurate value is K_0 = 9.0918095. See also footnotes in TableVI.

Table VIII. Ratio of variational K matrix elements to the accurate value for k = 0.15 au.[a]

n	AF	OAF	MNK	MNR	OMN	Schwinger
0	--	--	--	--	--	1.0124
2	1.0005	1.0009	1.0004	1.0010	1.0009	1.0000
4	1.0006	1.0004	1.0002	1.0005	1.0004	1.0000

[a]Accurate value is $K_0 = -1.7449393$. See also footnotes in TableVI.

The exactly soluble two-channel model problem used by Huck,[30] Nesbet,[25-26] and more recently by Harris,[27] is defined by the Hamiltonian $H = H_0 + V$ with

$$H_0 = |\chi_1 > \left(-\frac{1}{2}\frac{d^2}{dr^2}\right) < \chi_1| + |\chi_2 > \left(-\frac{1}{2}\frac{d^2}{dr^2} + \Delta E\right) < \chi_2| \quad (37)$$

and

$$V = \sum_{m \neq n}^{2} |\chi_m > V_{mn} < \chi_n| \quad (38)$$

where

$$V_{12} = V_{21} = \begin{cases} \frac{1}{2} C (r < a) \\ 0 \; (r > a) \end{cases} \quad (39)$$

and $< \chi_m|\chi_n > = \delta_{mn}$.

In terms of the regular eigenfunctions of H_0 ,

$$S_m(r_1,r_2) = \chi_m(r_1)\, k_m^{-\frac{1}{2}} \sin k_m r_2 \quad (m = 1,2) \quad (40)$$

the Schwinger variational functional for the K-matrix is given by

$$[K_{mn}] = \frac{< \Psi_m|U|S_n > < S_m|U|\Psi_n >}{< \Psi_m|U\, G_0\, U - U|\Psi_n >} \quad (41)$$

where $U = 2V$. In Eq.(41), the free-particle (standing wave) Green's function G_0 is

$$G_0(r_1r_2;r_1'r_2') = -\sum_{m=1}^{2} S_m(r_1,r_2)\, C_m(r_1',r_>), \tag{42}$$

where C_m is the irregular solution of H_0, and $r_> = \max(r_2,r_2')$ and $r_2 = \min(r_2,r_2')$. As usual, the wave function is expanded in terms of a basis set $\{\eta_i^m | m = 1,2,\ i = 1,\ldots,N\}$. From the stationary condition the K-matrix elements can be written as

$$K_{mn} = \sum_{a,b}^{2} \sum_{i,j}^{N} \langle S_m|U|\eta_i^a\rangle D_{ij}^{ab} \langle \eta_j^b|U|S_n\rangle, \tag{43}$$

where

$$\left(D^{-1}\right)_{ij}^{ab} = \langle \eta_i^a|(U\, G_0\, U - U)|\eta_j^b\rangle . \tag{44}$$

In order to compare with the previous results of Nesbet et al.[25-27] we also choose the potential parameters $a = 1.0$ and $C^2 = 10.0$, and the energies $E = 0.5$ and $\Delta E = 0.375$ ($k_1 = 1.0$ and $k_2 = 0.5$). In our calculations the basis set is composed of only the L^2 functions

$$\eta_i^m = |\chi_m\rangle\, r^i\, e^{-pr} \quad (i = 1,2,\ldots N), \tag{45}$$

since the total energy ($E = 0.5$) is sufficiently lower than the height of the potential ($\frac{1}{2} C \backsim 1.58$). The optimum value of the parameter p for η_i^m was determined to be 2.5 by Nesbet[25] and Harris[27] for their variational method.

In Table IX we show the deviation (ΔK) of the K-matrices obtained from the Schwinger variational principle from the exact results as a function of the basis set size N. The results in terms of two different basis sets with an exponent of $p = 2.5$ (the same as that for the other methods) and 0.9 (the optimized value for the Schwinger principle are listed. We also show the deviations of the K-matrices obtained by Nesbet with the Anomaly-Free (AF),[25] Optimized Anomaly-Free (OAF),[26a] and Restricted Interpolated Anomaly-Free (RIAF)[26b] methods and by Harris et al. with the Minimum Norm (MN)[27] method. Our results are quite clear and impressive. The rate of convergence of the Schwinger method with $p = 2.5$ is much faster than those of the other methods. For instance, the Schwinger K-matrix with two basis functions ($N = 2$) is already closer to the exact value than the K-matrices of the other methods using 25 basis functions. Furthermore, with $n = 6$, the Schwinger K-matrices have almost completely converged to the exact values, whereas the results of the other variational methods are still far from being converged.

Table IX. The deviation of the variationally determined K-matrices from the exact values.[a] (ΔK)

	N	AF[b]	MN[c]	OAF[d]	RIAF[e]	Schwinger (p^f=2.5)	Schwinger (p^f=0.9)
ΔK_{11}	N = 1	--	-18.72853	--	--	-19.09277	-0.14259
	2	--	-58.41920	--	--	- 0.26131	-1.53971
	4	-5.61743	- 5.69784	-4.54847	--	- 0.01000	-0.00004
	6	-2.99726	- 3.29061	-2.99989	-3.00448	0.0	0.0
	10	-1.39131	- 1.40472	-1.37881	-1.39040	0.0	0.0
	25	-0.36330	--	-0.33532	-0.29985	0.0	0.0
ΔK_{12}	N = 1	--	11.86603	--	--	12.10630	-0.09141
	2	--	37.84736	--	--	0.07861	0.96567
	4	3.57888	3.66193	2.89239	--	0.00600	0.00002
	6	1.91396	2.09620	1.91579	1.91839	0.0	0.0
	10	0.88909	0.89739	0.88115	0.88850	0.0	0.0
	25	0.23037	--	0.21363	0.19117	0.0	0.0
ΔK_{22}	N = 1	--	- 7.54469	--	--	- 7.57920	0.19492
	2	--	24.52584	--	--	0.01638	-0.60478
	4	-2.28298	- 2.35953	-1.84341	--	- 0.00355	-0.00002
	6	-1.22397	- 1.33719	-1.22525	-1.22670	0.0	0.0
	10	-0.56896	- 0.57410	-0.56393	-0.56858	0.0	0.0
	25	-0.14621	--	-0.13619	-0.12198	0.0	0.0

a) The exact values are: K_{11} = 21.76525, K_{12} = K_{21} = -14.12742 and K_{22} = 8.73385 (see Ref. 25).
b) Ref. 25, 26a. c) Ref. 27. d) Ref. 26. e) Ref. 26a. f) The exponent in the basis functions.

VI. CONCLUSIONS

In this article we have discussed both the development of an iterative approach to the Schwinger variational principle and the results of application of this method to several problems in electron-molecule collisions. These results show that the iterative Schwinger technique is a very promising approach to the solution of the electron-molecule scattering problem. The method has several desirable and practical features such as encouraging convergence characteristics for trial scattering wave functions containing small sets of discrete basis functions and a criterion for the systematic convergence of the procedure to the exact scattering solution for the potential chosen to describe the interation.

This work was supported by grant No. CHE79-15807 from the National Science Foundation and by an Institutional grant from the United States Department of Energy No. EY-76-G-03-1305. The research reported in this paper made use of the Dreyfus-NSF Theoretical Chemistry Computer which was funded through grants from the Camille and Henry Dreyfus Foundation, the National Science Foundation (Grant No. CHE78-20235), and the Sloan Fund of the California Institute of Technology.

REFERENCES

1. For earlier applications of the Schwinger principle to the scattering of electrons by the hydrogen atom without exchange see S. Altshuler, Phys. Rev. 89, 1278 (1953) and B. H. Brandsen and J. S. C. McKee, Proc. Phys. Soc A 70, 398 (1957).
2. J. R. Taylor, Scattering Theory (Wiley, New York, 1971).
3. See, for example, R. K. Nesbet, Phys. Rev. 175, 134 (1968).
4. R. R. Lucchese and V. McKoy, J. Phys. B 12, L421 (1979).
5. D. K. Watson, R. R. Lucchese, V. McKoy, and T. N. Rescigno, Phys. Rev. A 20, 1474 (1980).
6. D. K. Watson, T. N. Rescigno, and B. V. McKoy, "Schwinger variational calculations for electron scattering by polar molecules" J. Phys. B (1981) - accepted for publication.
7. R. R. Lucchese, D. K. Watson, and V. McKoy, Phys. Rev. A 22, 421 (1980).
8. R. R. Lucchese and V. McKoy, "Studies of the elastic scattering of electrons by CO_2 in the static-exchange approximation" Phys. Rev. A - submitted for publication.
9. R. R. Lucchese and V. McKoy, "Photoionization cross sections of CO_2," Phys. Rev. A - to be published.
10. R. R. Lucchese and V. McKoy, "Photoionization in Acetylene," J. Chem. Phys. - to be published.
11. L. Collins, D. Robb, and M. Morrison, private communication. Details are given by these authors in Phys. Rev. A 21,488(1980).

12. R. K. Nesbet, J. Chem. Phys. 40, 3619 (1964).
13. P. E. Cade, K. D. Sales, and A. C. Wahl, J. Chem. Phys. 44, 1973 (1966).
14. R. R. Lucchese and V. McKoy, "An iterative approach to the Schwinger variational principle applied to electron-molecular ion collisions," Phys. Rev. A (1981) - in press.
15. D. Robb and L. Collins, Phys. Rev. A 22, 2474 (1980).
16. T. Kato, Phys. Rev. 80, 475 (1950).
17. L. M. Delves, Adv. Nucl. Phys. 5, 1 (1972).
18. D. Thirumalai and D. G. Truhlar, Chem. Phys. Lett. 70, 330 (1980).
19. J. Callaway, Phys. Lett. 77A, 137 (1980).
20. For reviews, see a) D. G. Truhlar, J. Abdallah, Jr., and R. L. Smith, Adv. Chem. Phys. 25, 211 (1974); b) R. K. Nesbet, Adv. Atom. Mol. Phys. 13, 315 (1977); c) J. Callaway, Phys. Rev. 45, 89 (1978).
21. K. Takatsuka and V. McKoy, "A variational scattering theory using a functional of fractional form. I. General Theory," Phys. Rev. A - accepted for publication.
22. J. M. Blatt and J. D. Jackson, Phys. Rev. 76, 18 (1949).
23. T. Kato, Prog. Theor. Phys. 6, 295 (1951).
24. R. G. Newton, Scattering Theory of Waves and Particles (McGraw-Hill, New York, 1966) p. 321.
25. R. K. Nesbet, Phys. Rev. 179, 60 (1969).
26. a) R. K. Nesbet and R. S. Oberoi, Phys. Rev. A 6, 1855 (1972); b) R. K. Nesbet, Phys. Rev. A 18, 955 (1978).
27. F. E. Harris and H. H. MIchels, Phys. Rev. Lett. 22, 1036 (1969).
28. K. Takatsuka, R. Lucchese, and V. McKoy, "Relationship between the Schwinger and Kohn-type variational principles in scattering theory," Phys. Rev. A - to be published.
29. K. Takatsuka and V. McKoy, "An extension of the Schwinger variational principle beyond the static-exchange approximation," Phys. Rev. A - to be published.
30. R. J. Huck, Proc. Phys. Soc. (London) A 70, 369 (1957).
31. K. Takatsuka and V. McKoy, Phys. Rev. Lett. 45, 1396 (1980).

RECENT DEVELOPMENTS IN COMPLEX SCALING

T. N. Rescigno

Theoretical Atomic and Molecular Physics Group
Lawrence Livermore National Laboratory
University of California
Livermore, California 94550

I. INTRODUCTION

My objective in this talk will be to discuss some recent developments in the use of complex basis function techniques to study resonance, as well as certain types of non-resonant, scattering phenomena. Complex scaling techniques and other closely related methods have continued to attract the attention of computational physicists and chemists and have now reached a point of development where meaningful calculations on many-electron (more than two!) atoms and molecules are beginning to appear feasible. The field has evolved very rapidly over the past few years and I will not attempt anything like an exhaustive review of the subject. I can refer the interested reader to the review article by McCurdy[1] and to a volume of the International Journal of Quantum Chemistry[2] devoted entirely to the subject. These sources give an excellent summary of both mathematical and computational techniques in complex scaling through 1978. The scope of this talk will be limited to several developments since 1978 with which I have been connected.

The first of these developments is a discussion of direct methods for computing partial resonance widths.[3] The imaginary part of the complex resonance energy computed by any direct method gives only the total resonance width, but it can be shown that the eigenfunction associated with the complex resonance eigenvalue can be used to provide partial width information - a simple consequence of the separable nature of the S-matrix near a resonance.

The recent use of complex scaling methods for studying molecular resonances has stimulated considerable interest in generalizations which are applicable to other continuum processes.[4, 5] In Section III of this talk, I will review methods for computing resolvent matrix elements - and molecular photoionization cross sections in particular - within the framework of the Born-Oppenheimer approximation. This discussion will also serve to establish a connection between two methods proposed for treating molecular problems[6, 7] and will focus attention on certain numerical complications that must be dealt with.

Finally, in Section IV, I will review some very recent work on how complex scaling ideas can be used to adapt simple SCF theory to the study of negative ion shape resonances in atoms and molecules.

II. PARTIAL WIDTHS

Consider the scattering of a structureless particle by an idealized target with no rearrangements and with a potential which does not couple partial waves. These restrictions can be relaxed at the cost of substantially more cumbersome notation. The asymptotic form of the scattering wave function for partial-wave ℓ is given by

$$\Psi^{+}_{\alpha,k_\alpha,\ell}(r,x) \xrightarrow[r\to\infty]{} \frac{i}{(2\pi k_\alpha)^{1/2}} \sum_\beta \left\{ (-1)^\ell \delta_{\alpha\beta} \frac{e^{-ik_\alpha r}}{k_\alpha^{1/2}\, r} - S^\ell_{\beta,\alpha}(E) \frac{e^{ik_\beta r}}{k_\beta^{1/2}\, r} \right\} \Phi_\beta(x) \tag{1}$$

where $\phi_\beta(X)$ is the target wave function for channel β, and $S^\ell_{\beta\alpha}$(E is the partial wave S matrix. The channel momenta k_γ are related to the total energy E and target energy ε_γ by

$$k_\gamma = [\, 2\mu(E-\mathcal{E}_\gamma)\,]^{\frac{1}{2}} \tag{2}$$

At a resonance eigenvalue ε_R, all non-vanishing S matrix elements $S_{\beta\alpha}(E_R)$ have a pole. Thus at large r, the wave function with k^R_α given by

$$k^R_\alpha = \left[2\mu(\varepsilon_R - \mathcal{E}_\alpha)\right]^{\frac{1}{2}} \tag{3}$$

has only outgoing waves in each open channel and satisfies the asymptotic boundary condition

$$\Psi_{\alpha,k_\alpha^R,\ell}(r,x) \sim \sum_\beta D_\beta^\alpha \frac{e^{ik_\beta^R r}}{r} \Phi_\beta(x); \quad D_\beta^\alpha \equiv -S_{\beta,\alpha}^\ell / k_\beta^{1/2} \tag{4}$$

For a problem with M open channels, we use a trial variational function of the form

$$\psi_t = \sum_{i=1}^{N-M} C_i \phi_i(r,x) + \sum_\beta C_\beta \theta_\beta(r,x) \tag{5}$$

where the functions $\phi_i(r,x)$ are square-integrable basis functions and the sum over β is limited to open channels. The functions θ_β are given by

$$\theta_\beta = g(r) \frac{e^{ik_\beta r}}{r} \Phi_\beta(x) \tag{6}$$

where g(r) is a cutoff function satisfying

$$g(r) \underset{r \to 0}{\sim} r^{\ell+1} \quad ; \quad g(r) \xrightarrow[r \to \infty]{} 1 \tag{7}$$

The coefficients C_i and C_β are determined by making the functional

$$I[\Psi_t] = \iint \psi_t(r,x)(H_\ell(r,x)-E)\psi_t(r,x) r^2 dr dx \tag{8}$$

where $H_\ell(r,x)$ is the partial-wave Hamiltonian, stationary with respect to variation of the coefficients.

Bardsley and Junker[8] proposed an iterative scheme for solving a single-channel version of Eq. (8) for resonances by varying the (complex) values of k_α^R in the trial function until Eq. (3) is satisfied for a particular eigenvalue.

There is an important point to note about Eq. (8). The free-free elements of $(H_\ell - E)$, i.e., matrix elements between two functions of the form of θ_β, are in general not well defined for values of the channel momenta in the lower half-plane. This difficulty depends specifically on the form of the interaction potential since the definition of the channel momenta in Eq. (2) guarantees that the contribution at large r from the kinetic energy to the free-free matrix elements of $(H_\ell - E)$ does not lead to divergences. The matrix is therefore defined, for a particular choice of basis, as the analytic-continuation of that matrix from the upper half-k plane (physical sheet of the E plane). This analytic continuation can be performed using analytic formulas for the free-free matrix elements, if they are available, or numerically as discussed by Isaacson et al.[9]

The coefficients of the continuum functions, θ_β, in Eq. (5) are determined by diagonalizing the secular matrix obtained from Eq. (8). There is, in general, no way to arrange for these coefficients to coincide with the D^α_β chosen in Eq. (4). We denote the vector which connects the two sets of coefficients as $\vec{\alpha}$:

$$C_\beta = \sum_\alpha a_\alpha D^\alpha_\beta \tag{9}$$

A single diagonalization of a multichannel Hamiltonian in a set of basis functions which do not fix the asymptotic normalization of the wave function does not provide eigenvectors from which complete scattering information can be obtained in the general nonresonant situation.[10] The resonance case is considerably simpler; complete information, that is, partial as well as total widths, can be obtained without requiring the mixing coefficients $\vec{a}$.

We can demonstrate this very simply. The partial-wave S matrix is given in terms of the partial-wave T matrix by

$$S^\ell_{\beta\alpha} = \delta_{\beta\alpha} - 2\pi i \mu\, k^{\frac{1}{2}}_\alpha\, k^{\frac{1}{2}}_\beta\, T^\ell_{\beta\alpha} \tag{10}$$

Since we are considering a problem with no partial-wave coupling, we can write $T^\ell_{\beta\alpha}$ as a matrix element of the T operator between functions of the form $\phi_\alpha(x)X_\ell(k_\alpha\vec{r})$, where

$$X_\ell(k_\alpha\vec{r}) = \left(\frac{2}{\pi}\right)^{\frac{1}{2}} i^\ell \frac{\hat{j}_\ell(k_\alpha r)}{k_\alpha r} Y_{\ell m}(\hat{r}) \tag{11}$$

and $\hat{j}_\ell$ is the regular Ricatti-Bessel function

$$T^\ell_{\beta\alpha} = \langle \Phi_\beta(x)X_\ell(k_\beta\vec{r})\,|T(E)|\,\Phi_\alpha(x)X_\ell(k_\alpha\vec{r})\rangle \tag{12}$$

We can establish the behavior of $T^{\ell}_{\beta\alpha}$ near a resonance by recalling that T(E) satisfies the operator equation for all E,

$$T(E) = V + VG(E)V \tag{13}$$

where V is the full potential and G(E) is the full Green's function (operator) for the system. The central point of the proof hinges on the fact that the T operator is a separable operator at the complex resonance energy. We can define the resonance wave function $|\psi_R\rangle$ and the corresponding function in the dual space $\langle\tilde{\psi}_R|$ (not merely the complex conjugate) in terms of the residue of the Green's function at the resonance pole

$$|\Psi_R\rangle\langle\tilde{\Psi}_R| = \lim_{E\to\varepsilon_R} (E-\varepsilon_R)\, G(E) \tag{14}$$

More and Gerjuoy[11] have discussed the fact that Eq. (14) fixes the normalization of the resonance wave function and also derive the relationship between $|\psi_R\rangle$ and $|\tilde{\psi}_R\rangle$. With the definition in Eq. (13) we can easily see the residue of T(E) at a resonance pole is a separable operator:

$$\lim_{E\to\varepsilon_R} (E-\varepsilon_R)\, T(E) = V|\Psi_R\rangle\langle\tilde{\Psi}_R|V. \tag{15}$$

The definition of $S^{\ell}_{\beta\alpha}$ in terms of $T^{\ell}_{\beta\alpha}$ Eq. [(10)] and Eqs. (11) and (12) can be used to evaluate the singular term in $S^{\ell}_{\beta\alpha}$. Combining all other contributions to the S matrix into a background term $S^{bg,\ell}_{\beta\alpha}$, the form of $S^{\ell}_{\beta\alpha}$ near a resonance is

$$S^{\ell}_{\beta,\alpha} = S^{bg,\ell}_{\beta,a} - i\,\frac{\gamma_\beta\tilde{\gamma}^*_\alpha}{E-\varepsilon_R} \tag{16}$$

where the factors of the residue are

$$\gamma_\beta \equiv (2\pi\mu k^R_\beta)^{\frac{1}{2}} \langle\Phi_\beta X_\ell(k^R_\beta\vec{r})|V|\Psi_R\rangle \tag{17}$$

$$\tilde{\gamma}_\alpha \equiv (2\pi\mu k^R_\alpha)^{\frac{1}{2}*} \langle\Phi_\alpha X_\ell(k^R_\alpha\vec{r})|V|\tilde{\Psi}_R\rangle$$

If we now follow the usual argument for a narrow resonance,[12] which is based on the unitarity of $S^{\ell}_{\beta\alpha}$ for all real energies, we are led to the usual result for the partial widths for decay into open channels Γ_α that if Γ_α is defined by

$$\Gamma_\alpha = |\gamma_\alpha|^2 \tag{18}$$

and the total width is defined by

$$\varepsilon_R = E_R - i\Gamma/2 \tag{19}$$

where E_R is the real part of ε_R, the partial widths sum to the total width

$$\Gamma = \sum_\alpha \Gamma_\alpha \tag{20}$$

Although the identification of the total width as the sum of the partial widths ′Eq. 18) is exact only in the case of a narrow resonance, it is important to note that the resonance form expressed in Eq. (16) is exact and that an unambiguous definition of both the complex resonance energy ε_R and the residue factors is provided by the (analytically continued) matrix elements defined in Eq. (17).

It is not convenient to use the definition of γ_β in Eq. (16) to compute the partial widths, because we have not fixed the normalization of ψ_t to match that of ψ_R. However, because of the particular form of ψ_t chosen in Eq. (5) we can derive a simple prescription for the ratio of the partial widths in two channels. Recalling the definition in Eq. (9), it is easy to see that, with the form of the S matrix in Eq. (16), the coefficients C_β in the definition of ψ_t are proportional (in the limit of a complete basis) to the amplitudes γ_β according to

$$C_\beta = \frac{i\gamma_\beta}{k_\beta^{\frac{1}{2}}} \sum_\alpha a_\alpha \tilde{\gamma}_\alpha^{*} \tag{21}$$

Thus the ratio of partial widths for any pair of open channels is given by

$$\frac{\Gamma_\beta}{\Gamma_\alpha} = \left| \frac{k_\beta}{k_\alpha} \; \frac{C_\beta}{C_\alpha} \right|^2 \tag{22}$$

Since the partial widths must sum to the total width, and the Siegert eigenvalue computed together with ψ_t yields the total width, we can use Eqs. 19) and (22) together to determine all of the partial widths for decay into open channels.

The formalism outlined here was applied to a model three-channel square-well problem. The computational details can be found in reference 3. The results of this study confirmed the conclusions reached above and showed that the use of complex "Siegert" functions combined with purely L^2 functions provides a convenient method for obtaining quantitative information about partial as well as total resonance widths. The procedure we have outlined above is applicable to any direct method which provides an accurate representation of the eigenfunction associated with the resonance eigenvalue. However, if one were to employ a trial function without reference to a particular asymptotic form, as in the complex coordinate method for example, one would also have to project onto the resonance eigenfunction with the unperturbed states $\phi_\alpha(x)X_\ell(k_\alpha r)$, evaluated at complex coordinates, to extract the coefficients which would play the same role as C in Eq. (21). We refer the reader to the recent work of Noro and Taylor for further details.[13]

III. COMPLEX-COORDINATE PROCEDURES FOR MOLECULES AND MOLECULAR PHOTOIONIZATION

Two computational techniques have been proposed to date for extending complex scaling techniques to molecular resonance problems. In an earlier study, McCurdy and Rescigno[6] proposed the use of ordinary floating Gaussian basis functions which are made complex by simply scaling the orbital exponents by a phase factor. It was argued that such functions, when used to form a matrix representation of the Born-Oppenheimer Hamiltonian with a complex-valued scalar product, would effectively provide an asymptotic scaling of the electronic coordinates and that such a scaling would be sufficient to render a resonance eigenfunction L^2; this supposition was supported by several illustrative calculations. More recently, Moiseyev and Corcoran[7] have described a procedure which superficially appears to be identical to the transformation $r \rightarrow \theta r = \lambda r e^{i\phi}$ that can be applied to the electronic coordinates of a dilatation analytic Hamiltonian. This simple analogy is complicated, however, by the fact that the branch-point singularities in the electron-nuclear attraction terms of the Born-Oppenheimer Hamiltonian render it a nonanalytic function of electron coordinates. In addition to these numerical studies, the formal work of Simon[14] establishes that complex scaling in the Born-Oppenheimer picture can be put on firm mathematical ground through the use of what Simon calls the method of "exterior complex scaling" in which the magnitudes of all electronic coordinates are only scaled outside a sphere which is large enough to enclose all the nuclei.

A recent development in this area was the observation by McCurdy[4] that the Moiseyev-Corcoran procedure[7] could be formally related to Simon's exterior complex scaling. This observation also led to a procedure for calculating the matrix elements of the resolvent needed to evaluate photoionization cross sections.[4,] I will not repeat the rather lengthy arguments needed to establish the formal connections between the various methods. These can be found in references 4 and 5. I will content myself here with a brief comparison of the McCurdy-Rescigno (MR)[6] and Moiseyev-Corcoran (MC)[7] procedures and their use in photoionization calculations.

The MR procedure employs basis functions which effectively scale $\vec{r}$ by θ asymptotically while avoiding the nonanalyticity problem at the nuclear centers. This is accomplished by forming a matrix representation of the unscaled Born-Oppenheimer Hamiltonian in a basis set of complex Cartesian Gaussians of the form

$$\psi^{\theta}_{lmn}(\alpha, \vec{r}, \vec{A}) = N_{lmn}(x - A_x)^l (y - A_y)^m (z - A_z)^n \, e^{-\alpha\theta^{-2}(\vec{r} - \vec{A})^2} \tag{23}$$

We have aruged that using such a basis is equivalent to using rotated coordinates in the Hamiltonian asympototically.

It is instructive to compare the behavior of the analytically continued matrix elements in the MR and MC procedures, particularly the matrix elements of the nuclear attraction potential. We will consider a matrix element of one term of the nuclear potential between two S-type Gaussians [l=m=n=0 in Eq. (23)] both centered at position $\vec{A}$:

$$I = \left(\frac{2\alpha}{\pi}\right)^{3/4} \left(\frac{2\beta}{\pi}\right)^{3/4} \int d^3r \, e^{-\alpha(\vec{r} - \vec{A})^2} \frac{1}{\sqrt{(\vec{r} - \vec{R})^2}} e^{-\beta(\vec{r} - \vec{A})^2} \tag{24}$$

$$= \left(\frac{32}{\pi}\right)^{1/2} \frac{(\alpha\beta)^{3/4}}{(\alpha + \beta)} F_0((\alpha + \beta)(\vec{R} - \vec{A})^2)$$

where $F_0(z)$ is the entire function of z,

$$F_0(z) = 1/2 \sqrt{\frac{\pi}{z}} \; \mathrm{erf}\,(\sqrt{z}) \, .$$

The MC procedure replaces $(\vec{r} - \vec{R})^2$ with $(\vec{r}\theta - \vec{R})^2$ in Eq. (24) and obtains a formula for I by factoring θ^{-1} out of the integrand. Evaluating the result at complex θ gives

$$I^{MC} = \frac{e^{-i\phi}}{\lambda} \left(\frac{32}{\pi}\right)^{1/2} \frac{(\alpha\beta)^{3/4}}{(\alpha + \beta)} F_0\left[(\alpha + \beta)\left(\frac{\vec{R}e^{-i\phi}}{\lambda} - \vec{A}\right)^2\right] . \tag{25}$$

In the M procedure, we simply multiply α and β in Eq. (24) by θ^{-2}, giving

$$I^{MR} = \frac{e^{-i\phi}}{\lambda} \left(\frac{32}{\pi}\right)^{1/2} \frac{(\alpha\beta)^{3/4}}{(\alpha + \beta)} F_0\left[\frac{(\alpha + \beta)e^{-2i\phi}}{\lambda^2}(\vec{R} - \vec{A})^2\right] \tag{26}$$

It is now clear from comparing Eqs. (25 and (26) that the MR procedure can be related to the MC procedure; if the basis functions are simply shifted from center $\vec{A}$ to $\vec{A}\theta^{-1}$ in the latter, it reduces to the McCurdy-Rescigno prescription.

Note that as the orbital exponents α and β become small, I^{MC} and I^{MR} limit to the same numerical value since $\lim_{z\to 0}F_0(z)\to 1$. Furthermore, since the asymptotic behavior of the eigenfunctions is determined by the most diffuse functions, it is clear that both procedures should yield the same spectrum in the limit of a complete expansion. However, since the behavior of the matrix elements for large values of α and β is quantitatively very different in the two procedures, one may expect to find significant differences between the two methods in numerical applications. In fact, we expect to see substantial numerical differences between the two approaches for the following reason. In a complex-coordinate calculation, the cusps that appear in the wavefunction because of the nuclear sigularties in the Born-Oppenheimer Hamiltonian are moved to complex centers. Recalling that the M procedure can be derived from the MC prescription by translating the basis functions to complex centers $\theta^{-1}\vec{A}$, we see immediately that the MR method allows the complex cusps in the molecular wave functions to be approximated by Gaussians of large exponents on the cusp centers, while the MC method effectively centers basis functions elsewhere. For this reason we expect the MR procedure to have better convergence properties.

We have applied both the MR and MC procedures to the calculation of the photoionization cross section of H_2^+. We make use of the fact that the cross section $\sigma(\omega)$ can be expressed as a matrix element of the resolvent:[15]

$$\sigma(\omega) \underset{\varepsilon\to 0}{\overset{\lim}{=}} \frac{4\pi\omega}{c} \, \mathrm{Im} < \psi_0|\mu \frac{1}{H-E_0-\omega-i\varepsilon} \mu|\psi_0 > \tag{27}$$

where ψ_0 is the wave function for the target in its initial state with energy E_0, μ is the dipole operator, and ω is the photon frequency. Under complex-scaling, the continuous spectrum of the atomic Hamiltonian is rotated off the real axis. This makes it possible to obtain convergent approximations to Eq. (27) by inverting a finite matrix representation of the scaled Hamiltonian obtained over a set of normalizable functions, whereas such a representation could not be used directly at real energies in the continuum for the unscaled Hamiltonian.[16]

An approximation to Eq. (27) is obtained as (see references 4 and 5 for details):

$$\sigma(\omega) = 4\pi\omega/c \ \mathrm{Im}\ \theta^3 \ \vec{f}\,\underline{(E_0 + \omega - H)}_\theta^{-1} \ \vec{f}, \tag{28}$$

where the inverse $\underline{(E_0 + \omega - H)}_\theta^{-1}$ is simply formed from the analytically continued matrix elements obtained either from the MR or MC procedures and the elements of $\vec{f}$ are given by

$$f_\alpha = \theta^{-3} \sum_\beta d_\beta \int d^3r \chi_\beta(\vec{r})\mu(\vec{r}\theta)\chi_\alpha(\vec{r}) \tag{29}$$

The expansion coefficients d_β in Eq. (29) make up the eigenvector corresponding to ψ_0 and are determined by diagonalizing $H_{\alpha,\beta}(\theta)$ over basis functions $\{\chi_\alpha\}$ of the initial-state symmetry. The matrix inverse $\underline{(E_0 + \omega - H)}_\theta^{-1}$ is then constructed over a set of functions of opposite parity, $\{\chi_\alpha\}$.

To facilitate our comparisons, we have used large basis sets both to calculate the $1s\sigma_g$ ground-state eigenfunction and to represent the matrix $\underline{(E_0 + \omega - H)}_\theta$ for the σ_u continuum. In this way, we are able to use the same basis sets for both methods while avoiding the criticism that the basis set was optimum for one method but not for the other. After choosing the basis-set exponents, one may optimize the scaling parameter θ in a given basis set because the results are formally independent of θ when converged. The usual procedure in atomic calculations is to fix the magnitude of θ at unity and vary the argument to find the region of greatest stability. We did that using the MR prescription for computing $H_{\alpha,\beta}(\theta)$, making no attempt to optimize the magnitude of θ. Figure 1 shows a superposition of five plots of the photo-ionization cross sections obtained by setting $\theta = \exp(i\pi\phi/180.0)$ and varying ϕ in 2.5° increments over a 10° interval centered approximately at the most stable point in ϕ. The results are stable within a few percent and agree essentially exactly with the exact values of Bates and Öpik.[17]

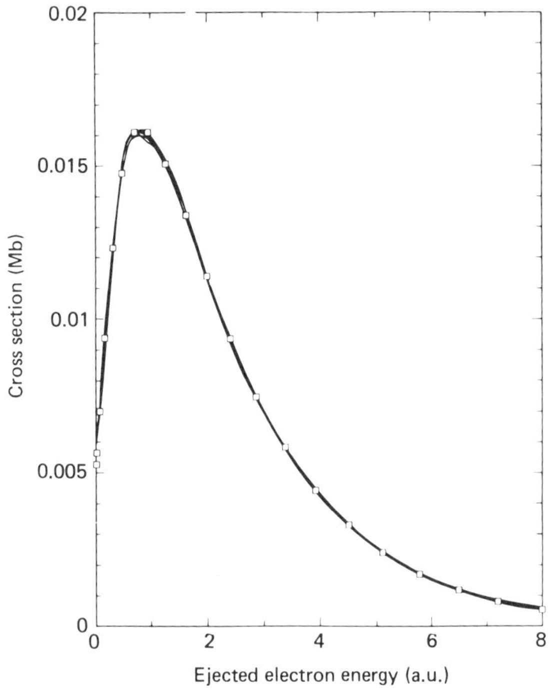

Fig. 1. Superposition of results for the parallel component of the photoionization cross section of H_2^+ computed using the method of Ref. 6 to analytically continue the Hamiltonian matrix elements. The superimposed curves are for ϕ varying over the most stable region (15° to 25°) in 2.5° increments.

We repeated this calculation with the procedure of Moiseyev and Corcoran. However, no recognizable region of stability could be found with $|\phi| = 1$, and we therefore varied $|\theta|$ to find a more optimum value. It is not surprising that this was necessary considering the behavior of nuclear attraction integrals with large exponents which Moiseyev and Corcoran point out and which we discussed earlier. We find the most favorable value of $|\theta|$ to be approximately 0.6, with little sensitivity for values varying between 0.5 and 0.8. Figure 2 shows a superposition of five plots for the photoionization cross sections obtained with $\theta = 0.6 \exp(i\phi/180.0)$ for values of ϕ in 2.5° increments over a 10 interval, again approximately centered in the region of greatest stability. Comparison of Fig. 1 and 2 shows that the Moiseyev-Corcoran technique is substantially less stable in this application than that of McCurdy and Rescigno.

have shown that accurate results can only be achieved when the orbitals referring to tightly bound core electrons are left unscaled and the outer valence electron orbitals made complex.[18] We have adopted a similar procedure for molecules, which I will illustrate in the next section. With this mix of real and complex orbitals, the two-electron integrals no longer scale simply and the computational details involved in implementing either method are comparable.

IV. COMPLEX SCF TECHNIQUES

A. Formulation

Almost all of the previous calculations using complex basis functions have relied on direct diagonalization of the Hamiltonian with many-electron basis states. In conventional structure calculations on bound states, a Hartree-Fock wavefunction is frequently generated as a starting point for more accurate treatments. It is logical to assume that the basic mathematics of SCF theory should also be applicable to the square-integrable resonance eigenstates of an analytically continued complex Hamiltonian. We have recently shown that, for the case of shape resonances in atoms and molecules, this is indeed the case.

The basic idea[19, 20] underlying our discussion of the SCF equations for resonances is that by beginning with the complex atomic Hamiltonian, $H(\{\theta r_j\})$, and following (essentially) the same variational arguments used to derive the bound state SCF equations, we can derive complex SCF equations appropriate for resonance states. This procedure is successful because the resonance eigenfunctions of the complex Hamiltonian are square-integrable. In that sense these eigenfunctions are sufficiently like bound state wavefunctions to allow treatment by the SCF methods currently employed in many bound state calculations.

We will specialize our discussion to the simple case of a shape resonance which corresponds to an electronic configuration with one electron outside a closed shell. The ^2P shape resonance encountered in e^- - beryllium scattering is an example of such a case; the $^2\Pi_g$ resonance state of N_2^- is a well known molecular example.

Starting with a trial variational wavefunction which is a single Slater determinant of spin orbitals

$$\Phi = \det|\psi_1\bar{\psi}_1\psi_2\bar{\psi}_2 \cdots \psi_k\bar{\psi}_k\psi_\mu| \cdot \tag{30}$$

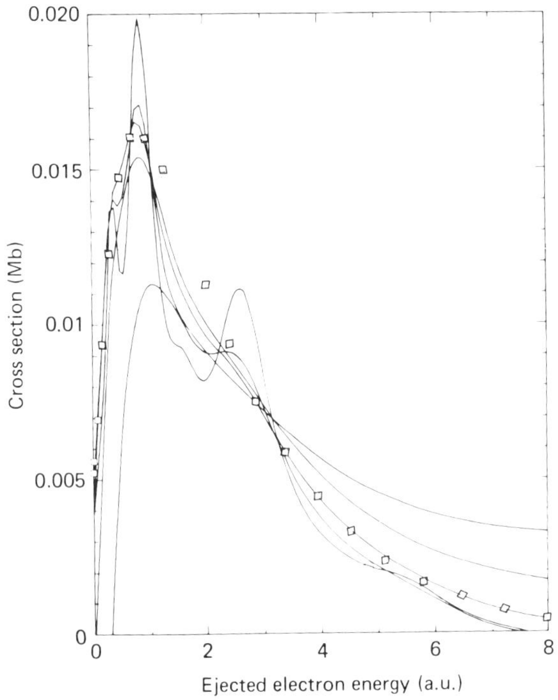

FIG. 2. As in Fig. 1, but using the method of Ref. 7 to analytically continue the Hamiltonian matrix elements. The magnitude of the scaling parameter is 0.6 and ϕ varies from 5° to 15° in 2.5° increments.

In comparing the two methods, we have concentrated on the behavior of the matrix elements of the nuclear attraction terms, since their non-analytic behavior has hindered the application of complex scaling techniques to molecular probelms. Moiseyev and Corcoran have also emphasized that the computation of two-electron matrix elements, which is the most time-consuming part of integral evaluation, is simpler with their procedure since the integrals are simply scaled by constants and need not be recomputed when the scale parameters are changed. It is noteworthy, however, that several applications of complex scaling techniques to atomic problems involving more than two electrons where each spin orbital, ψ_i, is the product of a spatial orbital ϕ_i (depending on the coordinates of one electron) and a spin function, and $\bar{\psi}_i$ denotes the spin orbital with opposite spin, we derive the SCF equations for the orbitals, ϕ_i, by setting to zero the first variations with respect to those orbitals of a particular functional. For the ordinary bound state problem that functional consists of the expectation value of the Hamiltonian with respect to Φ plus Lagrange multiplier terms which serve to apply the orthonormality constraint among the orbitals, ϕ_i. In our case we must generalize this expression slightly. Since the Hamiltonian, $H(\{\theta r_j\})$, is not hermitian, its eigenfunctions form a biorthogonal set , and the complex conjugate of the wavefunction which appears in the usual expectation value is not appropriate. Thus the functional we have chosen is

$$\begin{aligned} I = \int \Phi\, H(\{\vec{r}_j\theta\})\Phi d\tau_1 d\tau_2 \ldots d\tau_{2k+1} \\ - \sum_{i,j=1}^{k} \lambda_{ij} \int \phi_i(\vec{r})\phi_j(\vec{r}) d^3r \\ - 2\sum_{i=1}^{k} \lambda_{i\mu} \int \phi_i(\vec{r})\phi_\mu(\vec{r}) d^3r \\ - \varepsilon_\mu \int \phi_\mu(\vec{r})\phi_\mu(\vec{r}) d^3r \end{aligned} \tag{31}$$

where the matrix, λ_{ij}, of Lagrange multipliers is complex symmetric and the redundant terms in the first sum are for convenience only. As the scale factor, θ, approaches unity the SCF equations must approach the usual Hartree-Frock equations in which the orbital Hamiltonians are hermitian. The equations we drive using Eq. (31) have this property only if the orbitals, ϕ_i, in the trial wavefunction are real when $\theta=1$, because otherwise the Coulomb and exchange operators as defined below are nonhermitian. In practice, this simply means that we are constructing the correct analytic continuation of the working equations of the ordinary real valued SCF calculations of quantum chemistry. Note that in defining the functional in equation (31), we have in mind a class of problems which, in the $\theta\to 1$ limit, can be formulated in terms of real orbitals. If, for example, the orbitals of Eq. (31) were expressed as products of radial functions and spherical harmonics, the functional we would then use would have the complex conjugate of the angular variables of the left hand orbitals appearing in all the scalar products, but not the radial variables.

By setting the functional derivatives, $\delta I/\delta\phi_i$ and $\delta I/\delta\phi_u$, of equation (31) with respect to the open and closed shell orbitals to zero, we obtain the SCF equations. If we define the Fock operators by

$$\hat{F}_o = h_\theta + \theta^{-1} \sum_{j=1}^{k} (2\hat{J}_j - \hat{K}_j) + \theta^{-1}(\hat{J}_\mu - \tfrac{1}{2}\hat{K}_\mu)$$

$$\hat{F}_\mu = h_\theta + \theta^{-1} \sum_{j=1}^{k} (2\hat{J}_j - \hat{K}_j) \tag{32}$$

where the one electron Hamiltonian, h_∂, is (atomic units)

$$h_\theta = -\frac{\theta^{-2}}{2}\nabla^2 - \frac{Z\theta^{-1}}{r} \tag{33}$$

and the Coulomb and exchange operators are

$$\hat{J}_j = \int \frac{\phi_j(\vec{r}_2)\phi_j(\vec{r}_2)}{r_{12}} d^3r_2 \tag{34}$$

$$\hat{K}_j\phi_i = \phi_j(r_1)\int \frac{\phi_j(\vec{r}_2)\phi_i(\vec{r}_2)}{r_{12}} d^3r_2$$

the complex SCF equations can be written

$$2\hat{F}_o\phi_i = \sum_{j=1}^{k} \lambda_{ji}\phi_j + \lambda_{\mu i}\phi_\mu \tag{35}$$

$$\hat{F}_\mu\phi_\mu = \sum_{j=1}^{k} \lambda_{\mu j}\phi_j + \varepsilon_\mu\phi_\mu$$

$$\lambda_{ij} = \lambda_{ji}$$

The usual SCF procedure is to find a set of equivalent matrix equations by multiplying equation (35) by the orbitals and integrating. All the matrices appearing in these calculations are complex symmetric. Finally, we note that the value of the complex resonance energy from this procedure is the complex SCF energy, E_{CSCF}

$$
\begin{aligned}
E_{CSCF} &= \int \Phi H\,(\{\vec{r}_j\theta\})\Phi d\tau_1 \ldots d\tau_{2k+1} \\
&= \sum_{i=1}^{k} \int \phi_i (h_\theta + \hat{F}_o)\phi_i d^3r \\
&+ \tfrac{1}{2} \int \phi_\mu (h_\theta + \hat{F}_\mu)\phi_\mu d^3r
\end{aligned}
\qquad (36)
$$

B. Atomic Example

The ^2P shape resonance in e-beryllium scattering is an example of a resonance which the complex SCF approach should describe well. We have performed two sets of calculations on this system,[19] the first of which employs a (14s/16p) basis of real valued Gaussians chosen as follows. The complex SCF equations were solved in matrix form and E_{CSCF} from this calculation with $\theta = \exp(i\alpha)$ is plotted in Figure 3 for a range of α values between .3 and .5 radians. We were unable to converge the SCF equations for values of α less than .3. The trajectory of E_{CSCF} as a function of α is somewhat surprising in view of what has been observed in complex coordinate CI calculations.

In a configuration interaction calculation using the complex Hamiltonian, $H(\{\theta r_j\})$, there is a well known behavior to be expected of the complex resonance eigenvalue as a function of θ. That behavior, which can be shown to be a consequence of the complex version of the virial theorem,[21] is that for a given basis set the resonance eigenvalue, E_{res}, has a stationary point with respect to variation of θ so that, at some value of θ, $dE_{res}/d\theta$ vanishes. The stationary point does not necessarily occur with $|\theta| = 1$, but if a CI calculation is performed with $\theta = \exp(i\alpha)$ a sharp cusp in the complex value of E_{res} is often found as α is varied. The curve in Figure 3 corresponding to the (14s/16p) real basis function calculation does not show such a cusp, nor is E_{CSCF} particularly stable as α is varied. This is a troublesome point because the stationary value of E_{res} is usually taken to be the best approximation to the resonance energy.

The main curve in Figure 3 shows little evidence of stationarity with respect to variations in α. This problem is due entirely to an inadequate description of the orbitals which make up the Be core, particularly the tight 1s orbital. The problem has been discussed in detail elsewhere[18] and can be solved by using Gaussian basis functions with complex exponents given by $\theta\xi_i$ where ξ_i is real to expand the core orbitals (1s and 2s in this case). Thus we performed a second set of

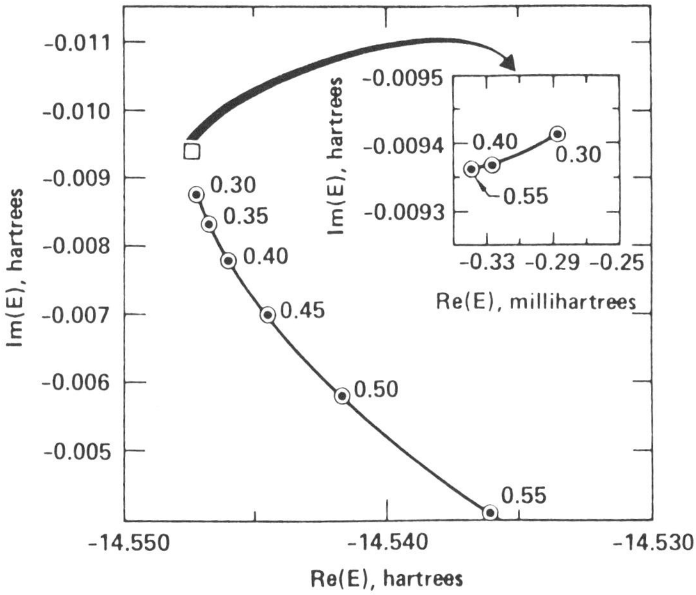

Figure 3. E_{CSCF} for Be^- as a function of α in the scaling parameter $\theta = \exp(i\alpha)$. The main cirve is calculated with real basis functions, and the insert is the complex basis function result for the same range of α. The real part of the energy scale in the insert is relative to -14.547 Hartrees.

calculations with only the s-functions scaled by θ. The results of this calculation are plotted in the inset to Figure 3. The stability of E_{CSCF} as a function of α is remarkable in this calculation, but still no sign of clear cusp behavior was observed. Fortunately E_{CSCF} is so stable that it is unnecessary to find a stationary point of E_{CSCF} in order to find the resonance energy. Choosing a value of α near the center of the range plotted in Figure 3 (α=0.4) we find E_{CSCF} = 14.54733 - .00937i. The value for the width of the resonance is therefore 0.51 eV or about half the value from a static exchange calculation.[18] To get an estimate of the position of the resonance it seems most reasonable to subtract the Hartree-Fock value of the ground state energy of Be from E_{CSCF} for Be^-, because one of the solutions of the complex SCF equations (a nonresonant, continuum solution) is the Hartree-Fock wavefunction for Be with the remaining electron in a continuum p orbital with (in a limiting sense) zero energy. The resonance position so obtained is 0.70 eV, only slightly lower that the static exchange value.[18]

There is a practical note which concerns the use of complex basis functions. In an earlier publication[18] we made use of the identity, written here for the radial part of a one electron problem,

$$\int \phi_i(r)H(r\theta)\phi_j(r)r^2dr = \theta^{-3}\int \phi_i(r\theta^{-1})H(r)\phi_j(r\theta^{-1})r^2dr \tag{37}$$

to perform a rotated coordinate calculation by diagonalizing the real Hamiltonian, $H(r)$, using complex basis functions, $\phi_i(r\theta^{-1})$. It is in fact more convenient to use the same device in complex SCF calculations. Thus the second of our calculations[19] on Be^- was actually performed with unscaled Fock operators and complex p basis functions of the form $\phi_i(r\theta^{-1})$ and real s basis functions, instead of scaled Fock operators with real p and complex s basis functions as described above. These two forms of the calculation are completely equivalent, but the prescription of unscaled Fock operators plus complex basis functions requires fewer complex two electron integrals and, more importantly, is a more convenient starting point for the extension to molecular problems.

B. Molecular Example

The $^2\Pi_g$ resonance state of N_2^- has been well characterized experimentally through extensive studies of vibrational excitation by electrons between 1 and 5 eV[22] and has also been the subject of numerous theoretical treatments.[23-27] We felt that this system would provide a good indication of whether the "method of complex basis functions" is an accurate and practical way to proceed with complex SCF studies on molecules.

The open-shell SCF equations for $N_2^-(^2\Pi_g)$, which has the electronic configuration $(1\sigma_g^2\ 1\sigma_u^2\ 2\sigma_g^2\ 2\sigma_u^2\ 3\sigma_g^2\ 1\pi_u^4\ 1\pi_g)\ ^2\Pi_g$, were solved in a mixed basis set of real and complex Cartesian Gaussian functions.[20] For these calculations, we use a real valued Hamiltonian again and Gaussian basis functions in which the exponents are scaled by θ^{-2}. The non-resonant core orbitals (all but $1\pi_g$) were again expanded solely in terms of real functions, for the same physical reasons which are discussed at length in reference 18. We also carried out several numerical tests to check the adequacy of the core-orbital basis. These are be discussed below.

The $1\pi_g$ orbital was expanded in terms of both real and complex Gaussians. The latter were chosen following the prescription of McCurdy and Rescigno (reference 6) by complex-scaling the orbital exponents of the Gaussian basis functions and keeping the associated nuclear centers real.

The core orbitals were expanded in a nuclear-centered (9s 5p)/[5s 3p] contracted basis, augmented with two d-polarization functions (α = 1.0 and .4). Several choices for the π_g orbital space were tested. A prelimenary set of calculations was done using a (4p) basis, augmented with additional diffuse d_π functions placed at the center-of-mass. All π_g functions were complex scaled. These calculations produced a complex total energy for $N_2^-(^2\Pi_g)$ which depended strongly on and varied monotonically with the rotation angle ϕ, the resonance width varying by roughly 40% over the range $15° \leq \phi \leq 25°$. A simple test was devised to demonstrate that this instability was due to the inadequacy of the π_g orbital basis and not the core orbital basis. The total SCF energy for $N_2^-(^2\Pi_g)$ can be written as $E = E_{core} + \varepsilon_{\Pi_g}$ where E_{core} is the core-orbital contribution to the total energy and ε_{Π_g} is the orbital energy of $1\pi_g$. A series of "static-exchange" calculations was performed at different angles with the core orbitals frozen. It was found that changes in ε_{Π_g} were precisely equal to the variations noted earlier in the SCF calculations.

The addition of several <u>real</u> p_π functions to the π_g orbital space greatly reduced the sensitivity of the energy to the rotation angle. The final π_g basis we used consisted of the (5p 2d)/[3p 2d] set of real functions, four nuclear-centered p_π functions with exponents .6, .26, .125, and .05 and six complex d_π functions placed at the center-of-mass with exponents ranging from .2 to .002 in a geometric series. The total energy as a function of rotation is shown in Figure 4 for an internuclear separation of 2.068 a.u., the equilibrium bond distance of N_2. The resonance width is found to vary by roughly % within a range of angles between $28° \leq \theta \leq 38°$. An average of the data over this range gives a resonance lifetime of .44 eV and an energy of 3.19 eV, when referenced to the SCF energy of ground state N_2 in the same basis. Table I compares these results to those of other recent theoretical calculations.

These preliminary results[20] are an encouraging indication that the complex SCF method, with properly chosen basis functions, can provide useful information about the lifetimes of certain types of molecular metastables and should provide a convenient starting point for further complex configuration-interaction studies.

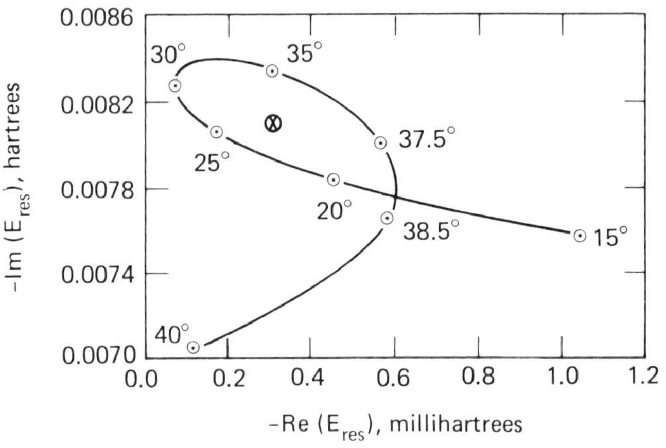

Figure 4. SCF Resonance energy of $N_2^-(^2\Pi_g)$ as a function of rotation angle. The real part of the energy scale is given relative to -108.8494 Hartrees. The internuclear separation is 2.068 a.u.

Table I. Comparison of electron resonance parameters for $N_2^-(^2\Pi_g)$. Energies and widths are in electron volts and R_o = 2.068 a.u.

	$\varepsilon_r(R_o)$	$\Gamma(R_o)$
Complex SCF[20]	3.19	.44
Krauss and Mies[24] (stabilization)	3.26	.8 .3
Schneider et al[25] (R-matrix)	2.15[a]	.34
Hazi et al[26]	3.23 (2.16[a])	.42
Levin and McKoy[27]	2.19[a]	.36

[a] Resonance energy relative to the energy of a fictitous N_2 neutral core made up with N_2^- orbitals.

It is worth noting that the simple SCF approach outlined above is not applicable in the case of a Feshbach resonance (for example, He (2s 2p) ^3P) because the complex SCF equations so derived have solutions which yield a real value for E_{CSCF} and consequently give no information about the resonance width. The decay of a Feshbach resonance is a correlation effect which cannot be described by a simple SCF treatment. MCSCF theory might offer an effective way to treat these cases.

V. REFERENCES

1. C. W. McCurdy in Electron-Molecule and Photon-Molecule Collisions, edited by T. Rescigno, V. McKoy and B. Schneider (Plenum, New York 1979).
2. Int. J. Quan. Chem 14, (4), 1978.
3. C. W. McCurdy and T. N. Rescigno, Phys. Rev. A 20, 2346 (1979).
4. C. W. McCurdy, Phys. Rev. A 21, 464 (1980).
5. C. W. McCurdy and T. N. Rescigno, Phys. Rev. A 21, 1499 (1980).
6. C. W. McCurdy and T. N. Rescigno, Phys. Rev. Letts. 41, 1364 (1978).
7. H. Moiseyev and C. T. Corcoran, Phys. Rev. A 20, 814 (1979).
8. J. N. Bardsley and B. R. Junker, J. Phys. B 5, L178 (1972).
9. A. D. Isaacson, C. W. McCurdy and W. H. Miller, Chem. Phys. 34, 311 (1978).
10. A. U. Hazi, Chem. Phys. Letts. 20, 251 ′1973).
11. R. M. More and E. Gerjuoy, Phys. Rev. A 7, 1288 (1973).
12. J. Taylor, Scattering Theory, (Wiley, New York, 1972), pp.411-413.
13. T. Noro and H. S. Taylor, J. Phys. B 13, L377 (1980).
14. B. Simon, Phys. Letts. 71A, 211 (1979).
15. P. W. Langhoff, S. T. Epstein and M. Karplus, Rev. Mod. Phys. 44, 602 (1972).
16. T. N. Rescigno and V. McKoy, Phys. Rev. A 12, 522 (1975).
17. D. R. Bates and I. Opik, J. Phys. B 1, 543 (1968).
18. T. N. Rescigno, C. W. McCurdy and A. E. Orel, Phys. Rev. A 17, 1931 (1978).
19. C. W. McCurdy, T. N. Rescigno, E. R. Davidson and J. G. Lauderdale, J. Chem. Phys. 73, 3268 (1980).
20. T. N. Rescigno, A. E. Orel and C. W. McCurdy, J. Chem. Phys. 73, (1980).
21. P. Froelich and E. Brandas, Phys. Rev. A 16, 2207 (1977); R. Yaris and P. Winkler, J. Phys. B 11, 1475 (1978).
22. G. J. Schultz in Principles of Laser Plasmas, edited by G. Bekefi (Wiley, New York 1976).

23. A. Temkin in Electron-Molecule and Photon-Molecule Collisions, edited by T. Rescigno, V. McKoy and B. Schneider (Plenum, New York 1979).
24. M. Krauss and F. H. Mies, Phys. Rev. A 1, 1592 (1970).
25. B. Schneider, M. LeDourneuf and Vo Ky Lan, Phys. Rev. Letts 43, 1926 (1979).
26. A. U. Hazi, T. N. Rescigno, and M. Kurila, Phys. Rev. A (submitted).
27. D. A. Levin and V. McKoy, Phys. Rev. A (submitted).

"Work performed under the auspices of the U.S. Department of Energy by the Lawrence Livermore Laboratory under contract number W-7405-ENG-48."

ALGEBRAIC VARIATIONAL METHOD CLOSE-COUPLING MODELS FOR ELECTRON EXCITATION OF ONE ELECTRON TARGETS

M. R. C. McDowell

Mathematics Dept., Royal Holloway College
(University of London)
Egham Hill, Egham, Surrey, England

ABSTRACT

Recent work by Morgan on a multichannel algebraic variational close coupling code is summarised and applications to excitation of He^+ and H are discussed. Particular attention is paid to numerical convergence of the method, in the energy range up to the n = 4 threshold where bound theorems apply. For He^+ comparison is made with experiment and with other theoretical results.

INTRODUCTION

This paper will summarise some recent advances in techniques for calculating electron impact excitation cross sections. I will concentrate on work done at Royal Holloway College. The main topic will be L.A. Morgan's development of Algebraic variational methods for close coupling by improving the treatment of the asymptotic region, and applications of her approach to scattering by one-electron targets.

THE ASYMPTOTIC ALGEBRAIC VARIATIONAL METHOD

We shall adopt the notation of Callaway (1978). Details of the analysis are given elsewhere (Morgan, 198o): for simplicity we here consider a two-electron system.

For scattering of an electron of energy $k^2_{\Gamma_a}$ rydbergs, the system being in channel Γ_a initially, the total wave function in that channel may be written such that

$$\lim_{r_2\to\infty} \Psi_{\Gamma_a}(\underline{x}_1,\underline{x}_2) = \sum_j G_{ja}(r_2)\tilde{\Psi}_{\Gamma_j}(\underline{x}_1,\underline{x}_2,\bar{r}_2) \tag{1}$$

where G_{ja} is a real function and $\tilde{\Psi}_{\Gamma_j}$ is independent of r_2, the sum being over all energetically accessible channels. The adoption of standing wave boundary conditions implies

$$r_2 G_{ja}(0) = 0$$

and

$$r_2 G_{ja}(r_2) \underset{r_2\to\infty}{\sim} k_j^{-1/2}[\delta_{ja} \sin H_j + K_{ja} \cos H_j] \qquad k_j^2 > 0$$

$$\underset{r_2\to\infty}{\sim} e^{-|k_j|r_2} \qquad k_j^2 < 0 \tag{2}$$

with

$$H_j = k_j r - 1/2\ell_j\pi + \frac{z}{k_j} \ell n z k_j r + \chi_j. \tag{3}$$

We are interested in the radial scattering function $G_{ja}(r)$ for the solution in channel j given an incident wave in channel a only. We denote the residual charge on the target by z (=0) for a neutral atom) so that

$$\chi_j = \arg \Gamma(\ell_j + 1 - \frac{iz}{k_j}) \tag{4}$$

is the Coulomb phase.

Then the essential scattering information is contained in the reactance matrix

$$\underline{K} = [K_{ja}] \tag{5}$$

which for real potentials is real symmetric.

The exact solution satisfies

$$\int \Psi^*_{\Gamma_j} (H - E) \Psi_{\Gamma_i} d\tau = 0, \qquad \forall\ i,j. \tag{6}$$

The solution is unknown: we proceed by constructing a trial solution, including a finite number (n_c) of channels, satisfying the correct boundary conditions in those channels, but with approximate K_{ja}. Then (6) reduces to, for systems with L-S coupling

$$I_{\alpha\alpha'}(LM_L SM_S) = \int \Psi_\alpha^* (H - E)\Psi_{\alpha'} d\tau = 0, \ \forall\alpha,\alpha' \tag{7}$$

where $\Gamma_i = \{\alpha_i, LM_L SM_S \Pi\}$. Allowing small arbitrary variations in the radial functions

$$r_2 \, \delta G_{\alpha\alpha'} \underset{r_2 \to \infty}{\sim} k_\alpha^{-1/2} \, \delta K_{\alpha\alpha'} \cos H_\alpha \tag{8}$$

we obtain the multichannel version of the Kohn variational principle

$$K_{\alpha\alpha'} = K_{\alpha\alpha'}(t) - I_{\alpha\alpha'}(t) \tag{9}$$

to second order in δG, where t indicates a quantity evaluated with the trial function.

The inverse Kohn method is obtained by choosing a trial function

$$r\, G'_{ja}(r) = (K^{-1})_{ja} f_{ja}^{(1)} + \delta_{ja} f_{ja}^{(2)} + \Phi_{ja} \tag{1o}$$

where $f^{(1)}$ is sin-like, $f^{(2)}$ cos-like asymptotically and Φ_{ja} is quadratically integrable. This yields

$$\underline{K}^{-1} = \underline{K}_t^{-1} + \underline{I}(t). \tag{11}$$

More generally we can choose

$$rG_{ja}(r) \underset{r \to \infty}{=} (\underline{\alpha}_1)_{ja} f_{ja}^{(1)} + (\underline{\alpha}_2)_{ja} f_{ja}^{(2)} + \Phi_{ja} \tag{12}$$

so that

$$\underline{K} = \underline{\alpha}_2 \, \underline{\alpha}_1^{-1} \tag{13}$$

and $\underline{\alpha}_1$, $\underline{\alpha}_2$ are arbitrarily chosen square matrices.

The total scattering function in channel j may then be written

$$\Psi_{\Gamma_j} = \sum_i \tilde{\Psi}_{\Gamma_i} (\bar{r}_j) \left[\sum_{\ell,s} \alpha_{j\ell}^{(s)} f_{\ell i}^s + \sum_\mu c_{ij}^{(\mu)} \eta_{i\mu} \right] \tag{14}$$

where s = 1 for sin-like terms, s = 0 for cos-like and the $\eta_{i\mu}$ are quadratically integrable one-electron radial functions. Rearranging, and writing

$$\phi_\nu \quad \tilde{\Psi}_i\ (\bar{r}_j)\ \eta_{i\mu}\ (r_j) \tag{15}$$

where η runs over all channels i and all orbitals μ used in that channel,

$$\Psi_{\Gamma_j} = \sum_{1,s} \alpha_{j1}^{(s)} \sum_i \tilde{\Psi}_{\Gamma_i}\ f_i^{(s)} + \sum_\nu C_{j\nu}\ \phi_\nu\ . \tag{16}$$

The stationary condition

$$\frac{\partial I_{ij}}{\partial C_{k\mu}} = 0,\ \forall\ i,j,k,\mu = 1,\ \ldots,n_c \tag{17}$$

gives a set of linear algebraic equations for the $C_{k\mu}$, in terms of the elements of $\underline{\alpha}^{(1)}$, $\underline{\alpha}^{(2)}$,

$$\sum_{\ell,s} \alpha_{j\ell}^{(s)} \sum_m <\phi_\mu|H-E|\tilde{\Psi}_m\ f_{\ell m}^{(s)}>$$

$$+ \sum_\mu C_j <\phi_\mu\ |H-E|\phi_\nu> = 0 \tag{18}$$

so that on solving these formally we obtain for our trial functional

$$I_{ij} = \sum_{k,\ell=1}^{n_c} \sum_{s,t=1}^{2} \alpha_{ik}^{(s)}\ M_{k\ell}^{st}\ \alpha_{j\ell}^{(t)} \tag{19}$$

in terms of tne M-matrix. This matrix may be expressed as

$$M = FF - \tilde{BF}(BB)^{-1}\ BF \tag{2o}$$

in which the free-free matrix (FF) is

$$FF_{k\ell}^{st} = \sum_{m,n} <\tilde{\Phi}_m\ f_{km}^{(s)}|H - E|\tilde{\phi}_n\ f_{\ell n}^{(t)}> \tag{21}$$

the bound-free matrix is

$$BF_{\ell\nu}^{t} = \sum_n <\phi_\nu|H-E|\tilde{\Phi}_n\ f_{\ell n}^{(r)}> \tag{22}$$

and

$$BB_{\mu\nu} = \langle\phi_\mu|H-E|\phi_\nu\rangle \tag{23}$$

is the bound-bound matrix.

Taking variations with respect to the elements of $\alpha^{(2)}$ allows us to express the Kohn principle (cf. Callaway 1978) as

$$K_{ij} = M^{11}_{ij} + \sum_{a,b} M^{21}_{ai} (M^{-1}_{22})_{ab} M^{21}_{bj} \tag{24}$$

and similarly the inverse Kohn principle becomes

$$(K^{-1})_{ij} = M^{22}_{ij} - \sum_{a,b} M^{12}_{ai} (M^{-1}_{11})_{ab} M^{12}_{bj} \tag{25}$$

We return to these expressions later.

Morgan's method then makes a specific choice of the trial scattering functions. Since there is some radius $r = R_o$ beyond which exchange is unimportant we can solve the non-exchange close-coupling equations for the finite n_c problem, without approximation, in the outer region, $r \geq R_o$. In the inner region $(0 \leq r \leq R_o)$ we write

$$f^{(s)}_{\ell i}(r) = \sum_t a^{st}_{\ell i}\, g^t_i(r), \qquad s = o,1 \tag{26}$$

where $g^{(t)}(r)$ are asymptotically sin-like and cos-like functions, regular at the origin. In the calculations to be reported here we took

$$g^{(1)}_i(r) = (1 - e^{-\gamma r})^{\ell_i} \sin k_i r$$

$$q^{(2)}_i(r) = (1 - e^{-\gamma r})^{\ell_i+1} \cos k_i r. \tag{27}$$

The coefficients $a^{st}_{\ell i}$ are determined by requiring that the total scattering function and its derivative in each channel are continuous at $r = R_o$. This is the principal difference from the R-matrix method.

In the outer region $(r \geq R_o)$ we solve

$$\frac{d^2}{dr^2} f^{(s)}_{ja}(0) = \sum_i \bar{V}_{ji} f^{(s)}_{ia}(0) \tag{28}$$

where

$$\bar{V}_{ji} = 2\,V_{ji} - \left(k_j^2 - \frac{\ell_j(\ell_j + 1)}{r^2}\right)\delta_{ij}$$

for n_c regular and n_c irregular solutions and their derivatives at R_o, for each of n_c values of j. Then $n^2{}_c$ 2 x 2 matrix inversions give the $a^{st}_{\ell i}$,

$$f^{(s)}_{ja}(0)\Big|_{R_o} = \sum_t a^{st}_{ja} \frac{dg_a}{dr}(R_o), \quad s = 1,2$$

$$\frac{df^{(s)}_{ja}}{dr}(0)\Big|_{R_o} = \sum_t a^{st}_{ja} \frac{dg_a}{dr}(R_o), \quad s = 1,2 \tag{3o}$$

$$j = 1,\ldots,n_c; \quad a = 1,\ldots,n_c.$$

The approach implied in (28) is a generalisation of that suggested by Oberoi and Nesbet (1973). They retained only the dominant $f^{(s)}_{jj}$ (s = 1,2; j = 1,111,n_c) components of the outer solutions, and applied their procedure to a two-channel model problem.

With our specific choice of the $g_i(r)$, we can evaluate the M matrix in this basis as soon as we have chosen target functions and correlation terms ϕ_ν. Notice that the evaluation of the bound-bound matrix is independent of the choice of matching condition, and can be evaluated for all impact energies at one blow, by adding the trivial energy-dependent term E $\langle\phi_\mu\phi_\nu\rangle$ at a later stage of the calculation. If we consider the bound-free matrix element

$$BF^t_{\ell\nu} = \sum_n \sum_s a^{ts}_{\ell n} \langle\phi_\nu|H-E|\tilde{\Phi}_n\, g^s_n\rangle \tag{31}$$

we see that $\underline{K}$ is known once the $a^{ts}_{\ell n}$ are determined, the matrix elements in (31) depending only on known functions.

The second major advance made by Morgan was to notice that the radial scattering functions occurring in the bound-free and bound-bound matrix elements always occur as

$$(H-E)\sum_{n,j} \Phi_n\, f^{(s)}_{ja}$$

and over the range $R_o \leqq r < +\infty$, this expression is identically zero, when exchange is neglected. Thus there is zero contribution to the direct outer bound-free and free-free integrals with this choice of outer solution. The bound-free exchange matrix elements involve for at least one electron, say co-ordinate r_p, a product of functions

$\eta_\nu(r_p) f_{ja}^{(s)}(r_r)$ which is exponentially small for $r_p \geqq R_o$: similarly for the free-free integrals. The outer bound-bound matrix elements are in any case exponentially small, so all outer integrals may be taken as zero. The model is completed by specifying the target states and the correlation terms.

In the investigations summarised here which are restricted to one electron targets, the one-electron Hamiltonian was diagonalised in a basis of Slater orbitals

$$\eta_\nu(r) = r^q e^{-\zeta r} \qquad (32)$$

with the parameters chosen to give the n = 1,2,3 states of the target exactly, and the n = 4 threshold to a precision of 1 in 10^4.

Table 1. Pseudostate thresholds in the 18 state basis for atomic hydrogen. Rydbergs above or below the ionisation threshold; all states $n \leqq 4$ being included exactly

	s	p	d	f	g
5	-0.172	0.0274	0.0419	0.0331	0.000
6	0.196	0.902	-	-	-
7	2.040	-	-	-	-

This produces eighteen states, of which we may treat 1s,...4s; 2p,111,4p; ed 4d; 4f as exact, the others being pseudostates which model the rest of the spectrum. The parameters are given by Hata et al (198ob), and yield the thresholds given in Table 1.

Clearly it would be more desirable to choose these thresholds to represent abscissae for a correct Gaussian integration on (-0.625,00), or to adopt proper moment generating techniques.

The correlation terms are again taken to be of the form (32) above, though the parameters are unrelated to those of the basis functions. Two choices were investigated.

Set A had q = 1, and the arguments of the exponentials in geometric progression (cf. Callaway 1978, p. 141) while for Set B, we fixed the argument at $\zeta = 2.4/z^2$, and let q run from 0 to a maximum value of 13 (this maximum geing constrained by the consequent size of the bound-bound matrix).

With these choices all matrix elements can be evaluated in closed form and the calculations are relatively rapid. The outer solutions depend on application of the Burke-Sehey expansion at some $R_A >> R_o$, and use the standard Norcross package (Norcross 1969) to obtain this. As a result there are difficulties very close to thresholds, and for large orbital angular momenta. The problems for large orbital angular momentum are only severe in the neutral atom case where they are exacerbated by the very diffuse nature of the target. In dealing with excitation to the n = 3 states, with electron density behaving as exp {-2r/3} the matching radius R_o is in the range 3o to 4o a_o. The radial functions $f_{ja}^{(s)}$ are sinusoidal near this boundary, but die rapidly inside the box; effectively the sinusoidal behaviour imposed must be suppressed by the cut-off factors $(1-e^{-\gamma r})$ and the correlation terms. The range of importance is out to the classical turning point

$$r_c \simeq \ell_j(\ell_j+j)/k_j^2 \tag{33}$$

where ℓ_j lies in the range (L-3, L+3). For example for the (3d,L+3) channel at the lowest energy considered (k^2_j = 0.033) r_c is outside the box for all L. The correlation functions should be chosen to allow an adequate representation of the physics in this region (R_o, r_c).

2.1 Choice of Variational Principle

Both the Kohn and inverse Kohn methods suffer from the occurrence of anomalies. In the Kohn case these arise in (24) when det $M^{22} \to 0$, and the inverse Kohn case (25) when det $M^{11} \to 0$. Both these conditions can occur simultaneously, and indeed these determinants vanish when the total wave function is exact. As pointed out by Nesbet (1969, 198o) there are no difficulties associated with eigenvalues of the bound-bound matrix, but both M^{11} and M^{22} have at least one zero eigenvalue between every pair of eigenvalues of the bound-bound matrix. In practice it is found that the width of these anomalies decreases rapidly as the model is improved. Seaton (1966) and Nesbet (1978) suggested an alternative approach, the Restricted Interpolated Anomaly Free (RIAF) method which obviates these difficulties. The idea is to apply a unitary transformation to the wave function, i.e. to M, so as to maximise whichever of $|M^{11}|$ or $|M^{22}|$ is smaller. Writing

$$M' = U^{\dagger}MU \tag{34}$$

with dimensions $2n_c \times 2n_c$, taking

$$\underline{U} = \begin{bmatrix} \cos \underline{\Delta} - \sin \Delta \\ \sin \underline{\Delta} - \cos \Delta \end{bmatrix} \tag{35}$$

and choosing

$$\underline{c} = \cos \Delta = \delta_{pq} \cos \phi_q, \; q = 1,\ldots,n_c$$

$$\underline{s} = \sin \Delta = \delta_{pq} \sin \phi_q, \; q = 1,\ldots,n_c \tag{36}$$

we vary the n_c real parameters ϕ_q such that

$$|M^{22'}| = \max_{\underline{\Delta}} \; M^{22}(\underline{\Delta}) = \max_{\underline{\Delta}} \; [\det T_{pq}] \tag{37}$$

with

$$T_{pq} = s_p s_q M^{11}_{pq} + s_p c_q M^{12}_{pq} + c_p s_q M^{21}_{pq} + c_p c_q M^{22}_{pq} \quad . \tag{38}$$

In practice this appears to remove all anomalies.

However, if channels associated with the pseudostates are open, difficulties arise from pseudoresonances lying below each pseudostate threshold (cf. Fon, Kingston and Burke 1979). For this reason we restrict ourselves to energies below the n = 4 threshold, and can thus assess our results by using the bound theorems, (Hahn et al 1964,b; Galitis 1965) that the singlet and the triplet eigenphase sums are bounded above.

3. RESULTS

Previous work by Callaway, McDowell and Morgan on excitation of the 1s → 1s, 2s, 2p transitions in atomic hydrogen is reviewed by Callaway (1978), and the results of the new code are in agreement. The new code has so far been applied to

(i) Excitation of the 1s → 2s, 2p transitions in He^+ and other one-electron positive ions just above the n = 2 threshold (Morgan 1979).

(ii) Excitation of all transitions between the n = 1,2,3 levels of atomic hydrogen at energies from the n = 3 to the n = 4 threshold (Hata et al 198o, a,b) and further applications to excitation of the n = 1,2,3,4 levels of atomic hydrogen at 35 and 54.4 eV are in progress. The next step is to add a structure package and extend the work to many-electron targets.

3.1 Excitation of He^+

Morgan (1979) showed that three-state results (1s, 2s, 2p) for (e,He^+) agreed with earlier work of Burke et al (1964). She then added pseudostates and additional correlation terms until the singlet and triplet eigenphase sums converged. Convergence was obtained with a (6s, 5p, 2d) basis, which included the n = 1,2,3 states exactly, and reproduced the exact static dipole polarisability of the target. Increasing to a (7s, 5p, 3d, 2f, 1g) basis changed the results by less than 1%. Morgan's results for the 1s - 2s transition are shown in Fig. 1 where they are compared with the three-state calculations (loc. cit.) and a calculation by Burke and Taylor (1969) with three-states and twenty Hylleras type correlation terms. The two lowest energy experimental points of Dolder and Peart (1973) are also shown, and lie well below the theoretical values.

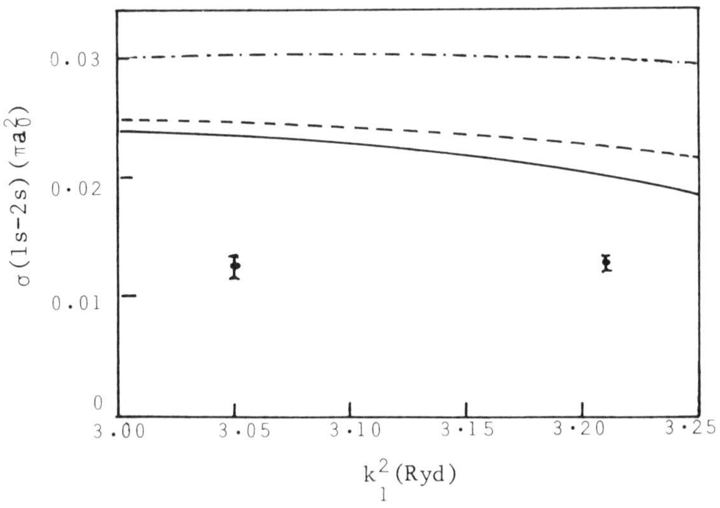

Fig. 1 Excitation of He^+ (1s - 2s). The chain curve is the three state result, the dashed curve the results of Burke and Taylor (1969) modified by Morgan, the full curve the algebraic variational method results of Morgan (see text). The circles with error bars are the two lowest energy measured values (Dolder and Peart, 1973). [From Morgan 1979 by permission of the Institute of Physics.]

It is interesting to note that a similar discrepancy exists for the resonance transition in Be^+ (2s → 2p), but not for ions of higher residual charge (Taylor et al 198o).

Very recently Wakid and Callaeay (198o) using a variant of our earlier approach with a (5s, 3p, 1d) basis with twenty correlation terms have investigated the same problem and obtained similar results to Burke and Taylor (1969) and Morgan (198o).

Wakid and Callaway (198ob) have considered the energy range from threshold to 1o2 eV, for the 1s . 2s transition only, extending earlier work by Henry and Matese (1976). They used a small basis (3s and 3p states) with up to fifteen correlation functions in each channel. For $L \geq 3$ a unitarised Coulomb Born Exchange procedure was adopted. The results decrease less rapidly above threshold than Morgan's but join smoothly to the Henry and Matese values and approach experiment from below at high energies. They suggest that the Dolder and Peart measurements may be unreliable up to 6o eV; but this suggestion is based on a calculation which cannot be tested by bound theorems.

Table 2. Electron impact excitation of the 1s → 2s and 1s → 2p transitions in He^+. (units of πa_0^2).

k_1^2	(a)	(b)	(c)
		1s - 2s	
3.08	0.0246	0.0242	0.0231
3.16	0.0239	0.0235	0.0217
3.24	0.0230	0.0219	0.0190
		1s - 2p	
3.08	0.0536	0.0578	0.0599
3.16	0.0556	0.0590	0.0588
3.24	0.0537	0.0571	0.0563

(a) Wakid and Callaway (198oa)
(b) Burke and Taylor (1969)
(c) Morgan (198o)

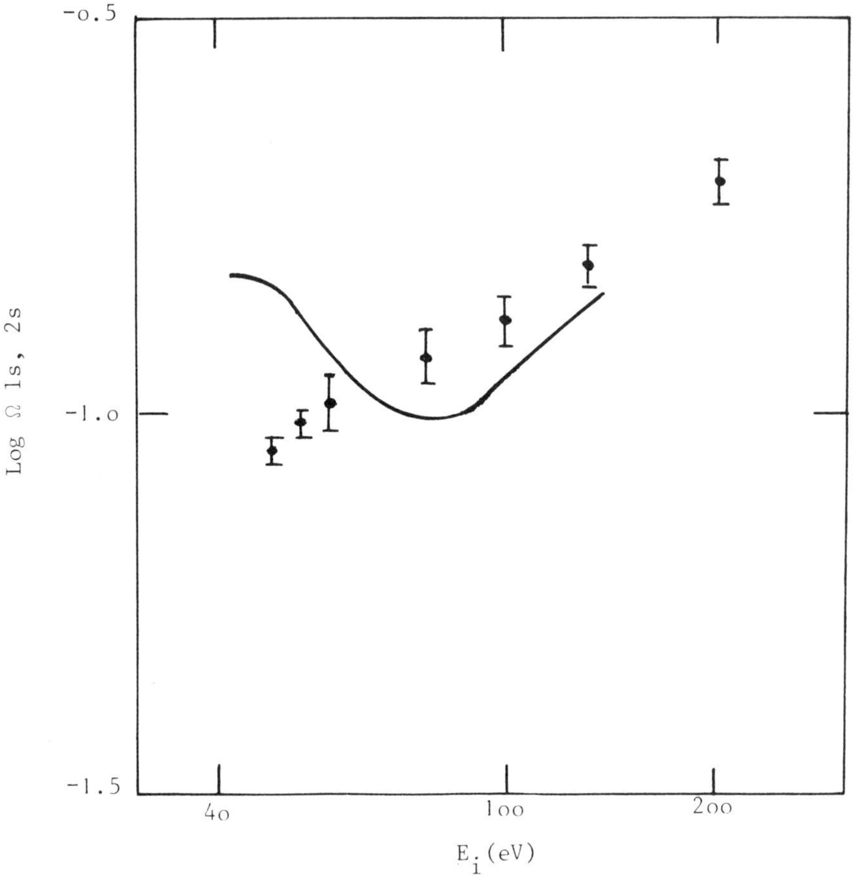

Fig. 2 The logarithm of the collision strength $\Omega_{1s,2s}$ for electron excitation of He^+. (From Wakid and Callaway 1980b).

Curve A Morgan 1979
P Wakid and Callaway 1980
Ɨ Dolder and Peart 1973

3.2 Excitation of atomic hydrogen

Hata et al (198o a,b) have used Morgan's algebraic variational method code to examine all transisions between the n = 1,2,3 levels of atomic hydrogen at energies between the n = 3 and n = 4 thresholds. Contributions from L > 6 were very small and could be neglected (see Table 3). They used an 18 state basis for L = 0,1 and reduced this to 16 states for L = 2 and 14 states for L $\geqq$ 3. The final calculations had thirteen correlation terms in each channel. Pseudostate thresholds are given in Table 1.

Excited states of neutral atoms are more difficult to tackle than are positive ions, because the system is so diffuse. Hata et al were satisfied with convergence of their eigenphase sums to 1o%.

Table 3. Convergence in total orbital angular momentum L for the 2s→n = 3 and 2p→n = 3 cross sections at $k_1 2$ = 0.90 ryd. (units of πa_o^2).

L	2s	2p
0	0.388	0.192
1	6.470	3.782
2	3.740	4.595
3	0.764	2.370
4	2.554	1.715
5	0.199	0.165
6	0.010	0.010
	14.125	12.830

Resonance structure in the L = 0,1 contributions has been discussed by Hata et al (198ob). As well as the expected series of Feshbach resonances, they found a 3_S shape resonance just above the n = 3 threshold. No such resonance is seen above n = 2.

The asymptotic coupling matrix for L = 0 and n = 2 channels open is

	3s,L	2p,L+1
2s,L	0	-6
2p,L + 1	-6	2

and diagonalising to uncouple the equations in the form

$$\left\{ \frac{d^2}{dr^2} + k^2 - \frac{\Lambda(\Lambda + 1)}{r^2} \right\} g_i \quad (r) = 0$$

the eigenvalues are

$$\Lambda_{21}^{(0)} = 7.08, \qquad \Lambda_{22}^{(0)} = -5.08.$$

The negative eigenvalue which is less than $-1/4$ corresponds to the single series of Feshbach resonances. The positive eigenvalue is large so the potential is strongly repulsive, and there is no shape resonance. For the n = 3 channels we have for the asymptotic coupling matrix

	3s L	3p,L+1	3d,L+2
3s L	0	- 1.47	0
3p,L+1	-1.47	2	-10.39
3d,L+2	0	-10.39	6

This has eigenvalues $(\Lambda_{nj}^{(L)})$,

$$\Lambda_{31}^{(0)} = -6.77 \qquad \Lambda_{32}^{(0)} = 0.131 \qquad \Lambda_{33}^{(0)} = 14.64$$

and again there is a single series of Feshbach resonances. The eigenvalue $\Lambda_{33}^{(0)}$ again gives strong repulsion and no resonances. However, the small positive eigenvalue $\Lambda_{32}^{(0)}$ produces a weakly repulsive r^{-2} potential which is dominated for some r by the attractive polarisation potential of the excited n = 3 state so that an L = 0 potential well capable of supporting one bound state is produced.

The stability of the results with respect to match point R_o and the starting point for the outer integration R_A is illustrated in Table 4. The comparison is for a six-state basis (1s 2s 2p 3s 3p 3d) with 11 correlation terms in each chennel. The results of Burke et at (1967) are given for comparison. We found that a value of R_o of at least 4o a_o was required. The agreement with Burke et al is satisfactory except for 1s - 1s and 2s - 3s.

Table 4. Results for L = 2 at k_1^2 = 0.90, in units of πa_o^2.

(a) R_o = 34 R_A = 150

(b) R_o = 40 R_A = 150

(c) R_o = 40 R_A = 120

(d) Burke et al.

Transition	a	b	c	d
1s 1s	0.239	0.257	0.258	0.162
1s 2s	0.061	0.059	0.059	0.06o
1s 3s	0.013	0.012	0.012	0.012
2s 3s	0.310	0.310	0.310	0.878
2s 3p	1.467	1.496	1.486	1.424

However, our six-state calculation, though not our eighteen-state, is sensitive to the number of correlation terms included. The results in Table 5 show the effect of going to 14 correlation terms in each channel in a six-state calculation, and compares these results with the eighteen-state values (with nine correlation terms).

Table 5. Results for L = 2, k_1^2 = 0.90 in

(a) 6-state, 11 correlation terms

(b) 6-state, 14 correlation terms

(c) 18-state, 9 correlation terms

Transition	a	b	c
1s 1s	0.258	0.115	0.343
1s 2s	0.059	0.063	0.062
1s 3s	0.012	0.012	0.017
2s 3s	0.310	0.342	0.247
2s 3p	1.486	1.343	1.416

Again the 1s - 1s and 2s - 3s values are the most sensitive, the elastic 1s - 1s result being extremely sensitive to correlation. However, within the six-state model the 14 correlation term values obtained in Kohn, Rabinow and RIAF variational methods are in much closer agreement than with only 11 correlation terms. No more than 9 correlation terms could be used in the eighteen-state model for $L \geqq 2$ without exceeding storage capacity for the bound-bound matrix.

Cross sections for each transition have been given by Hata et al (198ob) and a detailed discussion of the n = 2→n = 3 cross section is in Hata et al (198oa) and need not be repeated here.

ACKNOWLEDGEMENTS

The work described in this paper was carried out jointly with Dr. J. Hata and Dr. L.A. Morgan, using a code written by Dr. Morgan. The research was carried out under SRC Research Grant GR/A65o9.6.

REFERENCES

Burke, P. G., McVicar, D. D. and Smith, K., 1964, Proc. Phys. Soc. (London) A 83:397.
Burke, P. G., Ormonde, S. and Whitaker, W., 1967, Proc. Phys. Soc. (London) A 92:319.
Burke, P. G. and Taylor, A. F., 1969, J. Phys. B: Atom. Molec. Phys. 2:44.
Callaway, J., 1978, Physics Reports, 45:89.
Callaway, J., Morgan, L. A. and McDowell, M. R. C., 1975, J. Phys. B: Atom. Molec. Phys. 8:2181 and 9:2043, 1976.
Dolder, K. T. and Peart, B., 1973, J. Phys. B: Atom. Molec. Phys. 6:2415.
Fon, W. C., Berrington, K. A., Burke P. G. and Kingston, A. E., 1979, J. Phys. B: Atom. Molec. Phys. 11:1861.
Galitis, M., 1965, Sov. Phys. JETP, 20:107.
Hahn, Y., O'Malley, T. F. and Spruch, L., 1964, Phys. Rev. 134: 397 and 911.
Hata, J., Morgan, L. A. and McDowell, M. R. C., 1980a, J. Phys. B: Atom. Molec. Phys. 13: L347.
Hata, J., Morgan, L. A. and McDowell, M. R. C., 1980b, J. Phys. B: Atom. Molec. Phys., in press.
Henry, R. J. W. and Matese, J. J., 1976, Phys. Rev. A, 14:1368.
Morgan, L. A., 1979, J. Phys. B: Atom. Molec. Phys. 12:L735.
Morgan, L. A., 1980, J. Phys. B: Atom. Molec. Phys. 13:3703.
Morgan, L. A., McDowell, M. R. C. and Callaway, J., 1977, J. Phys. B: Atom. Molec. Phys. 10:3297.
Nesbet, R. K., 1969, Phys. Rev. A, 179:60.
Nesbet, R. K., 1978, Phys. Rev. A, 18:915.
Nesbet, R. K., 1980, "Variational Methods in Atomic Scattering", Plenum Press, N.Y.
Norcross, D. W., 1969, Comp. Phys. Comm. 1:88.
Oberoi, R. S. and Nesbet, R. K., 1973, J. Comp. Phys. 12:526.
Seaton, M. J., 1966, Proc. Phys. Soc. (London) A 89:469.
Taylor, P. O., Phaneuf, R. A. and Dunn, G. H., 1980, Phys. Rev. A, 22:435.
Wakid, S. and Callaway, J., 1980a, Phys. Lett. A 78:137.
Wakid, S. and Callaway, J., 1980b, submitted to J. Phys. B.

SCATTERING CALCULATIONS WITH L^2 BASIS FUNCTIONS: GAUSS QUADRATURE OF THE SPECTRAL DENSITY

John T. Broad

Fakultät für Chemie
Universität Bielefeld
48 Bielefeld, Germany

I. INTRODUCTION

In the past few years, several successful attempts[1-5] have been made to extend the use of square integrable (L^2) basis sets in atomic and molecular calculations from bound to scattering states. While all of these methods exploit the idea that the interaction region in a scattering experiment is short ranged, and hence well representable in a finite L^2 basis, they differ in how the boundary conditions at infinity appropriate to scattering are accounted for in the basis. It is my purpose here to examine exactly how these methods work, primarily on the simple, but useful, example of attractive Coulomb scattering.

Part II reviews the results of solving the Coulomb Schrödinger Equation in a particular infinite L^2 basis and that truncating to a finite basis generates a Gauss quadrature of the spectral density, with the Sturm sequence polynomials as the underlying orthogonal set. While this quadrature is the same as that generated by the moment technique[4] and is, in some sense, optimal for approximating matrix elements of smooth operator functions of the Coulomb Hamiltonian, a second set of functions, conjugate to the polynomials, must be introduced to represent singular integrals such as the matrix elements of the Green's function appearing in perturbation sums. Then, the explicit splitting of the Green's matrix into its quadrature approximation and a correction term displays how the rotation of the coordinate r in the complex plane[5] bounds the correction term and allows direct use of the quadrature approximations. Finally, the close formal analogies between the traditional Jost function[7] analysis in the coordinate, r, and the dependence of the quadrature on the size of the basis, n, are developed.

Part III presents applications of part II to problems of interest. First, the Coulomb Green's matrix in the basis set can be used directly to simplify the expressions for hydrogen atom properties such as the Bethe logarithm contribution to the Lamb shift[3]. More importantly, an arbitrary potential of finite rank in the basis can be added to the Coulomb Hamiltonian, thereby generating a Gauß quadrature of the new spectral density and yielding a smooth interpolation of the phase shift and a suggestive expression for the Fredholm determinant relative to pure Coulomb scattering. Next, this approach can be incorporated to account for one free electron in an L^2 close coupling calculation[9] in an infinite basis, with the other electrons in target pseudostates. This finally suggests that basis set approaches to the 3 particle problem might be understood in terms of multidimensional quadratures.

II. GAUSS QUADRATURE OF THE RADIAL COULOMB SPECTRUM

A. The Orthogonality of the Sturm Sequence Polynomials

Let us first review the solutions of the radial Schrödinger equation for the attractive Coulomb Hamiltonian for angular momentum l

$$H = \frac{1}{2}\frac{d^2}{dr^2} + \frac{l(l+1)}{2r^2} - \frac{1}{r}\,, \tag{1}$$

in atomic units. The solution regular at the origin takes the form[7,10]

$$\Psi(r,k) = \left(\frac{2}{\pi k}\right)^{\frac{1}{2}} \frac{\Gamma(l+1-it)(2kr)^{l+1}}{\Gamma(2l+2)e^{-\pi t/2}}\, e^{ikr}\Phi(l+1-it;2l+2;-2ikr), \tag{2}$$

where k is the wave vector, with $E = \frac{k^2}{2}$, and $t = \frac{1}{k}$.

The normalizable bound states appear where the confluent hypergeometric function Φ truncates to a polynomial and $\Gamma(l+1-it)$ has simple poles at

$$it_b = n_b + l + 1 \quad \text{where } (n_b > 0), \tag{3}$$

while the set of solutions at all energies, bound and continuum, is complete:

$$\delta(r-r') = \int_0^\infty dE\Psi(r,k)\Psi^*(r',k) + \sum_b \Psi_b(r)\Psi_b^*(r')$$
$$\equiv \sum\!\!\!\!\!\!\int dE\Psi(r,k)\Psi^*(r',k)\,, \tag{4}$$

where the terms in the bound state sum are $-2\pi i$ times the residue of the integrand at the bound state energies of Eq. (3). Yamani and Fishman[2] introduced the complete L^2 basis in r,

$$\Phi_n(r) = (\lambda r)^{l+1} e^{-\lambda r/2} L_n^{2l+1}(\lambda r) \ , \tag{5}$$

where L_n^{2l+1} is a Laguerre polynomial[10] and λ is a real, positive scaling parameter. Although this basis is not orthogonal, the overlap is tridiagonal[2,6] i.e.

$$S_{nn'} = \langle\Phi_n|\Phi_{n'}\rangle \text{ is non zero only if } n = n', \text{ or } n' \pm 1, \tag{6}$$

and the relation to the biorthogonal functions, $\bar{\Phi}_n$, with $\langle\bar{\Phi}_n|\Phi_{n'}\rangle = \sigma_{nn'}$ is particularly simple: $\bar{\Phi}_n(r) = \Phi_n(r)/rN_n$ where N_n is a normalization constant[2,6].

What makes the basis special is that the matrix elements of the radial Coulomb Hamiltonian (Eq. 1) are also tridiagonal[2,6]:

$$H_{nn'} = \langle\Phi_n|H|\Phi_{n'}\rangle \text{ is non zero only if } n = n', \text{ or } n \pm 1. \tag{7}$$

This means that if the scattering wave function is expanded in the basis set as:

$$\Psi(r,k) = \sum_{n=0}^{\infty} \Phi_n(r)\Psi_n(k) \ , \tag{8}$$

the infinite matrix Schrödinger equation for the coefficients

$$\sum_{n'=0}^{\infty} (H_{nn'} - ES_{nn'})\Psi_{n'}(k) = 0 \tag{9}$$

is really only a three term recursion relation connecting $\Psi_n(k)$ with $\Psi_{n\pm1}(k)$. (Clearly the expansion of a non L^2 function, $\Psi(r,k,)$, in L^2 functions has its meaning as a distribution, i.e. when integrated over r with some L^2 function, and converges best at small r^l). It was found that conformally mapping the energy from $(0,\infty)$ to $(-1,1)$ by defining

$$x(E) = \frac{E - \lambda^2/8}{E + \lambda^2/8} \tag{10}$$

allows the energy dependence of the off diagonal terms to be factored out as[2,6]

$$H_{nn'} - ES_{nn'} = 2(E + \lambda^2/8)(h_{nn'} - x\delta_{nn'}/\eta_{n'}) \ . \tag{11}$$

The Schrödinger Eq. (4) then becomes the three term recursion relation

$$h_{nn+1}\Psi_{n+1}(k) + (h_{nn} - x(E)/\eta_n)\Psi_n(k) + h_{nn-1}\Psi_{n-1}(k) = 0 \tag{12}$$

obeying the initial condition

$$h_{01}\Psi_1(k) + (h_{00} - x(E)/\eta_0)\Psi_0(k) = 0 \ . \tag{12a}$$

Furthermore, truncating these equations after N terms and writing the remainder as an inhomogeneity as:

$$\begin{bmatrix} h_{00}-x/\eta_0 & h_{01} & & & \\ h_{10} & \cdot & \cdot & \cdot & \cdot \\ & \cdot & \cdot & \cdot & \\ & & \cdot & \cdot & \cdot \\ & & \cdot & \cdot & \cdot h_{N-1,N-1}-x/\eta_{N-1} \end{bmatrix} \begin{bmatrix} \Psi_0(k) \\ \Psi_1(k) \\ \cdot \\ \cdot \\ \Psi_{N-1}(k) \end{bmatrix} = \begin{bmatrix} 0 \\ 0 \\ \cdot \\ \cdot \\ -h_{N-1,}(k) \end{bmatrix}$$

reveals that $\Psi_N(k)$ has zeros exactly at the eigenvalues $x_j^{(N)}$ of the truncated matrix $(\eta_n^{-1/2} h_{nn'} \eta_{n'}^{-1/2})$. On the other hand, we already know another set of functions of x obeying the same recursion relations (12) and solely determined by these zeros, namely the Sturm sequence polynomials $p_n(x)$ starting with $p_o(x) = 1$. Evidently, the coefficients $\Psi_n(K)$ are the Sturm sequence polynomials multiplied by a function independent of n

$$\Psi_n(k) = \Psi_o(k) p_n(x) \ . \tag{13}$$

That the Sturm sequence polynomials are orthogonal to each other followc immediately from taking matrix elements of the completeness relation (4) in the basis, giving

$$\not{\int} dx \rho(x) p_n(x) p_{n'}(x) = \eta_n \delta_{nn'} \ , \tag{14}$$

where the weight function is

$$\rho(x) = 2(E + \lambda^2/8) \quad |\Psi_0(k)|^2/\lambda \ . \tag{15}$$

Since $\Psi_0(k)$ contains the factor $\Gamma(1 + 1 - it)$, the weight function has simple poles at all the bound states; the sum in Eq. (14) indicates $-2\pi i$ times the sum of the residues at these poles, just as in the completeness integral, Eq. (4). Incidentally, Yamani[2,3] found the polynomials generated by the Hamiltonian Eq. (1) in the basis (5) to be a logical extension of the allowed range of the parameter values of certain polynomials examined by Pollaczek in 1950[10]. They are expressed as F(-n,b,c,z), where b and c depend on 1 and E while z depends only on the energy[3,6].

B. Gauß Quadrature of the Spectral Density

Once we know that the eigenvalues of the truncated Hamiltonian matrix fall at the zeros of polynomials orthogonal on a weight function with the right singularities at the bound states, we can formulate a Gauß quadrature tailor made to the radial Coulomb Hamiltonian. Furthermore, since this quadrature is generated merely by introducing the basis set, there must be a relation between the quadrature and the eigenvectors of the truncated Hamiltonian, or pseudostates. If we denote a pseudostate in r by

$$\Psi_j^{(N)}(r) = \sum_{n=0}^{N-1} \Phi_n(r)\Psi_{nj}^{(N)} , \qquad (16)$$

where $\Psi_{nj}^{(N)}$ is the nth component of the jth eigenvector, then normalizing these eigenvectors to 1 and exploiting the relation of the Sturm sequence polynomials to the eigenvectors[10] gives

$$\Psi_{nj}^{(N)} = c_j p_n(x_j^{(N)}), \qquad (17)$$

where

$$c_j^{-2} = \sum_{n=0}^{N-1} p_n^2(x_j^{(N)})/\eta_n^2(1-x_j^{(N)})/\lambda$$

A look at the sum in the expression for the constants c_j recalls the Christoffel formula[10] for the quadrature weights $w_j^{(N)}$ giving finally with Eq. (13) and (15) that

$$\Psi_{nj}^{(N)}\Psi_{nj}^{(N)*} = \frac{dE}{dx}\Big]_{x_j^{(N)}} w_j^{(N)}\Psi_n(k_j^{(N)})\Psi_{n'}^*(k_j^{(N)})/\rho(x_j^{(N)}) , \qquad (18)$$

where $k_j^{(N)}$ is related to $x_j^{(N)}$ by Eq. (10). This bears out the experience of Hazi and Taylor[1] that out to the radii where the basis set expansion is good the actual scattering wave functions evaluated at a pseudostate eigenvalue differ from the pseudostate eigenvector only by a normalization factor.

Applying this idea to approximate a matrix element of an operator function Op(H) between initial and final states <i and f> is revealing. Using the spectral representation gives

$$<i|Op(H)|f> = \sum\!\!\!\!\!\!\int dE <i|\Psi(k)> O(E) <\Psi(k)|f> ,$$

while introducing the basis set (5), truncating to N terms, and using the expansion in Eq. (8) yields

$$<i|Op(H)|f> \simeq \sum_{n=0}^{N-1}\sum_{n'=0}^{N-1} <i|\Phi_n> \sum\!\!\!\!\!\!\int dE\Psi_n(k)O(E)\Psi_{n'}(k) <\Phi_{n'}|f> .$$

Using Eqs. (3) and (5) then sets the $\sum\!\!\!\!\!\!\int$ up for a quadrature of degree N, which, using Eq. (18) gives

$$<i|Op(H)|f> \approx \sum_{n=0}^{N-1}\sum_{n'=0}^{N-1} <i|\Phi_n> \sum_{j=1}^{N} \Psi_{nj}^{(N)} O(E_j^{(N)})\Psi_{n'j}^{(N)} <\Phi_{n'}|f>$$

which in turn using the definition of the pseudostates in Eq. (16)

simplifies to

$$<i|Op(H)|f> \approx \sum_{j=1}^{N} <i|\Psi_j^{(N)}> O(E_j^{(N)}) < \Psi_j^{(N)}|f> .$$

Thus approximating the spectral resolution of the identity by a sum over pseudostates is not a bad approximation after all; it has the quality of a Gauß quadrature tailor made to the spectral density of H. Evidently, it only requires that the states $<i|$and$|f>$ be well represented by the finite basis and that O(E) be a smooth function throughout the range of integration.

C. A Second Solution and the Green's Matrix

From the above discussion, it is clear that a quadrature approximation will not suffice for the most interesting operator function of H, the Green's function; it would replace the infinite number of bound state poles and branch cut along the positive energy axis by a set of N poles and zeros. To get the correct behavior we need something analogous to the irregular Coulomb wave, yet an attempt to expand the irregular Coulomb wave in the basis set (5) of regular functions is sure to fail. The next idea is to go directly to the recursion relation (12) satisfied by Ψ_n and p_n, but to start with another boundary condition at n = 0. Designating the 2nd function by $q_n(K)$, it should satisfy

$$h_{nn+1}q_{n+1}(k) + (h_{nn} - x/\eta_n)q_n(k) + h_{nn-1}q_{n-1}(k) = 0 \qquad (19)$$

for $n \geq 1$

with the initial condition

$$h_{01}q(k) + (h_{00} - x/\eta_0)q(k) = 1 . \qquad (19a)$$

Using the recursion relation[12] for the polynomials and their orthogonality, one can immediately write down an integral representation of the q_n as

$$q_n(k) = \not{\int} dx'\rho(x')/(x' - x). \qquad (20)$$

For the radial Coulomb Hamiltonian, the $q_n(k)$ can be shown to be the other branch of the ${}_2F_1$ of the modified Pollaczek polynomials, which might be called modified Pollaczek functions of the second kind.

Before using the second function to construct the Green's matrix, it is fruitful to apply the Gauß quadrature carefully to the integral representation (20) of $q_n(k)$.

Subtracting and adding the residue at the singularity:

$$q_n(k) = \mathcal{I}dx'\rho(x')[p_n(x') - p_n(x) + p_n(x)]/(x' - x)$$

removes the singularity to allow a quadrature to be done at least on the first two terms. Indeed, since $[p_n(x') - p_n(x)]/(x' - x)$ is a polynomial of degree n-1, a quadrature of degree n is surely exact[10], with the $p_n(x')$ vanishing by construction at the abscissas to give

$$q_n(k) = p_n(x)\{\mathcal{I}dx'\rho(x')/(x' - x) - \sum_{j=1}^{n} w_j^{(n)}/(x_j^{(n)} - x)\} \quad (21)$$

Note that the first term in Eq. (21) is $q_o(k)$, while the second term is its quadrature approximation, which in the limit that k goes to the positive real axis from above becomes

$$q_n(k + io) = p\mathcal{I}dx'\rho(x')p_n(x')/(x' - x) + i\pi\rho(x)p_n(x). \quad (22)$$

Evidently the q_o term alone in Eq. (21) accounts for the multivalued nature of q_n.

After constructing and examining the second function, it is straight forward to use the recursion relations (12) and (19) to express the matrix inverse of (ES - H), the Green's matrix, as

$$G_{nn'}(E) = -2(1 - x)p_{n_<}(x)q_{n_>}(x)/\lambda, \quad (23)$$

where $n_>$ and $n_<$ are the greater and the lesser of n and n', or in the spectral representation as

$$G_{nn'}(E) = +2(1 - x)/\lambda\ \mathcal{I}dx'\rho(x')p_n(x')/(x - x') . \quad (24)$$

While splitting off the residue at the singularity in the integrand in terms of $n_<$ and $n_>$ and using the orthogonality of $p_{n_<}(x')$ to $(p_{n_<}(x') - p_{n_>}(x))/(x' - x)$ merely reproduces Eq. (23), it is more useful to examine a quadrature of some degree N > n and n'. Splitting off the singularity as

$$G_{nn'}(E) = + \frac{2(1 - x)}{\lambda} \mathcal{I} \frac{dx'\rho(x')}{(x - x')} [p_n(x')p_{n'}(x') - p_n(x)p_{n'}(x) + p_n(x)p_{n'}(x)]$$

allows an exact quadrature of degree N on the first two terms, giving

$$G_{nn'}(E) = -\frac{2(1-x)}{\lambda}\left\{\sum_{j=1}^{N}\frac{W_j^{(N)}p_n(x_j^{(N)})p_{n'}(x_j^{(N)})}{x_j^{(N)}-x}\right.$$

$$\left.+\,p_n(x)p_{n'}(x)\left[\int\frac{dx'p(x')}{x'-x}-\sum_{j=1}^{N}\frac{W_j^{(N)}}{x_j^{(N)}-x}\right]\right\},$$

since the integrand of the first two terms contains the weight function times a polynomial of the degree $n+n'-1 \leq 2N-1$ in x'. Now, using Eqs. (10), (13), (15) and (18) to simplify the first term and Eq. (21) for the second term yields

$$G_{nn'}(E) = \sum_{j=1}^{N}\frac{\Psi_{nj}^{(N)}\Psi_{n'j}^{(N)}}{E-E_j^{(N)}} - \frac{2(1-x)p_n(x)p_{n'}(x)q_n(k)}{\lambda p_n(x)}. \quad (25)$$

Thus each element of the Green's matrix in the first N basis functions splits into two terms: the pseudostate representation of $(ES - H)^{-1}$ as an N×N matrix and a correction term, which since $p_N(x_j^{(N)}) = 0$, exactly removes the pseudostate poles and contains the multivalued analytic structure of G.

D. Relation to the Rotated Coordinate Method

With the explicit splitting of the inner block (n and n' < N) of the Green's matrix into the sum over pseudostates and the correction term, we can expose how the rotated coordinate method works. When the coordinate, r, is replaced by $re^{i\theta}$, where θ is a constant angle between 0 and $\pi/2$, the continuous spectrum of the hydrogen atom is rotated down by 2θ into the complex energy plane, while the bound states remain fixed at the Rydberg values in Eq. (3). Experience shows[11] that the pseudostates try to mimic this behavior, with the lowest few eigenvalues remaining close to the lowest few Rydberg states, while the highest few eigenvalues lie close to the rotated branch cut. This means that, when Eq. (25) for the Green's matrix is evaluated at real positive energies away from the rotated pseudostates, the correction term becomes much less important. Indeed, using Eq. (21), the correction term contains the difference between $q_o(k)$ and its rational fraction approximation, which is known[10] to converge away from the poles.

E. Formal Analogy to Jost Function Theory

Before turning to applications to more complicated systems, it is useful to expose the analogy between Jost Function theory[7] in r and the development here in n. Clearly the polynomials $p_n(x)$, with $p_o = 1$, are in a sense regular in n the way the regular Jost solution is in r. What plays the role here of the irregular solution and of the Jost function?

Using Eq. (22) for real k, one can write the polynomials as

$$p_n(x) = [q_n(k) - q_n^*(k)]/2\pi\rho(x) , \tag{26}$$

which means, using Eq. (13) and (15), that the coefficient of the regular physical solution can be expressed as

$$\Psi_n(x) = \frac{2(1 - x)(q_n(k) - q_n^*(k))}{\lambda\Psi_o^* 2i} = \frac{i}{2} (e_n^* - \frac{\Psi_o}{\Psi_o^*} e_n), \tag{27}$$

where

$$e_n(k) \equiv 2(1 - x)q_n(k)/\lambda\Psi_o ;$$

hence, $e_n(k)$ plays the role of the irregular solution and $1/\Psi_o(k)$ that of the Jost function. This is more evident if we use Eqs. (26) - (28) to write (29) $p_n(x) = e_n(k)/\Psi_o^* - e_n^*(k)/\Psi_o)/2i\pi$ and remember that $1/\Psi_o(k)$ has zeros at all the bound states.

The relation between the Jost function and the limiting behavior in the irregular solution as r goes to zero must be translated here into behavior in e_n as n goes to zero, which follows from Eqs. (11) and (28) and the recursion relation, Eq. (19)

$$(H_{o1} - ES_o)e_1 + (H - ES_{oo})e_o = 1/\Psi_o \tag{30}$$

On the other hand, the recursion relations (2) and (9) also can be used to create the analogy in n of the Jost function as a Wronskian

$$(H_{nn+1} - ES_{nn+1})(e_{n+1}e_n^* - e_n e_{n+1}^*) = 2i$$

while

$$(H_{nn+1} - ES_{nn+1})(e_{n+1}p_n - e_n p_{n+1}) = 1/\Psi_o$$

Moreover, the role of the inverse square of the Jost function as the spectral density of the regular solution could not be more clearly expressed than by the orthogonality of the polynomials with respect to the weight function in Eqs. (4) and (5).

Finally, one can ask for which function the irregular solutions, $e_n(k)$, serve as expansions coefficients in the basis:

$$\xi(r,k) \doteq \sum_{n\geq 0}^{\infty} \Phi_n(r)e_n(k) \tag{31}$$

From the recursion relation (30), it follows that $\xi(r,k)$ satisfies an inhomogeneous Schrödinger equation

$$(H - E)\xi = \bar{\Phi}_o(r)/\Psi_o \quad , \tag{32}$$

and has an asymptotic form at large r

$$\xi(r,k) \underset{r\to\infty}{\sim} \sqrt{2\pi/k}\, \exp[i(kr + t\ln 2kr - l\pi/2 + \sigma_1)] \tag{33}$$

identical with the irregular Coulomb wave, which incidentally makes it, together with its regularity at its origin, the natural candidate for use as the second regularized Coulomb function in Kohn principle applications.

III. APPLICATIONS AND EXTENSIONS

A. H-atom properties

The explicit form, Eq. (23), of the matrix elements of the Coulomb Green's function in the L^2 basis (5) finds immediate application in calculating properties of the hydrogen atom. For instance, in calculations of the Lamb shift, it is necessary to evaluate the Bethe logarithm[12], which contains an energy integral of the matrix element $\langle\Psi_b|\vec{p}G(E)\vec{p}|\Psi_b\rangle$, where Ψ_b is the bound state whose shift is to be calculated and $\vec{p}$ the linear momentum. Since an appropriate choice of the scaling parameter, λ, causes one of the basis functions (5) to coincide with Ψ_b, and in a partial wave decomposition the operator $\vec{p}$ only couples neighboring l and n values, the matrix element reduces to a linear combination of just a few matrix elements[6] of G, all known analytically. Incidentally, at the energy E_b, since the Coulomb potential is diagonal in the basis (5), the representation of the Green's function collapses to the diagonal Sturm sum[12].

B. The J-Matrix Method[2,6]

Adding a potential of finite rank in the basis, i.e.

$$\tilde{V}_{nn'} = \langle\Phi_n|V|\Phi_{n'}\rangle \tag{34}$$

non zero only for n and n'<N changes the analysis above for the pure Coulomb potential very little.

The new Hamiltonian

$$\tilde{H} = H + \tilde{V} \tag{35}$$

can be made tridiagonal, or of Jacobi, or J-Matrix, form in N steps by a Householder transformation, thereby leading to the three term recursion relations analogous to Eq. (2) and the orthogonality of the Sturm sequence polynomials, p_n, with respect to the square of the first expansion coefficient, $\tilde{\Psi}_o$. Moreover, the finite range of V in n implies the boundary condition

$$\tilde{e}_n = e_n \quad \text{for all } n \geq N , \tag{36}$$

and gives, after some algebra, the phase shift relative to the Coulomb phase as

$$e^{2i\delta} = \frac{\tilde{p}_N(x)q^*_{N-1}(k) - \tilde{p}_{N-1}(x)q^*_N(k)}{\tilde{p}_N(x)q_{N-1}(k) - \tilde{p}_{N-1}(x)q_N(k)}$$

in terms of the Sturm sequence polynomials, $\tilde{p}_n$, of $\tilde{H}$ and the second functions, q_n, of the pure Coulomb H. On the real k axis, δ is a smooth function of k, equal to the phase of the complex quantity, $q^*_N(k_j^{(N)})$, at the pseudostate, or Harris[2], eigenvalues, with no spurious singularities, such as appear in the Kohn method. Furthermore, since V is of finite rank, the Fredholm determinant,

$$D(k) = \det[G(k)(E - H)] = \det[(E - \tilde{H})/(E - H)] , \tag{38}$$

can be evaluated explicitly as

$$D(k) = \Psi_o/\tilde{\Psi}_o = h_{N-1}(q_N\tilde{p}_{N-1} - q_{N-1}\tilde{p}_N) . \tag{39}$$

This displays the significance of the square of the absolute value of the Fredholm as a ratio of spectral densities and gives a more sophisticated alternative to the quantum defect method in the location of the bound states at its zeros. In addition, note, using Eq. (22) for the imaginary part of D takes the form of a reference function times a polynomial as proposed by Langhoff[4] in his Stieltjes immaging work. Unfortunately, the finite rank approximation to the potential gives neither a variationally stable, phase shift, nor a correct high energy Born term. The Kato correction[2] provides an immediate remedy, while the Schwinger ansatz[13] using the matrix elements of the Coulomb Green's function provides a more radical cure.

C. Electron-Ion Scattering

Generalization of the potential scattering J-Matrix formulations to traditional multichannel scattering in the close-coupling sense with a different basis (5) for each target state is straightforward and has been carried out[9] and applied with success to the photoelectron detachment from H^-[9]. An electron atom calculation would explicitly use the Coulomb Green's Matrix developed here. To what extent using target positive energy pseudostates in the close coupling calculation correctly accounts for ionization of the target, at least within the model that the fast and slow electrons see different boundary conditions[9], is open to speculation, and whether introducing a multidimensional quadrature is a useful computational way of formulating the 3 particle continuum is a tantalizing, but unproven, idea for further work.

REFERENCES

1. A. Hazi and H. S. Taylor, Phys. Rev. A1,1109 (1970).
2. E. J. Heller and H. A. Yamani, Phys. Rev. A9,1201,1209 (1974); H. A. Yamani and L. Fishman, J.Math.Phys. 16,410 (1975); E. J. Heller, Phys.Rev. A12,1222 (1975).
3. E. J. Heller, W. P. Reinhardt and H. A. Yamani, J.Comp.Phys. 13,536 (1973); E. J. Heller, T. N. Rescigno and W. P. Reinhardt, Phys.Rev. A8,2946 (1973); H. A. Yamani and W. P. Reimhardt, Phys. Rev. A11,1156 (1975).
4. P. W. Langhoff in Electron-Molecule and Photon-Molecule Collisions, T. N. Rescigno, B. V. Mckoy and P. W. Langhoff, B. Schneider, ed., A. U. Hazi, ibid; A. Givwer, C. Asaro, B. V. Mckoy and P. W. Langhoff, J.Chem.Phys. 72, 713 (1980) and references therein.
5. C. W. McCurdy and T. N. Rescigno, Phys.Rev. A21,1499 (1980) and references therein.
6. J. T. Broad, Phys. Rev. A18,1012 (1978); J. T. Broad in Springer Lecture Notes in Mathematics, Proceedings of the ZIF Conference on Eigenvalue Problems, J. Hinze ed., to be published in 1981.
7. R. g. Newton, Scattering Theory of Waves and Particles (McGraw-Hill, New York, 1966), Chapter 12 and Section 14.3.
8. A. Macquet, Phys.Rev. A15,1088 (1977).
9. J. T. Broad and W. P. Reinhardt, Phys.Rev. A14, 1976.
10. Higher Transcendetal Functions, A. Erdelyi, ed (McGraw-Hill, New York, 1953), especially Chapter 10.
11. S.-I. Chu and W. P. Reinhardt, Phys.Rev.Lett. 39,1195 (1977); C. Cerjan et al, Int.J.Quantum Chem. XIV,393 (1978) and references therein.
12. H.A. Bethe and E. E. Salpeter, Quantum Mechanichs of One and Two Electron Atoms (Springer, Berlin, 1957); S. Klarsfeld and A. Macquet, Phys.Lett. 43B,201 (1973).
13. D. K. Watson et al, Phys.Rev. A21,738 (1980) and references therein.

MOLECULAR RESONANCE PHENOMENA*

A. U. Hazi

Theoretical Atomic and Molecular Physics Group
Lawrence Livermore National Laboratory
University of California
Livermore, California 94550

I. INTRODUCTION

Resonant autoionizing states of molecules and autodetaching states of molecular negative ions play important roles in many collision phenomena involving low energy electrons, e.g. dissociative attachment, associative and Penning ionization, dissociative recombination, etc.[1,2] Although the formal theory[2,3] describing resonant scattering of electrons from molecules has been developed during the past 20 years, only a few quantitative studies of molecular resonance phenomena had appeared in the literature until recently.[2,4] During the last three years, significant progress has been made toward the development of reliable and practical methods for calculating the decay width of molecular resonances within the Born-Oppenheimer approximation. The method of complex scaling[5,6] and the Stieltjes-moment-theory technique[7,8] are the two ab initio methods which have received most of the attention. In his talk, T. Rescigno will review recent work on the method of complex scaling as it applies to calculating molecular resonance parameters and photoionization cross sections. In my talk, I will focus on the Stieltjes-moment-theory technique and its recent applications to resonance phenomena in diatomic molecules.

Originally, Langhoff and co-workers developed the Stieltjes-moment-theory technique to treat the photoionization of molecules.[9] However, they quickly recognized[10] that the method is applicable to a variety of phenomena involving continuous spectra. In the context of resonances occuring in electron-atom or electron-molecule collisions, the method is based on projection-operator techniques and the golden-rule definition of the resonance width, $\Gamma(E)$.[11] Square-integrable, L^2, basis functions are used to describe both the resonant and the non-resonant parts of the scattering wave function. Stieltjes-moment-theory is employed to extract a continuous approximation for the width $\Gamma(E)$ from a discrete representation of the background continuum. Since the method avoids the explicit construction of molecular continuum or scattering wave functions by using L^2 basis functions exclusively, it has several advantages. Its numerical implementation requires only existing atomic and molecular electronic structure codes. Many-electron effects, such as correlation and target polarization, can be incorporated into the calculation of the width via configuration interaction (CI) techniques. Since the method produces a width $\Gamma(E)$ which is a function of the scattering energy, the corresponding "shift" $\Delta(E)$ can also be obtained by directly evaluating the principal value integral (Hilbert transform) relating $\Delta(E)$ and $\Gamma(E)$.[11] On the other hand, since scattering boundary conditions are not explicitly enforced, the Stieltjes-moment-theory technique does not normally provide partial widths or branching ratios for resonances which decay into more than one open channel.

Since the Asilomar workshop[8] in 1978 little evolution of the Stieltjes-moment-theory, as applied to molecular resonances, has taken place. Most of the recent work has focused on numerical applications to selected resonance states of diatomic molecules, in order to demonstrate the method's reliability and to assess the accuracy of the calculated widths. Specifically, we have used the Stieltjes technique, along with configuration interaction wave functions, to calculate energies and widths for both Feshbach resonances, such as the $^1\Sigma_g^+$, $^1\Sigma_u^+$ and $^1\Pi_u$ doubly-excited, autoionizing states of H_2, and low-energy shape-resonances, such as the well-known $^2\Pi_g$ state of N_2^- and the $^2\Sigma_u^+$ state of F_2^-. In all of these cases, we have studied the dependence of the electronic resonance parameters on the internuclear distance. Similar calculations are now in progress on some triatomic molecules, such as H_3, HCO and CH_2^-. In addition, the method has been applied to calculating ionization rates for collisions between negative mesons and simple atoms.[12] I believe the variety of these applications clearly demonstrates that, at the present time, the Stieltjes-moment-theory technique is the most useful ab initio method for calculating the widths of molecular resonances.

In the remainder of the talk, I will review the Stieltjes-moment-theory technique (Section II), and discuss its applications to the $^1\Sigma_u^+$ autoionizing state H_2 and the $^2\Pi_g$

shape-resonance of N_2^- (Section III). I will give a brief summary in Section IV.

II. THEORY

A. General Remarks

In Feshbach's theory of resonances,[11] the width of an isolated resonance which decays into a single open channel is defined according to the "golden-rule" formula:

$$\Gamma(E) = 2\pi \left| \langle \phi_r (H - E) \psi_E^{\pm} \rangle \right|^2 \qquad (1)$$

Here ϕ_r is a localized, L^2 function describing the resonance state. The function $\psi_E^{\pm}$ is an energy-normalized scattering function which represents the non-resonant, background continuum at energy E. The operator H is the full, many-electron Hamiltonian. For narrow resonances, the physical width is $\Gamma(E_r)$, where E_r is the resonance energy.

For molecular resonances, the wave function ϕ_r can be computed using well-established, quantum chemical techniques, such as the multiconfiguration-self-consistent field (MCSCF) or configuration interaction (CI) procedures, formulated within the stabilization method.[13] However, the direct calculation of the non-resonant scattering function $\psi_E^{\pm}$ is quite difficult because of the non-spherical and non-local nature of the electron-molecule interaction potential. In the Stieltjes-moment-theory method, this difficulty is avoided by expanding $\psi_E^{\pm}$ in terms of square-integrable, L^2, many-electron basis function which are constructed to be orthogonal to the resonance wave function ϕ_r through the use of projection operators. This allows for an accurate representation of the background continuum without sacrificing the computational simplicity inherent in the exclusive use of electronic structure codes.

As presently formulated, the Stieltjes method uses CI techniques for constructing the wave functions and it involves three major computational steps: (i) The total space of many-electron configurations (CI-space) is partitioned into resonant and non-resonant parts using suitably defined projection operators. The resonance wave function ϕ_r and the discrete, L^2, approximations to the background solutions are obtained by diagonalizing the full Hamiltonian in these two respective subspaces. (ii) The width matrix-elements connecting the two subspaces are calculated from the "golden-rule" formula, Eq. (1). (iii) Stieltjes-moment-theory techniques[9,10] are employed to extract correctly normalized widths from the discrete representation of the background continuum.

B. Calculation of the Resonant and Non-Resonant Wave Functions

For a given molecule, one starts by selecting a suitable, orthonormal set of one-electron basis functions. The choice of the molecular orbitals is governed by the electronic structure of the target and the expected nature of the resonance under consideration. The basis must contain two groups of orbitals: those which are required to describe, at some level of approximation, the resonance and the molecule left behind by the scattered electron, and those which can approximate the wave function of the scattering electron itself. Since both the resonance and the target states are "bound" states, the molecular orbitals in the first group are essentially the same as one would use in any electronic structure calculation of the ground, or low-lying excited states. That is, the first group should include the orbitals occupied in the the single configuration (e.g., Hartree-Fock) wave functions, as well as the so-called "correlating" orbitals which are required to describe electron correlation and configuration mixing in the resonance and the target states. On the other hand, the job of the second group of orbitals is to approximate continuum wave functions, and therefore they do not normally occur in bound-state calculations. This group must include numerous diffuse basis functions, but only in the symmetry that corresponds to the open-channel into which resonance decays. To illustrate these features, I will give specific examples of basis sets in Section III, where application to resonances of H_2 and N_2^- are discussed.

To define the total CI space, $P_o + Q_o$, one constructs from the one-electron basis set all the many-electron configurations which are required to describe accurately the structure of the resonance, its energy and its decay width. In general, internal consistency between the levels of approximation used for the resonant (Q_o) and non-resonant (P_o) subspaces is more important than absolute convergence with respect to the number of particle-hole excitations included in the CI calculation. The subspace P_o is constructed from those configurations which, to first order, describe the open channel, i.e., target + scattered electron. In other words, the non-resonant background is represented by all possible antisymmetrized, Kroenecker products of the essential target configurations and the molecular orbitals chosen to represent the scattered electron (the second group of orbitals mentioned in the previous paragraph). All other configurations are included in the subspace Q_o. The resonance wave function ϕ_r and the corresponding "unshifted" resonance energy ε_r are obtained by diagonalizing Q_oHQ_o. If the choice of P_o is physically reasonable, i.e., if P_o space is constructed from all the configurations which are essential for describing the target molecule or molecular ion, the lowest eigenstate of Q_oHQ_o will correspond to the resonance state. In the case of Feshbach resonances, the definition of P_o is straightforward, and it is usually dictated by symmetry, as the example given in Section IIIA will show. Shape-resonances are more difficult to treat because the resonance is configurationally identical to the decay channel. For example, in the case of N_2^-, the resonance has the electron configuration $1\sigma_g^2 \ldots 1\pi_u^4\ 3\sigma_g^2\ 1\pi_g$, whereas the background continuum is given by $1\sigma_g^2 \ldots\ 1\pi_u^4\ 3\sigma_g^2\ k\pi_g$ where $k\pi_g$ is a scattering orbital. In such cases, more ambiguous criteria, such as the localization of the resonance orbital,[13,14] must be used to obtain a suitable definition of P_o.

Once Q_oHQ_o is diagonalized, two new projection operators P and Q are introduced in order to obtain an accurate representation of the non-resonant continuum. The operators P and Q are defined as

$$Q = \sum_i \phi_{ri}><\phi_{ri} \tag{2a}$$

and

$$P = 1 - Q = P_o + (Q_o - Q) \tag{2b}$$

In Eq. (2a) the index i runs over all the resonances of a given symmetry under consideration, e.g., a Rydberg series of autoionizing states. It is important to note that P space contains not only those configurations which approximate the decay channel (originally in subspace P_o) but also the higher, non-resonant solutions of Q_oHQ_o. I have found that this repartitioning of the total CI space is essential to incorporate, as fully as possible, many-electron correlation and polarization effects in the description of the non-resonant continuum. In the case of many-electron targets, such effects must be accounted for if accurate widths are to be obtained.

Once the projection operator P is defined, the discrete representation of the non-resonant continuum is obtained by diagonalizing PHP in the basis of all L^2 configurations which make-up the total space $P_o + Q_o$. The resulting eigenfunctions satisfy

$$<\chi_n |PHP| \chi_m> = \delta_{nm}\, \varepsilon_n \tag{3}$$

and

$$<\chi_n |\chi_m> = \delta_{nm} \tag{4}$$

All the solutions with non-zero eigenvalues are orthogonal to the resonance functions $\{\phi_{ri}\}$ by construction.

C. Calculation of the Resonance Width

It is now recognized that the eigenvalues and the eigenfunctions which result from diagonalizing the Hamiltonian of a scattering problem in an L^2 basis will form a discrete representation of the scattering continuum. For sufficiently large basis sets, each solution χ_n approximates $\psi_E^{\pm}$ with $E = \varepsilon_n$ in a region near the nuclei, except for an overall normalization factor. Since the "golden-rule" formula in Eq. (1) contains the L^2 function ϕ_r which is localized near the nuclei, the matrix-element has no significant contribution from the asymptotic region where χ_n fails to approximate $\psi_E^{\pm}$.[15] Consequently, it is possible[16] to use $\{\chi_n\}$ to calculate the width matrix-elements:

$$\gamma_n = 2\pi |<\phi_r H \chi_n>|^2 \tag{5}$$

However, χ_n cannot be used directly to approximate $\Gamma(\varepsilon_n)$ because χ_n is unit-normalized, whereas $\psi_{\varepsilon_n}^+$ is energy normalized. Some way must be found to determine the normalization constant relating χ_n and $\psi_{\varepsilon_n}^+$.

The same problem occurs when one employs L^2 basis functions to calculate continuum oscillator strengths or photoionization cross sections which also describe transitions from bound initial states to continuum final states. Langhoff and co-workers have shown that accurate photoionization cross sections can be computed from variationally constructed, discrete pseudospectra employing Stieltjes moment theory.[9,10] In the present context, the same technique is used to extract appropriately normalized resonance widths from the pseudospectrum $\{\gamma_n \varepsilon_n\}$ defined in Eqs. (3)-(5).

It is useful to consider the so-called cumulative function F(E) defined according to the equations:

$$F(E) = \int^{E} dE' \quad \Gamma(E') \tag{6a}$$

$$\Gamma(E) = dF/dE \tag{6b}$$

The pseudospectrum $\{\gamma_n \varepsilon_n\}$ associated with the non-resonant solutions of PHP defines a histogram approximation to F(E):

$$\tilde{F}(E) = 0, \qquad 0 \leqslant E \leqslant \varepsilon_1 \tag{7a}$$

$$\tilde{F}(E) = \sum_{n=1}^{k} \gamma_n = \sum_{n=1}^{k} 2\pi \, |<\phi_r H \chi_n>|^2 \qquad \varepsilon_k < E < \varepsilon_{k+1} \tag{7b}$$

At the rise points, $E = \varepsilon_k$, the cumulative function is approximated by

$$\tilde{F}(E) = 1/2[\tilde{F}(\varepsilon_k -) + F(\varepsilon_k +)]$$
$$= \sum_{n=1}^{k-1} \gamma_n + 1/2\,\gamma_k \qquad E = \varepsilon_k \tag{8}$$

To gain insight into the meaning of Eq. (7), let us consider the result that one would obtain by evaluating the integral in Eq. (6a) using numerical quadrature, i.e.,

$$F(E) \approx 2\pi \sum_n \omega_n \,|\langle \phi_r H \psi^{\pm}_{E_n} \rangle|^2, \tag{9}$$

where E_n and ω_n are the quadrature points and weights. A comparison of Eq. (7b) and (9) shows that, in the Stieltjes development, the non-resonant eigenvalues ε_n are used as the points associated with an equivalent quadrature.[17] Furthermore, the normalization constants relating χ_n and $\psi^{\pm}_{\varepsilon_n}$ $n = 1, 2, \ldots$ determine implicitly the quadrature weights, ω_n.

The Stieltjes derivative of the histogram approximation to F(E) is obtained as the slope of a straight line connecting the values of F at two neighboring rise points [Eq. (8)]. The resulting histogram approximation for $\Gamma(E)$ has the form

$$\tilde{\Gamma}(E_k) = \frac{\gamma_k + \gamma_{k+1}}{2(\varepsilon_{k+1} - \varepsilon_k)} \qquad k = 1, 2, \ldots \tag{10}$$

where E_k are the half-way points, $E_k = 1/2(\varepsilon_k + \varepsilon_{k+1})$. The result in Eq. (10) shows that the "correct" normalization constants associated with the discrete, non-resonant eigenfunctions χ_n are determined by the density of eigenvalues representing the continuous spectrum of PHP.

In actual computations employing Stieltjes-theory, one does not work directly with the pseudospectrum $\{\gamma_n, \varepsilon_n, n = 1 \ldots N\}$ that results from the CI calculations. Instead, one performs a moment analysis to obtain a "smoothed" spectrum

$$\{\bar{\gamma}_n(M), \bar{\varepsilon}_n(M), n = 1 \ldots M\}$$

whose elements are uniquely determined by the first 2M inverse power moments of the original pseudospectrum provided $M \leqslant N$. Usually, the calculations must be repeated for several values of M ($M \ll N$) until mutually consistent results are obtained. The details of the moment analysis are given in the literature,[9] so I will not discuss them here.

It should be noted that the Stieltjes procedure approximates $\Gamma(E)$ only at the discrete energies E_k. Since the resonance energy E_r does not usually correspond to a given E_k, one must interpolate the widths given by Eq. (10) in order to calculate the physical width $\Gamma(E_r)$. One alternative approach is to fit the approximate cumulative function given by Eq. (8) to an appropriate analytic function of E, which can be differentiated at any energy to yield $\Gamma(E)$. This approach has the advantage that cumulative functions resulting from different moment analyses, i.e. those corresponding to different values of M, can be fitted simultaneously.

D. Calculation of the Resonance Shift

In Feshbach's theory of resonances,[11] the energy shift, $\Delta(E)$, of an isolated resonance is given by the principal value integral:

$$\Delta(E) = (2\pi)^{-1} \mathcal{P} \int \frac{\Gamma(E')dE'}{E - E'} \qquad (11)$$

For narrow resonances, the resonance energy E_r satisfies the equation:

$$E_r = \varepsilon_r + \Delta(E_r) \qquad (12)$$

where ε_r is the eigenvalue of Q_0HQ_0. The Stieltjes-moment-theory technique allows for calculating the shift, $\Delta(E)$, since it provides an approximation for not only the physical width, but also the width as a function of energy, i.e., $\Gamma(E)$.

III. APPLICATIONS TO H_2^{**} AND N_2^-

A. $^1\Sigma_u^+(1\sigma_u\,2\sigma_g)$ Autoionizing State of H_2

The calculation of resonance widths using the Stieltjes-moment-theory technique is conceptually the simplest in the case of Feshbach resonances. As an illustration, I will discuss the $^1\Sigma_u^+(1\sigma_u 2\sigma_g)$ doubly-excited autoionizing state of H_2 for which several other theoretical results are available for comparison.

As I discussed in Section IIB, the one-electron basis set must contain two groups of orbitals: one containing those required to describe the resonance and the target states, and the other containing those required to approximate the scattered electron. In the present case, the first group consisted of $6\sigma_g$, $6\sigma_u$, $4\pi_g$ and $4\pi_u$ molecular orbitals which were linear combinations of contracted Gaussian basis functions located on the two nuclei.[18,19] The $1\sigma_g$ and $1\sigma_u$ orbitals corresponded to the two lowest states of H_2^+, whereas the other orbitals were chosen to represent the $(1\sigma_u\ 2\sigma_g)$ resonance state and the other excited states of H_2^+. Since the latter are Rydberg states, the first group already contained some diffuse, two-center $s\sigma$ and $p\pi$ basis functions. Since the $(1\sigma_u\ 2\sigma_g)$ resonance decays to the $^2\Sigma_g^+$ state of H_2^+, the ejected electron has σ_u symmetry. Consequently, we chose 12 $p\sigma$ Gaussian functions, placed on the center of the molecule, as the second group of orbitals. These Gaussian functions had a geometric sequence of exponents (0.32, 0.16, 0.08, ... 0.00031, 0.000155) and were Schmidt-orthogonalized sequentially to the other six σ_u orbitals.

For H_2 the total CI space, $P_o + Q_o$, contained all 124 $^1\Sigma_u^+$ configurations which can be constructed from the full basis of $6\sigma_g$, $18\sigma_u$, $4\pi_g$ and $4\pi_u$ orbitals. Since the open channel corresponds to H_2^+ $(^2\Sigma_g^+)$ + $e(k\sigma_u)$, the configurations $(1\sigma_g\ n\sigma_u)$, n = 1, ... 18, defined the non-resonant subspace P_o. The $(1\sigma_u\ 2\sigma_g)$ resonance was obtained as the lowest eigenstate of the 106x106 Q_oHQ_o-matrix. Since the above basis set provided reasonably good descriptions of the three lowest members of the $^1\Sigma_u^+$ autoionizing Rydberg series converging to the $^2\Sigma_u^+$ state of H_2^+, we chose all three resonances to define the resonant subspace Q [see Eq. (2a)]. The fully correlated non-resonant scattering solutions were obtained by diagonalizing the Hamiltonian in the 121 dimensional subspace orthogonal to Q.

Table 1

Comparison of calculated widths for the $^1\Sigma_u^+(1\sigma_u\ 2\sigma_g)$ autoionizing state of H_2

R(au)	E_r(eV)[a]	Γ(eV) Stieltjes	Coulomb[b]	2-state[c]
1.0	39.46	0.22	0.17	
1.4	30.71	0.40	0.29	0.40
1.8	25.54	0.65	0.38	
2.0	23.71	0.74	0.43	0.73
2.5	20.55	0.89	0.56	0.87
3.0	18.67	1.05	0.70	

a Relative to the ground state of H_2.
b Taken from Reference 19.
c D. Robb, (unpublished).

The Stieltjes calculations were performed at six internuclear distances between 1.0 and 3.0 a_o. The width of the $^1\Sigma_u^+(1\sigma_u\ 2\sigma_g)$ state was determined by evaluating Γ(E) at an energy corresponding to the lowest eigenvalue of Q_oHQ_o.

Table 1 compares the resonance widths obtained with the Stieltjes technique to other theoretical results. Previously, Kirby, Guberman and Dalgarno calculated the width of the $^1\Sigma_u^+(1\sigma_u\ 2\sigma_g)$ autoionizing state using a resonance wave function which was determined in the same $(6\sigma_g, 6\sigma_u, 4\pi_g, 4\pi_u)$, basis set as used in the Stieltjes calculations. However, they employed antisymmetrized products of the $1\sigma_g$ orbital and simple Coulomb functions to approximate the non-resonant wave functions appearing in the golden-rule formula, Eq. (1). As Table 1 shows, the use of simple Coulomb functions to describe the ejected electron seriously underestimates the resonance width, especially at larger internuclear separations. The last column of Table 1 gives the widths obtained very recently by D. Robb in a fixed-nuclei, 2-state close-coupling calculation of e - H_2^+ scattering. The lowest $^2\Sigma_g^+$ and $^2\Sigma_u^+$ states of H_2^+ were included in the basis, and the width was computed by fitting the energy dependence of the σ_u eigenphase sum to the Breit-Wigner formula. The agreement between the Stieltjes and close-coupling results is very good. Finally, it is interesting to note that the width of the $^1\Sigma_u^+(1\sigma_u\ 2\sigma_g)$ resonance increases significantly as the internuclear distance increases, which is precisely the behavior I predicted some time ago[20] based on the separated-atom limits of the relevant molecular orbitals.

B. $^2\Pi_g$ Shape-Resonance of N_2^-

At the Asilomar conference,[8] I reported widths for the well-known, low-energy $^2\Pi_g$ resonance of N_2^- which were obtained with the Stieltjes-moment-theory technique using a frozen-core description of the resonance state. Since target polarization was neglected in that work, both the computed resonance energy and width were quite similar to those extracted from various "static-exchange" scattering calculations, but they were significantly larger than the experimental values. Consequently, the Stieltjes-moment-theory calculations were repeated using a CI description of $N_2^-(^2\Pi_g)$ to account for core-polarization.

To build an orbital basis for N_2, we started with the [5s, 3p] contracted Gaussian basis set of Dunning,[21] to which we added a $p\pi$ Rydberg function ($\zeta = 0.03$), as well as one $d\sigma(\zeta = 0.8)$ and two $d\pi(\zeta = 1.0, 0.4)$ polarization functions. A self-consistent-field (SCF) calculation of the ground, $^1\Sigma_g^+$ state in this basis yielded $9\sigma_g$, $9\sigma_u$, $6\pi_g$ and $6\pi_u$ orbitals which formed the first group of orbitals defined in Section IIB. The scattering basis set contained 13 additional π_g orbitals which were constructed from the primitive Gaussians given in Table 2. As usual, a geometric sequence of exponents were used for the diffuse functions located at the center of the molecule.

Table 2

Exponents of the π_g Gaussian basis functions used to describe the scattered electron in $N_2^-(^2\Pi_g)$

Two-center[a] p_x:	0.08
Two-center[a] d_{xz}:	0.15, 0.06
One-center[b] d_{xz}:	0.36, 0.18, 0.09, 0.045, 0.0225, 0.011, 0.0055, 0.00275, 0.00137, 0.00068

[a] Functions located on the two-nuclei
[b] Functions located on the center of the molecule.

In the case of shape-resonances, the definition of resonant and non-resonant subspaces (P_o and Q_o) is somewhat ambigious. We ran several test calculations to identify which of the 19 π_g orbitals in the basis were essential to obtain a reasonable description of $N_2^-(^2\Pi_g)$. Since this resonance has the electron configuration $1\sigma_g^2\ 1\sigma_u^2\ 2\sigma_g^2\ 2\sigma_u^2\ 1\pi_u^4\ 3\sigma_g^2\ 1\pi_g$, where $1\pi_g$ is basically a valence orbital, we found that the $p\pi_g(0.03)$ Rydberg function included originally in the first group of orbitals was unnecessary. Consequently, we shifted this function to the "scattering" set and used the resulting $14\pi_g$ orbitals, along with the SCF wave function for $N_2(X^1\Sigma_g^+)$, to define the non-resonant subspace P_o. The configurations forming Q_o were generated by taking all possible single-replacements of the valence orbitals occupied in the three reference configurations $1\sigma_g^2 \ldots 3\sigma_g^2\ n\pi_g$, n = 1, 2, 3. The nitrogen 1s core orbitals ($1\sigma_g$ and $1\sigma_u$) remained doubly-occupied in all configurations, and the single-excitations were generated within the $9\sigma_g$, $9\sigma_u$, $5\pi_g$, $6\pi_u$ basis. The resulting total CI space had 193 configurations. It should be noted that the primary purpose of the above mentioned single-replacements is to describe the short-range distortion or polarization of the N_2 target while the scattered electron is captured in the $1\pi_g$ resonant orbital.

With the above choice of P_o, the $^2\Pi_g$ resonance corresponded to the lowest eigenstate of Q_oHQ_o, which was then used to define the projection operator Q. (See Section IIB.) The 18 discrete, non-resonant background solutions were obtained by diagonalizing the Hamiltonian in the 192 dimensional subspace orthogonal to Q. The Stieltjes calculations were performed for nine different internuclear separations between 1.744 and 2.391 a_o.

Figure 1 shows the dependence of the calculated width on both the internuclear distance, R, and the kinetic energy of the ejected electron, E. The width $\Gamma(E)$ increases quite rapidly with decreasing internuclear distance. For a fixed value of R, $\Gamma(E)$ increases rapidly between 0 and 3 eV, and it reaches a broad maximum between 4.5 and 5.5 eV. Since in the low energy region $d\Gamma/dE$ is of the order of 0.1-1.0, we also estimated the energy shift $\Delta(E)$ by using the raw Stieltjes data to approximate the principal value integral.[8,22] We calculated the resonance enregies E_r from Eq. (12) and then used these values to obtain the physical width $\Gamma(E_r)$ as a function of internuclear distance.

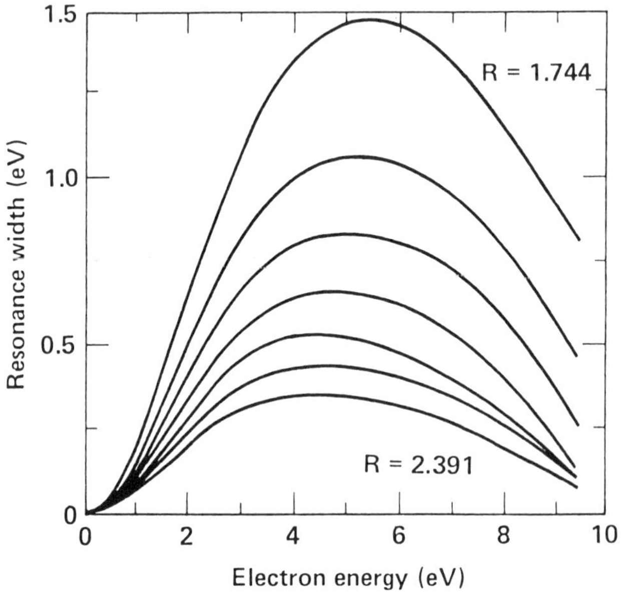

FIG. 1. Dependence of the resonance width $\Gamma(E)$ on electron energy and internuclear distance, as calculated with the CI-Stieltjes-moment-theory technique for $N_2^-(^2\Pi_g)$. The different curves correspond to internuclear distances of 1.744, 1.868, 1.968, 2.068, 2.168, 2.268 and 2.391 a_o, and Γ decreases monotonically with increasing R.

Figure 2 compares the potential energy curve of $N_2^-(^2\Pi_g)$ calculated with the CI wave function to our earlier results obtained with a frozen-core, static-exchange approximation,[8] as well as to the curve extracted from the recent R-matrix calculations of Schneider, LeDourneuf and Vo Ky Lan.[23] Figure 2 also shows the computed, SCF, potential energy curves of $N_2(X\ ^1\Sigma_g^+)$. Figure 3 compares the widths of the $N_2^-(^2\Pi_g)$ resonance which have been determined in several recent theoretical studies. The results included in Figures 2 and 3 show that neglecting the polarization of N_2 during the collision leads to resonance energies and widths which are significantly larger than those implied by the experimental vibrational excitation cross sections for N_2.[24] On the other hand, all of the recent ab initio calculations[23,25,26] which accounted for short-range distortions of the N_2 molecule produced resonance parameters in good agreement with each other. This agreement is especially encouraging since the ab initio treatments employed different techniques to describe target polarization. The latest Stieltjes work relied on single-excitation CI, whereas the R-matrix[23] and T-matrix[26] treatments constructed the electron-molecule interaction potential from molecular orbitals which were determined in "stabilization-self-consistent field" calculations of $N_2^-(^2\Pi_g)$.[27]

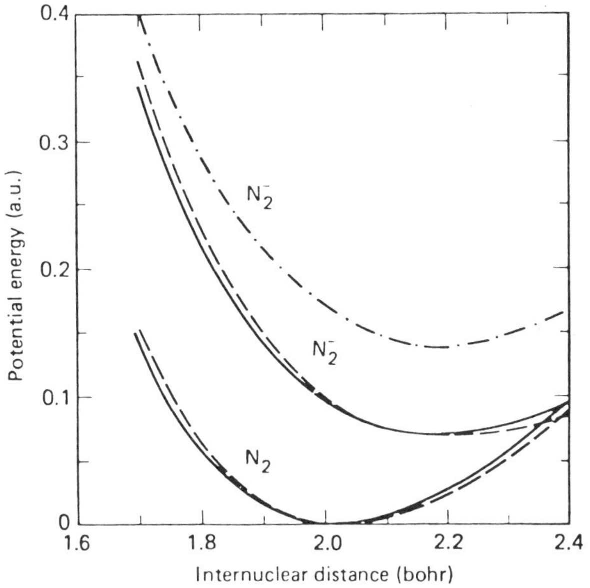

FIG. 2. Potential energy curves of $N_2(^1\Sigma_g^+)$ and $N_2^-(^2\Pi_g)$. Solid lines: R-matrix calculation with polarized core from (Ref. 23). Dot-dashed line: Stieltjes calculation with frozen core (from Ref. 8). Dashed lines: Stieltjes calculation with polarized core (from Ref. 25).

The accuracy of the fixed-nuclei resonance parameters of $N_2^-(^2\Pi_g)$ can be judged from the vibrational excitation cross sections which have been calculated for N_2 using these parameters in different models to describe the dynamics of the nuclei. The R-matrix calculation of Schneider and co-workers[23] produced, for the first time, ab initio vibrational excitation cross sections which had both the correct magnitude and the proper structure in comparison to the experimental data.[24] T. Rescigno, M. Kurilla and myself used the resonance parameters

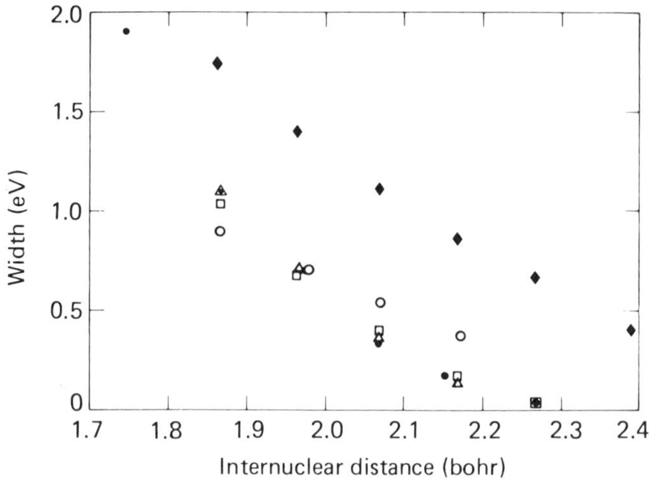

FIG. 3. Electronic resonance width of $N_2^-(^2\Pi_g)$ as a function of internuclear distance. ●: R-matrix calculation with polarized core (from Ref. 23). ■: Stieltjes calculation with frozen core (from Ref. 8). □: Stieltjes calculation with polarized core (from Ref. 25). △: T-matrix calculation with polarized core (from Ref. 26). ○: semiempirical results of Dube and Herzenberg (from Ref. 28).

obtained with both the R-matrix and the Stieltjes-moment-theory techniques to calculate vibrational excitation cross sections for N_2 with the complex-potential or "boomerang" model.[28] Again, we found reasonable agreement between the experimental[24] and the theoretical results.[25] In our work, we also examined carefully the dependence of the computed cross sections on the quality of the fixed-nuclei, electronic resonance parameters employed in the "boomerang" model. We found that, as expected, the inclusion of target polarization is absolutely essential, and that the positions and the relative heights of the peaks in the vibrational excitation cross sections are rather sensitive to the potential energy curve of N_2^-. A forthcoming paper[25] describes our study of the resonant vibrational excitation of N_2 by electron impact, and I refer you to it for additional details.

Finally, it is worth noting in Figure 3 that the width derived by Dube and Herzenberg in their semiempirical study[28] differs from those obtained in the various ab initio calculations in two respects: it is about 50% larger at R = 2.068 bohr, and it has substantially smaller, negative slope. Nevertheless both the semi-empirical and the ab initio resonance parameters produce realistic vibrational excitation cross sections when used in the "boomerang" model. This suggests that there may not be a unique set of fixed-nuclei resonance parameters which give correct cross sections, and that caution should be used in extracting molecular resonance parameters directly from experimental data.

IV. SUMMARY

In this talk I have attempted to show that Stieltjes-moment-theory provides a practical and a reasonably accurate method for calculating the widths of molecular resonances. The method seems to possess a number of advantages for molecular applications, since it avoids the explicit construction of continuum wavefunctions. It is very simple to implement the technique numerically, because it requires only existing bound-state electronic structure codes. Through the use of configuration interaction techniques, many-electron correlation and polarization effects can be included in the description of both the resonance and the non-resonant background continuum.

To illustrate the utility and the accuracy of the Stieltjes-moment-theory technique, used in conjunction with configuration interaction (CI) wave functions, I have discussed recent applications to the $^1\Sigma_u(1\sigma_u\ 2\sigma_g)$ autoionizing resonance state of H_2 and the well known $^2\Pi_g$ state of N_2^-. I have described in detail the choices of the one-electron basis sets and the types of many-electron configurations appropriate for these two cases. I have also given guidelines for the selection of the projection operators defining the resonant and non-resonant subspaces in the case of both Feshbach and shape-resonances.

The numerical results indicate that the Stieltjes-moment-theory technique, which employs L^2 basis functions exclusively, produces as accurate resonance parameters as can be extracted from direct electron-molecule scattering calculations, provided approximately the same approximations are used to describe important physical effects such as target polarization. Furthermore, as our recent work on N_2 shows, the method provides sufficiently accurate fixed-nuclei electronic resonance parameters to be used in ab initio calculation of resonant vibrational excitation cross sections. Similar calculations are currently in progress for low-energy electron-F_2 collisions,[29] where we intend to study both vibrational excitation and dissociative attachment via the $^2\Sigma_u^+$ resonance of F_2^-.

Finally, I should mention that additional work is still required to refine the method and to gain more computational experience. Applicability of the technique to resonances occurring in electron-polar molecule collisions needs to be investigated. The criteria for choosing Gaussian-type basis sets which can adequately describe the background, non-resonant scattering continuum in triatomic molecules have not yet been established. Nor has the method been applied to Feshbach resonances occurring in negative ions of complex diatomic molecules. I am confident that the theoretical study of molecular resonances will continue to be an active area of research and that we will witness significant progress during the next few years.

V. REFERENCES

1. G. J. Schulz, Rev. Mod. Phys. 45, 423 (1973).
2. J. N. Bardsley, in Electron-Molecule and Photon-Molecule Collisions, edited by T. Rescigno, V. McKoy and B. Schneider (Plenum, New York, 1979) p. 267.
3. T. F. O'Malley, Phys. Rev. 150, 14 (1966).
4. F. Fiquet-Fayard, Vacuum 24, 533 (1974).
5. C. W. McCurdy, in Electron-Molecule and Photon-Molecule Collisions, edited by T. Rescigno, V. McKoy and B. Schneider (Plenum, New York, 1979) p. 299.
6. B. R. Junker, Phys. Rev. Lett. 44, 1487 (1980).
7. A. U. Hazi, J. Phys. B. 11, L259 (1978).
8. A. U. Hazi in Electron-Molecule and Photon-Molecule Collisions, edited by T. Rescigno, V. McKoy and B. Schneider (Plenum, New York, 1979) p. 281.
9. P. W. Langhoff, in Electron-Molecule and Photon-Molecule Collisions, edited by T. Rescigno, V. McKoy and B. Schneider (Plenum, New York, 1979) p. 183.
10. P. W. Langhoff, Int. J. Quant. Chem. Symp. 8, 347 (1974).
11. H. Feshbach, Ann. Phys. (New York) 19, 287 (1962).
12. J. S. Cohen, R. L. Martin and W. R. Wadt, Phys. Rev., to be published.
13. H. S. Taylor, Adv. Chem. Phys. 18, 91 (1970).
14. H. S. Taylor and A. U. Hazi, Phys. Rev. A14, 2071 (1976).
15. W. H. Miller, Chem. Phys. Lett. 4, 627 (1970).
16. A. P. Hickman, A. D. Isaacson and W. H. Miller, J. Chem. Phys. 66, 1483 (1977).
17. E. J. Heller, W. P Reinhardt and M. A. Yamani, J. Comput. Phys. 13, 536 (1973).
18. S. Guberman, to be published.
19. K. Kirby, S. Guberman and A. Dalgarno, J. Chem. Phys. 70, 4635 (1979).
20. A. Hazi, Chem. Phys. Lett. 25, 259 (1974).
21. T. H. Dunning, J. Chem. Phys. 53, 2823 (1970).
22. P. W. Langhoff and W. P Reinhardt, Chem. Phys. Lett. 24, 495 (1974).
23. B. I. Schneider, M. LeDourneuf, and Vo Ky Lan, Phys. Rev. Lett. 43, 1926 (1979).
24. G. J. Schulz, in Principles of Laser Plasmas, edited by G. Bekefi (Interscience, New York 1976).

25. A. U. Hazi, T. Rescigno and M. Kurilla, Phys. Rev. A23, 1089 (1981).
26. D. A. Levin and V. McKoy, unpublished.
27. M. Krauss and F. H. Mies, Phys. Rev. A1, 1592 (1970).
28. L. Dube and A. Herzenberg, Phys. Rev. A20, 194 (1979).
29. A. U. Hazi, T. Rescigno and A. E. Orel, to be published.

*Work performed under the auspices of the U.S. Department of Energy by the Lawrence Livermore National Laboratory under contract No. W-7405-ENG-48.

THE R-MATRIX THEORY OF ELECTRON-MOLECULE SCATTERING

Barry I. Schneider

Los Alamos Scientific Laboratory
University of California
Los Alamos, New Mexico 87545

ABSTRACT

The R-matrix theory of electron scattering from diatomic molecules is extended to include the effects of nuclear excitation. The theory is derived in a very general form and used to establish the connection between the R-matrix method, the Kapur-Peierls states, the Siegert resonances and the Feshbach projection method. The power of this approach as an analytic and computational tool is demonstrated by considering the vibrational excitation of a homonuclear diatomic molecule dominated by an isolated electronic resonance state.

INTRODUCTION

The R-matrix method has proven to be a very useful technique in the calculation of atomic and molecular collision cross sections[1]. My intention in this article is to present a more general derivation of the theory including the effects of nuclear as well as electronic motion[2]. The reason for this is not purely formal for as we shall see there is a close connection between the R-matrix method, the Kapur-Peierls technique, the Siegert states and the Feshbach projection operator approach to resonant scattering processes. Originally much of this material was part of a paper written to clarify the role of the Born-Oppenheimer approximation in scattering processes dominated by an electronic resonant state[3]. The theory was resurrected a few years later but in a much more diluted form by Phil Burke, Maryvonne Le Dournef and myself and applied very successfully to the resonant vibrational excitation of N_2 by low energy electrons. The result of that calculation, performed in collaboration with Maryvonne and Vo Ky Lan[4] is described elsewhere in this volume. For those interested in the applications of the theory to electron-molecule collisions I have included a list of references at the end of the article.

THEORY

We require a solution of the Schrödinger equation

$$(H - E)|\Psi_E) = (H_e^{N+1} + T_R - E)|\Psi_E) = 0 \tag{1}$$

subject to scattering boundary conditions. Following Bloch we place both the electronic co-ordinate of the incident electron, r_{N+1} and the nuclear co-ordinate R in a "box" of radius (a_e, a_R). The appropriate asymptotic form of the wavefunction is

$$\lim_{r_{N+1}\to a_e} \Psi_E(1\ldots N+1,R) = \sum_{j,\alpha} \Psi_{j\alpha}^{AB}(1\ldots a_e,R) \tag{2a}$$

where

$$\Psi_{j\alpha}^{AB}(1\ldots N,a_e,R) = \Psi_j^{AB}(1\ldots N,R)\chi_{j\alpha}^{AB}(R)\Phi_{j\alpha}^{AB}(a_e) \tag{2b}$$

and

$$\lim_{R\to a_R} \Psi_E(1\ldots N+1,R) = \Psi^{A-}(1\ldots N_j)\Psi^B(N_{j+1},\ldots N+1)\chi^{AB-}(a_R) \tag{2c}$$

$$\Psi_j^{AB}(1\ldots N,R) = j^{th} \text{ electronic state of target} \tag{2d}$$

$$\Psi_{j\alpha}^{AB}(R) = \alpha^{th} \text{ vibrational state of } j^{th} \text{ electronic state of target} \tag{2e}$$

$$\Psi^{A-}(1\ldots N_j)\Psi^B(N_{j+1},\ldots N+1) = \text{electronic wave function of negative ion.} \tag{2f}$$

Thus equations (2b) and (2c) refer to electronic-vibrational excitation and dissociative attachment respectively. We exclude from consideration the three-body process which would lead to direct dissociation of the molecule. The Bloch L operator[5] may be defined as

$$L = \frac{1}{2} \sum_{j,\alpha} |\Psi_j^{AB}\chi_{j\alpha}^{AB})\delta(r_{N+1} - a_e)(\frac{\partial}{\partial r_{N+1}} - L_{j\alpha})(\Psi_j^{AB}\chi_{j\alpha}|$$

$$+ \frac{1}{2m_N} |\Psi^{A-}\Psi^{B})\delta(R-a_R)(\frac{\partial}{\partial R} - L_{AB-})(\Psi^{A-}\Psi^{B}| \tag{3}$$

with

$$L_{j\alpha} = \frac{\partial}{\partial r} \ln(\Phi_{j\alpha}^{AB})^{+}|R_{N+1} = a_e \tag{4a}$$

$$L_{AB^-} = \frac{\partial}{\partial R} \ln(\chi^{AB^-})^{+}|R = a_R \tag{4b}$$

where $\Phi_{j\alpha}^{AB}$ and χ^{AB^-} have been separated into an incoming (-) and outgoing (+) wave asymptotically for all open channels. For closed channels ($E_{j\alpha} = -k_{j\alpha}^2$) we write

$$L_{j\alpha} = k_{j\alpha}. \tag{4c}$$

The solution of the scattering problem within the box may be written as

$$|\Psi_E) = (H + L - E)^{-1} L|\Psi_E) . \tag{5a}$$

By separating $|\Psi_E)$ into incoming and outgoing waves asymptotically we have

$$|\Psi_E) = (H + L - E)^{-1} L|\Psi_E^-) \tag{5b}$$

where we have used the property of L that

$$L|\Psi_E) = L|\Psi_E^-) . \tag{5c}$$

To effect a solution of equation (5b) we introduce a complete system

$$H_e^{N+1}|\Psi_{E_i}^0) = E_i^0(R)|\Psi_{E_i}^0) \tag{6a}$$

$$(T_R + E_i^0(R)|\theta_{iq}) = \varepsilon_{iq}|\theta_{iq}) \tag{6b}$$

subject to the boundary conditions

$$\frac{\partial}{\partial r_{N+1}} \Psi_{E_i}^0 (1...N+1,R)|_{r_{N+1}=a_e} = b_e\Psi_{E_i}^0 (1...a_e,R) \tag{6c}$$

$$\frac{\partial}{\partial R}\,\theta_{iq}(R)\big|_{R=a_R} = b_N\theta_{iq}(a_R) \tag{6d}$$

these states would satisfy the equation

$$(H_e^{N+1} + T_R + L_b)|\Psi^0_{E_i}\theta_{iq}) = \varepsilon_{iq}|\Psi^0_{E_i}\theta_{iq}) \tag{7}$$

where L_b is identical to L except for the replacement of $L_{j\alpha}$ with b_e and L_{AB}^- with b_N, if all derivations of electronic wavefunctions with respect to the nuclear coordinate R were neglected. To keep the theory exact we divide the Hamiltonian into an unperturbed and perturbed part. In the unperturbed part we keep all terms diagonal in the electronic state. Thus we define

$$H_0 = \sum_{i,q} |\Psi^0_{E_i}\theta_{iq})(\Psi^0_{E_i}\theta_{iq}|H_e^{N+1} + T_R + L_b|\Psi^0_{E_i}\theta_{iq})(\Psi^0_{E_i}\theta_{iq}| \tag{8a}$$

$$H' = \sum_{\substack{i\neq j\\ q,r}} |\Psi^0_{E_i}\theta_{iq})(\Psi^0_{E_i}\theta_{iq}|H_e^{N+1} + T_R + L_b|\Psi^0_{E_j}\theta_{jr})(\Psi^0_{E_j}\theta_{jr}| + L - L_b \;. \tag{8b}$$

The inverse operator defined in equation (5a) satisfies the equation

$$(H_0 + L_b - E)G = I - H'G \;. \tag{9}$$

It is important to realize that H' is not a Hermitian operator due to presence of the logarithmic derivatives of the open channel wavefunctions in the definition of L. We expand the operator G as

$$G = \sum_{iq,jr} |\Psi^0_{E_i}\theta_{iq})G_{iq,jr}(\Psi^0_{E_j}\theta_{jr}| \tag{10}$$

and get an equation for $G_{iq,jr}$ by substituting into equation (9)

$$(\varepsilon_{iq} - E)G_{iq,jr} = \delta_{ij}\delta_{qr} - \sum_{k,s}(\Psi^0_{E_i}\theta_{iq}\; H'|\Psi^0_{E_k}\theta_{ks})G_{ks,jr} \tag{11}$$

where

$$(\Psi^0_{E_i}\theta_{iq}|H'|\Psi^0_{E_k}\theta_{ks}) = (\Psi^0_{E_i}\theta_{iq}|H_e^{N+1} + T_R + L_b|\Psi^0_{E_k}\theta_{ks}) + (\Psi^0_{E_i}\theta_{iq}|L - L_b|\Psi^0_{E_k}\theta_{ks}) \;. \tag{12a}$$

Explicitly

$$(\Psi^0_{E_i}\theta_{iq}|L - L_b|\Psi^0_{E_k}\theta_{ks})$$

$$= \frac{1}{2}\sum_{j,\alpha}(\chi^{AB}_{j\alpha}|g^j_{E_i}(a_e,R)|\theta_{iq})(L_{j\alpha} - b_e)(\theta_{ks}|g^j_{E_k}(a_e,R)|\chi^{AB}_{j\alpha})$$

$$+ \frac{1}{2m_N}g^{AB^-}_{E_i}(a_R)\theta_{iq}(a_R)(L_{AB^-} - b_N)g^{AB^-}_{E_k}(a_R)\theta_{ks}(a_R) \quad (12b)$$

where

$$g^j_{E_i}(a_e,R) = (\Psi^{AB}_j|\ {}^0_{E_i})_{r_1 \cdots r_N, r_{N+1}=a_e} \quad (12c)$$

$$g^{AB^-}_{E_i}(a_R) = (\Psi^{A^-}\Psi^B|\ {}^0_{E_i})_{r_1 \cdots r_{N+1}, R=a_R} \quad . \quad (12d)$$

It is important to note that $g^j_{E_i}$ depends only parametrically on the nuclear co-ordinate and may be obtained from an electronic R-matrix calculation at fixed geometry of the nuclei. Since the perturbation has both a real and imaginary part, it is possible to define a level shift and width matrix as

$$\Delta E_{iq,ks} = (\Psi^0_{E_i}\theta_{iq}|H^{N+1}_e + T_R + L_b|\Psi^0_{E_k}\theta_{ks})$$

$$+ \mathrm{Re}(\Psi^0_{E_i}\theta_{iq}|L - L_b|\Psi^0_{E_k}\theta_{ks}) \quad (13a)$$

$$\Gamma_{iq,ks} = -\mathrm{Im}(\Psi^0_{E_i}\theta_{iq}|L - L_b|\Psi^0_{E_k}\theta_{ks})$$

$$= -\mathrm{Im}\{\frac{1}{2}\sum_{j,\alpha}(\chi^{AB}_{j\alpha}|g^j_{E_i}(a_e,R)|\theta_{iq}) \times (L_{j\alpha} - b)(\theta_{ks}|g^j_{E_k}(a_e,R)|\chi^{AB}_{j\alpha})$$

$$+ \frac{1}{2m_N}g^{AB^-}_{E_i}(a_R)\theta_{iq}(a_R)(L_{AB^-} - b_N)g^{AB^-}_{E_k}(a_R)\theta_{ks}(a_R)\} \quad . \quad (13b)$$

Thus we may cast equation (11) in the following form:

$$(\varepsilon_{iq} - E)G_{iq,jr} = \delta_{ij}\delta_{qr} - \sum_{k,s}[\Delta E_{iq,ks} - i\Gamma_{iq,ks}]G_{ks,jr} \quad (14a)$$

$$\sum_{ks} [(\varepsilon_{iq} - E)\delta_{ik}\delta_{qs} + \Delta E_{iq,ks} - i\Gamma_{iq,ks}]G_{ks,jr} = \delta_{ij}\delta_{qr} \ . \quad (14b)$$

A formal solution of equation (5b) is therefore

$$|\Psi_E) = \sum_{iq,jr} |\Psi^0_{E_i}\theta_{iq})(\Psi^0_{E_i}\theta_{iq}|(\varepsilon + \Delta E - i\Gamma - E)^{-1}|\Psi^0_{E_j}\theta_{jr})$$

$$\times (\Psi^0_{E_j}\theta_{jr}|L|\Psi^-_E) \ . \quad (15)$$

The salient features of equation (15) are the presence of a complex level shift matrix and <u>only</u> the incoming wave on the right hand side of the equality. By projecting equation (15) onto the asymptotic states of the target we may derive appropriate expressions for the S-matrix elements of physical interest. Thus we have

$$(\Psi^{AB}_p \chi^{AB}_{p\alpha}|\Psi_E) = F^E_{p\alpha}(r)$$

$$= \sum_{iq,jr} (\chi^{AB}_{p\alpha}|F^p_{E_i}(r,R)|\theta_{iq})(\Psi^0_{E_i}\theta_{iq}|(\varepsilon + \Delta E - i\Gamma - E)^{-1}|\Psi^0_{E_j}\theta_{jr})$$

$$\times (\Psi^0_{E_j}\theta_{jr}|L|\Psi^-_E) \quad (16a)$$

$$(\Psi^{A^-}\Psi^B|\Psi_E) = F^E_{AB^-}(R)$$

$$= \sum_{iq,jr} F^{AB^-}_{E_i}(R)|\theta_{iq})(\Psi^0_{E_i}\theta_{iq}|(\varepsilon + \Delta E - i\Gamma - E)^{-1}|\Psi^0_{E_j}\theta_{jr})$$

$$\times (\Psi^0_{E_j}\theta_{jr}|L|\Psi^-_E) \ . \quad (16b)$$

Asymptotically

$$F^E_{p\alpha}(a_e) = \sum_{iq,jr} (\chi^{AB}_{p\alpha}|g^p_{E_i}(a_e,R)|\theta_{iq})$$

$$\times (\Psi^0_{E_i}\theta_{iq}|(\varepsilon + \Delta E - i\Gamma - E)^{-1}|\Psi^0_{E_j}\theta_{jr})(\Psi^0_{E_j}\theta_{jr}|L|\Psi^-_E) \quad (17a)$$

$$F^{E}_{AB^-}(a_R) = \sum_{iq,jr} g^{AB^-}_{E_i}(a_R)\theta_{iq}(a_R)$$

$$\times\ (\Psi^{0}_{E_i}\theta_{iq}|(\varepsilon + \Delta E - i\Gamma - E)^{-1}|\Psi^{0}_{E_j}\theta_{jr})(\Psi^{0}_{E_j}\theta_{jr}|L|\Psi^{-}_{E}). \qquad (17b)$$

The matrix element on the right hand side involving the operator L may be written as

$$(\Psi^{0}_{E_j}\theta_{jr}|L|\Psi^{-}_{E}) = \frac{1}{2}\sum_{k,\beta}(\chi^{AB}_{k\beta}|g^{k}_{E_j}(a_e,R)|\theta_{jr})\left(\frac{\partial F^{E}_{k\beta}}{\partial r} - L_{k\beta}F^{E}_{k\beta}\right)_{a_e}$$

$$+ \frac{1}{2m_N} g^{AB^-}_{E_j}(a_R)\theta_{jr}(a_R)\left(\frac{\partial F^{E}_{AB^-}}{\partial R} - L_{AB^-}F^{E}_{AB^-}\right)_{a_R} \qquad (18a)$$

$$(\Psi^{0}_{E_j}\theta_{jr}|L|\Psi^{-}_{E}) = \frac{1}{2}\sum_{k,\beta}(\chi^{AB}_{k\beta}|g^{k}_{E_j}(a_e,R)|\theta_{jr})\left(\frac{\partial F^{-E}_{k\beta}}{\partial r} - L_{k\beta}F^{-E}_{k\beta}\right)_{a_e} \qquad (18b)$$

because there is no incident wave in the dissociative channel. Finally we have

$$F^{E}_{p\alpha}(a_e) = \sum_{iq,jr,k,} (\chi^{AB}_{p\alpha}|g^{p}_{E_i}(a_e,R)|\theta_{iq})$$

$$\times\ (\chi^{0}_{E_i}\theta_{iq}|(\varepsilon + \Delta E - i\Gamma - E)^{-1}|\Psi^{0}_{E_j}\theta_{jr})$$

$$\frac{1}{2}(\chi^{AB}_{k\beta}|g^{k}_{E_j}(a_e,R)|\theta_{jr})\left(\frac{\partial F^{-E}_{k}}{\partial r} - L_{k\beta}F^{-E}_{k\beta}\right)_{a_e} \qquad (19a)$$

$$F^{E}_{AB^-}(a_R) = \sum_{\substack{iq,jr\\k,\beta}} g^{AB^-}_{E_i}(a_R)\theta_{iq}(a_R)$$

$$\times\ (\Psi^{0}_{E_i}\theta_{iq}|(\varepsilon + \Delta E - i\Gamma - E)^{-1}|\Psi^{0}_{E_j}\theta_{jr})$$

$$\times\ \frac{1}{2}(\chi^{AB}_{k\beta}|g^{k}_{E_j}(a_e,R)|\theta_{jr})\left(\frac{\partial F^{-E}_{k\beta}}{\partial r} - L_{k\beta}F^{-E}_{k\beta}\right)_{a_e} \qquad (19b)$$

By defining the two operators

$$G_{p,k}(R|R') = \sum_{iq,jr} g^{p}_{E_i}(a_e,R)\theta_{iq}(R) \times (\Psi^{0}_{E_i}\theta_{iq}|(\varepsilon + \Delta E - i\Gamma - E)^{-1}|\Psi^{0}_{E_j}\theta_{jr}) \times \theta_{jr}(R')g^{k}_{E_j}(a_e,R') \tag{20a}$$

$$G_{AB^-,k}(a_R|R) = \sum_{iq,jr} g^{AB^-}_{E_i}(a_R)\theta_{iq}(a_R) \times (\Psi^{0}_{E_i}\theta_{iq}|(\varepsilon + \Delta E + i\Gamma - E)^{-1}|\Psi^{0}_{E_j}\theta_{jr}) \times \theta_{jr}(R)g^{k}_{E_j}(a_e,R) \tag{20b}$$

we obtain

$$F^{E}_{p\alpha}(a_e) = \sum_{k,\beta}(\chi^{AB}_{p\alpha}|G_{pk}(R|R')|\chi^{AB}_{k\beta})\,\frac{1}{2}\,\left(\frac{\partial F^{-E}_{k}}{\partial r} - L_{k\beta}F^{-E}_{k\beta}\right)_{a_e} \tag{21a}$$

$$F^{E}_{AB^-}(a_R) = \sum_{k,\beta}(G_{AB^-,k}(a_R|R)|\chi^{AB}_{k\beta})\,\frac{1}{2}\,\left(\frac{\partial F^{-E}_{k\beta}}{\partial r} - L_{k\beta}F^{-E}_{k\beta}\right)_{a_e} \tag{21b}$$

or equivalently

$$F^{+E}_{p\alpha}(a_e) = -F^{-E}_{p\alpha}(a_e) + \sum_{k,\beta}\frac{1}{2}\,(\chi^{AB}_{p\alpha}|G_{pk}(R|R')|\chi^{AB}_{k\beta})\left(\frac{\partial F^{-E}_{k\beta}}{\partial r} - L_{k\beta}F^{-E}_{k\beta}\right)_{a_e} \tag{21c}$$

$$F^{+E}_{AB^-}(a_e) = \sum_{k,\beta}\frac{1}{2}\,(G_{AB^-,k}(a_R|R)|\chi^{AB}_{k\beta})\left(\frac{\partial F^{-E}_{k\beta}}{\partial r} - L_{k\beta}F^{-E}_{k\beta}\right)_{a_e} \tag{21d}$$

Clearly the whole theory rests on the ability to evaluate the Green's functions appearing in equations (21). By returning to our original equation (10) it is simple matter to show

$$G_{p,k}(R|R') = \int \Psi^{O*}_{E_p}(r_1 \ldots r_N, R) G(R_1 \ldots r_N a_e R | r'_1 \ldots r'_N a_e R')$$

$$\times\ \Psi^{O}_{E_k}(r'_1 \ldots r'_N, R')\ d\vec{r}_1 \ldots d\vec{r}_N d\vec{r}'_1 \ldots d\vec{r}'_N \quad (22a)$$

$$G_{A\bar{B},k}(a_R|R') = \int \Psi^{\bar{A}^*}(r_1 \ldots r_j)\Psi^{B^*}(r_{j+1} \ldots r_{N+1})$$

$$\times\ G(r_1 \ldots r_{N+1} a_R | r'_1 \ldots r'_N a_e R')$$

$$\times\ \Psi^{O}_{E_k}(r'_1 \ldots r'_N, R')\ d\vec{r}_1 \ldots d\vec{r}_{N+1} d\vec{r}'_1 \ldots d\vec{r}'_N \ . \quad (22b)$$

There are a number of points worth noting about equation (11). The presence of the complex level shift and width matrix implies that the solutions of the eigenvalue problem

$$H|\Psi_\lambda) = E_\lambda(E)|\Psi_\lambda) \quad (23a)$$

$$H^+|\chi_\lambda) = E^*_\lambda(E)|\chi_\lambda) \quad (23b)$$

form a bi-orthogonal set. This basis known as the Kapur-Peierls basis[5] has an implicit dependence on the incident energy due to the presence of the channel energies in the definition of L. Because of the boundary condition the eigenfunctions of equation (23a) contain only outgoing waves while those of (23b) only incoming. It is easy to show that all the eigenvalues of equation (23a) lie in the lower half of the complex plane by realizing that Γ is a positive semi-definite matrix. The bi-orthogonality relationship implied by equations (23) allow a simple spectral resolution of G

$$G = \sum_\lambda \frac{|\Psi_\lambda)(\chi_\lambda|}{E_\lambda(E) - E} \ . \quad (24)$$

Thus in a mathematical sense the Kapur-Peierls basis is the natural basis for the problem. At first glance, it might appear that those $E_\lambda(E)$ lying near the real axis would be responsible for any resonances in the collisions processes. In point of fact, however, resonant poles are an intrinsic property of the system and do not depend on the incident energy. The resonant poles may be located by examining equation (24) as a function of energy and determing those energies, ε_n satisfying the condition

$$E_\lambda(\varepsilon_n) = \varepsilon_n \quad . \tag{25}$$

These values, which are called the Siegert eigenvalues, correspond to boundary conditions which replace the actual energy by the resonance energy in the definition of L. By expanding

$$\begin{aligned} E_\lambda(E) &= E_\lambda(\varepsilon_n) + \left.\frac{\partial E_\lambda}{\partial E}\right|_{E=\varepsilon_n} (E - \varepsilon_n) + \ldots \\ &= \varepsilon_n + \gamma_{\lambda n}(E - \varepsilon_n) + \ldots \end{aligned} \tag{26}$$

we get

$$G = \sum_\lambda \frac{|\Psi_\lambda)(\chi_\lambda|}{(\varepsilon_n - E)(1 + \gamma_{\lambda n} + \ldots)} \quad . \tag{27}$$

The nth residue of this equation is

$$R_n = \sum_\lambda \left.\frac{|\Psi_\lambda)(\chi_\lambda|}{(1 + \gamma_{\lambda n})}\right|_{E=\varepsilon_n} \tag{28}$$

and allows us to expand G as

$$G = \sum_n \frac{R_n}{\varepsilon_n - E} \tag{29}$$

which shows that the Kapur-Peierls and Siegert states are related. Clearly both the Kapur-Peierls and Siegert eigenvalues depend on the nuclear co-ordinates. This dependence is due to the boundary conditions imposed on the eigenfunctions. The R-matrix basis does not suffer from this difficulty as the boundary conditions are entirely arbitrary and energy independent. Any nonadiabatic effects appear through the coupling of different electronic R-matrix levels in the inverse operator of equations (20). In many cases of physical interest the captured electron is moving much more rapidly than the nuclei in the internal region and the coupling is negligible. However, should it be necessary to account for nonadiabatic effects this can be done as in spectroscopic applications by computing the coupling matrix element implied by equation (20). There are several practical points to be emphasized about the inversion or diagonalization of the complex Hamiltonian of the Kapur-Peierls and Siegert theories. From the point of view of the Bloch operator formalism there is no need whatsoever to include complex basis functions in the formation of the matrix. The outgoing wave boundary conditions are replaced by an operator L. By introducing an arbitrary basis set and diagonalizing (or inverting) this modified operator we are insured of converging to the correct eigenvalues as the basis set is increased. Secondly, the matrix elements of the operator L are very simple to evaluate as they are surface not volume integrals. Thus the level shift and width operators arise from the flux of particles entering and leaving the interaction volume. In the light of these remarks it may well be unnecessary to modify existing integral programs to handle complex basis functions. One must be a little cautious however about questions of basis set convergence if a purely real expansion is used.

In order to illustrate how the R-matrix approach can be used as a computational tool and to provide physical insight into a simple resonant process, let us consider the reaction

$$e + A_2(\nu=0) \rightarrow A_2^{*-} \rightarrow e + A_2(\nu=n) \ . \tag{30}$$

Since no inelastic electronic channels appear the relevant equations are

$$F_\alpha^E(a_e) = \sum_\beta (\chi_\alpha^{A_2}|G(R|R')|\chi_\beta^{A_2}) \frac{1}{2} \left(\frac{\partial F_\beta^{-E}}{\partial r} - L_\beta F_\beta^{-E}\right)_{a_e} \tag{31a}$$

$$\begin{aligned} G(R|R') = & \sum_{iq,jr} g_{E_i}(a_e,R)\theta_{iq}(R) \\ & \times (\Psi_{E_i}^0\,\theta_{iq}|(\varepsilon + \Delta E - i\Gamma - E)^{-1}|\Psi_{E_j}^0\,\theta_{jr}) \\ & \times \theta_{jr}(R')g_{E_j}(a_e,R') \ . \end{aligned} \tag{31b}$$

We now assume that there is negligible nonadiabatic coupling between the R-matrix levels so that

$$G(R|R') = \sum_{i,q,r} g_{E_i}(a_e,R)\theta_{iq}(R)$$

$$\times\ (\Psi^0_{E_i}\theta_{iq}|(\varepsilon + \Delta E - i\Gamma - E)^{-1}|\Psi^0_{E_i}\theta_{ir})$$

$$\times\ \theta_{ir}(R')g_{E_i}(a_e,R)$$

$$= \sum_i g_{E_i}(a_e,R)G_i(R|R')g_{E_i}(a_e,R') \tag{32}$$

where

$$(T_R + \Delta E_i(R|R') - i\Gamma_i(R|R') - E)G_i(R|R') = \delta(R - R') \ . \tag{33}$$

Note that even in this simple case the nuclear motion must be described in terms of a nonlocal, complex potential. A more detailed examination of equations (13a,b) shows that both ΔE_i and Γ_i depend on energy as well. Equation (33) is identical in form to the equation derived using the Feshbach projection operator technique for an isolated electronic resonance. Finally we obtain

$$F^E_\alpha(a_e) = \sum_{\beta,i} (\chi^{A_2}_\alpha|g_{E_i}(a_e,R)G_i(R|R')g_{E_i}(a_e,R')|\chi^{A_2}_\beta) \times \frac{1}{2}\left(\frac{\partial F^{-E}_\beta}{\partial r} - L_\beta F^{-E}_\beta\right)_{a_e} \ . \tag{34}$$

From a computational point of view the essential difficulty is the calculation of $g_{E_i}(a_e,R)$. These may be obtained using previously developed fixed-nuclei electronic R-matrix programs at the required internuclear distances needed to perform the integration over R. The calculation of the Green's function $G_i(R|R')$ is essentially trivial at least for diatomic molecules. It may be obtained by numerical integration or by spectral decomposition. The appropriate states are, of course, the Kapur-Peierls biorthogonal set.

$$F^E_\alpha(a_e) \cong \sum_{\beta,i,n} \frac{(\chi^{A_2}_\alpha|g_{E_i}(a_e,R)|H_{in})(I_{in}|g_{E_i}(a_e,R)|\chi^{A_2}_\beta)}{E_{in}(E) - E} \times \frac{1}{2}\left(\frac{\partial F^{-E}_\beta}{\partial r} - L_\beta F^{-E}_\beta\right)_{a_e} \tag{36}$$

where $\hat{g}_{E_i}$ is some average $g_{E_i}(a_e,R)$. Thus the strength of the transition is governed by two factors, a Franck-Condon overlap of nuclear factors involving the target and compound state vibrational wavefunctions and an energy denominator which is singular at the compound state nuclear vibrational levels. This physical picture of vibrational resonances has been used by experimentalists for many years and has been derived using a wide variety of theoretical models[6]. The R-matrix approach has the advantage of being able to calculate the necessary quantities, ab initio, for both resonant and nonresonant collisions in a consistent fashion as well as provide physical insight into the nature of the scattering process.

REFERENCES

1. B. I. Schneider, "Electron and Photon Molecule Collisions," ed. by T. N. Rescigno, B. V. Mckoy and B. I. Schneider, Plenum Press, New York (1979).
2. B. I. Schneider, M. LeDourneuf and P. G. Burke, J. Phys. B 12: L365 (1979).
3. B. I. Schneider, Phys. Rev. A14:1823 (1976).
4. B. I. Schneider, M. LeDournef, and Voky Lan, Phys. Rev. Lett. 43:1926 (1979); Invited paper Symposium on Electron-Molecule Collisions, ed. I. Shimamura and M. Matsuzawa, University of Tokyo Press (1979).
5. C. Bloch, Nucl. Phys. 4:503 (1957).
6. D. T. Birtwistle and A. Herzenberg, Phys. Rev. A1:1592 (1970).

RECENT DEVELOPMENTS OF THE FRAME TRANSFORMATION THEORY

OF ELECTRON MOLECULE PROCESSES

I. R-MATRIX DESCRIPTION OF SHORT RANGE ELECTRONUCLEAR CORRELATIONS IN RESONANT VIBRATIONAL EXCITATION AND DISSOCIATIVE ATTACHMENT

M. Le Dourneuf, Vo Ky Lan and B.I. Schneider*

Observatiore de Paris
92190 Meudon
France

A computationally efficient technique for an *ab initio* description of electron-molecule processes is investigated within the framework of frame transformations. In the first part, we concentrate on resonant processes, determinant in vibrational excitation and dissociative attachment. The key tool of the frame transformations is then the R-matrix method which allows separate resolution of the Schrödinger equation in the close and distant interaction regions, where completely different physical conditions prevail: formation of a complex AB^- at short distances, electronic or nuclear dissociation ($e + AB$, $A^- + B$) at large distances. The implementation of the method is illustrated on the well known $N_2^-(^2\Pi g)$ vibrational resonance, showing that the *electronic correlations* are efficiently described by a simple Hartree-Fock representation of the AB^- complex and the *electronuclear correlations* are also efficiently described by a simple Born-Oppenheimer factorisation of the electronic and nuclear motions of the AB^- complex. The connection with alternative approaches, specially the widely successful Boomerang model, closes the discussion.

*During his 1978-79 sabbatical leave from his permanent institution, Los Alamos Scientific Laboratory, Los Alamos, New Mexico.

I. INTRODUCTION

The complexity of electron-molecule dynamics in its whole is well established. On the one hand, the electrostatic interaction is considerably more anisotropic and of longer range than in atoms, due to the presence of two or more uncentered nuclear singularities. On the other hand, the additional degrees of freedom associated to the nuclear motion leads to a formidable number of closely spaced rovibrational excitation chammels, as well as rearrangement processes ($e + AB \to A + B^-$). The huge difference in mass between the electrons and the nuclei in most cases allows one to neglect the nuclear motion during the electronic collision and therefore to obtain rotational and vibrational excitation cross sections by a mere frame transformation on the fixed nuclei scattering matrix ("impulse approximation" for the nuclei, more often referred to as the "adiabatic nuclei approximation" for the electron[1]). However, the most interesting situations of strong nuclear excitations - huge rotational and vibrational cross sections near threshold for polar molecules, 10^3 fold enhancement of the vibrational cross sections near shape resonances - result from strong and long-lasting electronuclear correlations which must be properly described.

None of the standard methods gives a uniformly satisfactory description of the electron-molecule dynamics. Only recently, the *traditional close coupling* has succeeded in providing converged fixed nuclei results for a variety of typical diatomics (homonuclear neutrals H_2, N_2, polar CO, HF, HCl, LiH, LiF, ionic H_2^+, N_2^+, CH^+) in the static exchange or model exchange approximations, with or without model polarisation[2,3]. Its extension to nuclear excitations - rotational[4], vibrational or hybrid close coupling (CC)[5]-, although straightforward in principle, has failed to produce converged results. At last, its extension to dissociative processes leads to the usual formal problems of rearrangement collisions. Interesting information has been provided by *pure L^2 methods*. The stabilisation method has given valuable information about negative ion resonances[6] but its weakness in the description of long range effects may limit its predictive value in some cases.[7] The pseudo bound state approach to electron scattering[8] can be considered as the precursor of the T-matrix and R-matrix methods, which belong to the *new generation of methods, hybrid between collision and bound state techniques*.

Indeed, the complexity of the electron-molecule scattering problem is most efficiently overcome by *taking advantage of the qualitatively different features of the interaction at short and large distances*, as suggested by Chang and Fano (1972) in their original paper on the Frame Transformation (FT) theory and summarized of Fig. 1. Roughly speaking, one must describe, for each degree of freedom, the *transition* from a purely decoupled situation at infinity - collisional electron decoupled from *N electrons* in a *Born-*

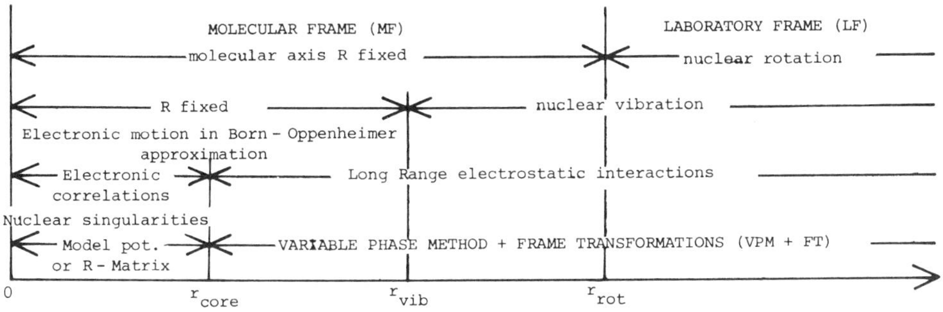

Fig. 1. The physical regions of the electronic configuration space in electron molecule scattering.

Oppenheimer (BO) core state - to the formation of a short range (R) molecular compound of *N+1 equivalent electrons,* strongly coupled to oneanother and to the nuclei forming *a Born-Oppenheimer compound state.* The practical implementation of the FT assumes a careful initial estimation of physical simplifications, the development of adapted tools to describe them accurately and the ultimate test of the calculation and physical assumptions by experiment. This review aims at describing substantial progresses made recently along these lines.

In this paper, we will show how the R-matrix (RM) formalism allows completely independent descriptions of the reaction zone (AB^- complex formation, most efficiently described by bound state techniques) and the decoupling zones (e + AB, A^- + B, most efficiently described by collisional techniques), providing the simplest description of resonant and dissociative processes (§II). Its practical implementation and efficiency to describe the SR electronuclear compound, will be illustrated on the widely studied problem of the $^2\Pi_g$ shape resonance in the vibrational excitation of N_2 around 2eV (§III). Finally the connection with methods akin to the R-matrix formalism, especially the resonant scattering theory, will be discussed (§IV).

The next paper will then describe two recent contributions: the single center (SC) collisional approach combined with the FT ideas. Expected is the contribution of the variable phase method (VPM) and related asymptotic techniques to a general implementation of the FT description of the long range (LR) electro nuclear correlations near threshold for rotational (vibrational) excitation. Less expected is the contribution of a newly developed adiabatic formulation of the SC close coupling as an efficient, complementary tool of the R-matrix in a physically transparent description of SR correlations.

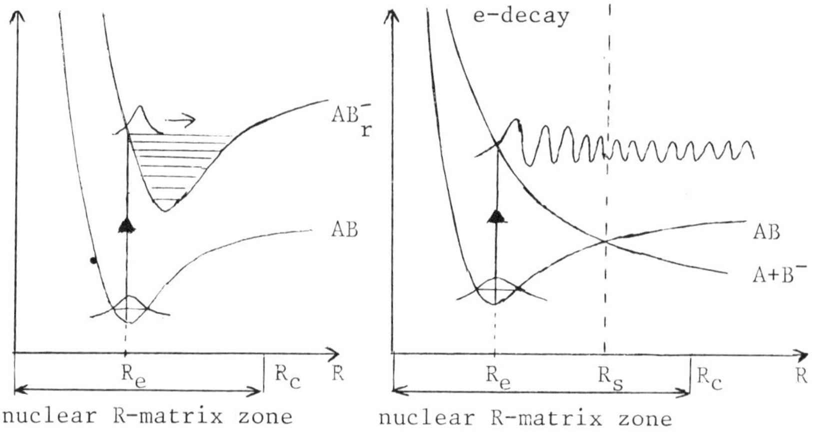

Fig. 2. The physical regions of the electronuclear configuration space.

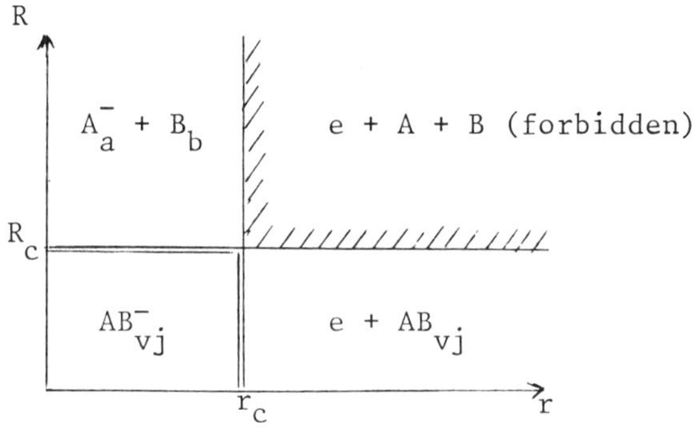

Fig. 3. a) Vibrational excitation b) Dissociative attachment

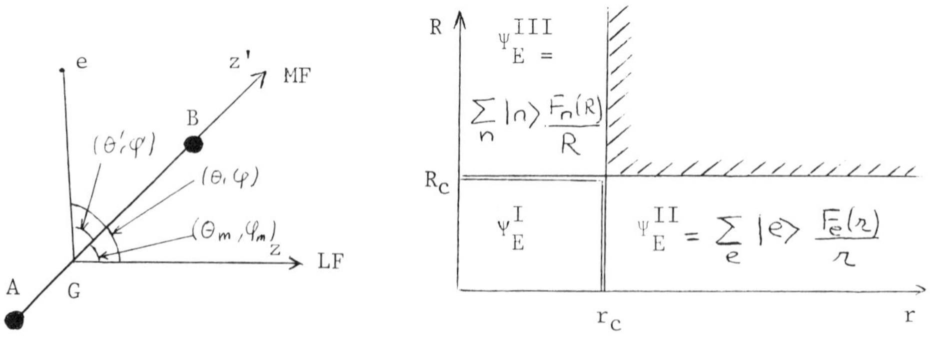

Fig. 4. a) The laboratory frame and molecular frame b) The different expansion of the wavefunction

Our final aim is to "prove" that the SR interactions correspond to the formation of a strongly correlated electronuclear compound, in which all the electrons follow adiabatically the nuclear motion, according to the usual Born Oppenheimer situation in bound states.

II. THE R-MATRIX DESCRIPTION OF THE ELECTRONUCLEAR REACTION ZONE

A. The R-Matrix Reaction Zone

The R-matrix electronuclear hypersphere should include the zone of strong electronuclear coupling.[9] The simplest estimate is provided by a rectangle in electronuclear radial coordinates (Fig. 2). The maximum nuclear excursion R_c is easily estimated in resonant vibrational excitation from classical arguments (Fig. 3a); and in dissociative attachment, it should usually go beyond the classical stabilisation point (Fig. 3b). The electronic compound limit r_c is then estimated for the largest internuclear distance R_c as the minimum radius of a sphere including most of the charge distribution of the target AB and the resonant compound $AB_{\bar{r}}$, inside which exchange and other strong electronic and electronuclear correlations are expected to be important. This simple choice obviously excludes three-body dissociation, which would require a substantial extension of the formalism.

B. The R-Matrix Hamiltonian

Following Bloch,[10] the dynamics inside the finite R-matrix zone is fully characterized by a modified Hamiltonian, later on referred to as the R-matrix Hamiltonian, which selects the Hermitian part of the original Hamiltonian for arbitrary functions, by adding convenient surface operators:

$$\tilde{H} = H_{el} + T_{nuc} + L^{b}_{el} + L^{B}_{nuc} \tag{1}$$

where L^{b}_{el} es the monoelectronic ejection surface operator

$$L^{b}_{el} = \sum_{e} |e\rangle \delta(r-r_c)\frac{1}{2}\left|\frac{1}{r}\frac{\delta}{\delta r} r-b\right|\langle e| \tag{2a}$$

on all possible electronic ejection channels

$$|e\rangle = \left|Y_{1\lambda}(\hat{r}'_{N+1})\phi^{BO}_{i\Lambda}(N,R)X^{N}_{v}(R)\right|_{MF}\left(\frac{2\bar{J}+1}{4\pi}\right)^{1/2} D^{\bar{J}*}_{\bar{M}\bar{\Lambda}}(\hat{R}) \tag{2b}$$

describing the escape of a collisional electron in partial wave 1, leaving the molecular target in electronic state $\phi^{BO}_{i\Lambda}$ (with projection Λ of the electronic angular momentum on the internuclear axis), and vibrational state $X^{N}_{v}(R)$, the whole system following the molecular rotation (molecular frame transformation defined in Fig. 4a). This assumption, following Chang and Fano's[11] suggestion that the rotational decoupling occurs at much larger distances, will be discussed at length in the accompanying paper.

L^{B}_{nuc} is the nuclear fragmentation surface operator

$$L^{B}_{nuc} = \sum_{n} |n\rangle \delta(R-R_b) \frac{1}{2\mu} \left| \frac{1}{R} \frac{d}{dR} R - B \right| \langle n| \tag{3a}$$

on all possible nuclear fragmentation channels

$$|n\rangle = \phi^{BO}_{i\bar{\Lambda}}(N+1,R) \left(\frac{2\bar{J}+1}{4\pi}\right)^{1/2} D^{\bar{J}*}_{\bar{M}\bar{\Lambda}}(\hat{R}) \tag{3b}$$

describing the partial wave $\bar{J}\bar{M}$ of the relative motion of two atomic fragments whose electronic state remains conveniently described by a molecular BO description.

b of Eq. (2a) and B of Eq. (3a) are two arbitrary constants of the R-matrix formalism, usually assumed to be zero in numerical applications.

C. The R-Matrix Basis

The basis solution Ψ_k of the Schrödinger eigenvalue equation

$$\{H_{el} + L^{b}_{el} + T_{nuc} + L^{B}_{nuc}\}\Psi_k = E_k \Psi_k \tag{4}$$

forms a discrete orthonormal set, selected by the boundary conditions associated with the singular Bloch operators

$$L^{b}_{el}\ \Psi_k = 0 \quad \Longleftrightarrow \tag{5a}$$

$$\frac{d}{dr} F_{ek} - bF_{ek} \Big|_{r=r_c} = 0 \qquad \text{for} \qquad \frac{F_{ek}}{r} = \langle ek|\Psi_k\rangle \tag{5b}$$

$$L^{B}_{nuc}\ \Psi_k = 0 \quad \Longleftrightarrow \tag{6a}$$

$$\frac{d}{dR} F_{nk} - BF_{nk} \Big|_{R=R_c} = 0 \qquad \text{for} \qquad \frac{F_{nk}(R)}{R} = \langle n|\Psi_k\rangle \tag{6b}$$

In contrast to the usual atomic R-matrix method, the molecular

R-matrix basis has up to now been obtained variationally, using the same analytical basis as in usual bound state codes (Slater,[12] Gaussian[13]).

The Born Oppenheimer adiabatic factorisation of the compound electronic and nuclear motion is, in most circumstances, expected to be a reasonably good approximation, in accord with the physical intuition that, within the strongly correlated compound, all the electrons are similarly accelerated by the nuclear Coulombic field and adapt themselves similarly to the nuclear motion.

The eigenstate determination splits then in two parts:

i) the R-matrix electronic eigenstates at fixed R,

$$\{H_{el}(N+1,R) + L^{b}_{el} - E_k(R)\}\phi_{k\bar{\Lambda}}(N+1,R) = 0 \qquad (7a)$$

which provide a discrete set of R-matrix electronic states and potential curves $E_k(R)$.

ii) the R-matrix vibrational sublevels for each electronic potential

$$\{T_{nuc} + L^{B}_{nuc} + E_k(R) - E_{k\nu}\}\chi^{N+1}_{k\nu}(R) = 0. \qquad (7b)$$

More insight into the practical significance of the Born-Oppenheimer approximation for the short range compound is gained by the use of the typical R-matrix potential curves of the N_2^- complex (Fig. 6). Remembering that the surface boundary condition (5) leads to quantized values of the wave number, which are inversely proportional to the box size r_c for a free particle, one expects that the corresponding quantisation of the compound electronic motion within the R-matrix zone will provide wide R-matrix electronic splittings and slowly R-dependent R-matrix electronic eigenstates.

"The Born-Oppenheimer diabatic" approach may be more convenient in exceptional cases of weakly avoided crossings. A generalized RM formalism proceeds then in three steps:

i) definition of a diabatic basis $\varphi_\kappa(N+1,R)$ of RM electronic states which are the $\varphi_\kappa(N+1,R)$ solutions of a truncated electronic Hamiltonian $\mathcal{H}_{el}$

$$\{\mathcal{H}_{el}(N+1,R) + L^{b}_{el} - \varepsilon_\kappa(R)\}\varphi_\kappa(N+1,R) = 0. \qquad (8a)$$

ii) determination of the corresponding nuclear basis

$$\{T_{nuc}(R) + L^{B}_{nuc} + \varepsilon_\kappa(R) - \varepsilon_{\kappa\nu}\}\chi^{N+1}_{\kappa\nu}(R) = 0. \qquad (8b)$$

iii) calculation of the final RM basis by diagonalization of the Hamiltonian on the diabatic basis

$$\psi_k = \sum_{\kappa\nu} c_{\kappa\nu}^{(k)} \varphi_\kappa (N+1,R) \chi_{\kappa\nu}^{N+1}(R). \tag{8c}$$

D. The R-Matrix Hypersphere

Within the R-matrix hypersphere, the solution of the exact Schrödinger equation at energy E is obtained formally

$$(H-E)\psi_E^I = 0 \Longleftrightarrow (H-E)\psi_E^I = (L_{el}^b + L_{nuc}^B)\psi_E^I \tag{9}$$

in terms of the RM Green function G and the surface amplitude

$$\psi_E^I = \underbrace{\sum_n \frac{|\psi_n\rangle\langle\psi_n|}{E_n - E}}_{G\,(E)} \underbrace{[L_{el}^b + L_{nuc}^B]\,\psi_E^I}_{\text{surface amplitude}} \tag{10}$$

E. The R-Matrix Boundary Conditions on the Hypersphere

The R-matrix boundary conditions on the hypersphere, which constitute the only information necessary to a scattering calculation, are obtained by projecting eq. 10 on the hypersurface basis

$$\begin{bmatrix} F_e(r_c) \\ \\ F_n(R_c) \end{bmatrix} = \begin{bmatrix} R_{ee} & R_{en} \\ \\ R_{ne} & R_{nn} \end{bmatrix} \begin{bmatrix} F'_e(r_c) \\ \\ F'_n(R_c) \end{bmatrix} . \tag{11}$$

F_e, F_n are the electronic and nuclear channel functions associated with the collisional expansions appropriate to the electronic and nuclear decoupling zones (Fig. 4b).

$$\psi_E^{II} = \sum_e |e\rangle \frac{F_e(r)}{r} \tag{12a}$$

$$\psi_E^{III} = \sum_n |n\rangle \frac{F_n(R)}{R} \tag{12b}$$

The R-matrix is defined as the surface projection of the inner zone Green function

$$\begin{aligned} R_{ee} &= \langle e\delta(r-r_c)|G|e\delta(r-r_c)\rangle \\ R_{en} &= \langle e\delta(r-r_c)|G|n\delta(R-R_c)\rangle \\ R_{nn} &= \langle n\delta(R-R_c)|G|n\delta(R-R_c)\rangle . \end{aligned} \tag{13}$$

Eq. (11) can be recast in a more convenient form, connecting the electronic and nuclear boundary conditions

$$\begin{bmatrix} F_e(r_c) \\ \\ F'_e(r_c) \end{bmatrix} = \begin{bmatrix} G^c_{en} & G^c_{e'n} \\ \\ G^c_{en'} & G^c_{e'n'} \end{bmatrix} \begin{bmatrix} F_n(R_c) \\ \\ F'_n(R_c) \end{bmatrix} \tag{14}$$

F. Propagation in the External Zone and Final Matching

Although we defer the reader to the next paper for a discussion of the most appropriate techniques to solve the CC equations accounting for residual LR potentials in the electronic and nuclear decoupling zones (Fig. 4b), it is clear that the propagation of a complete set of independent solutions provides the formal connection to the dynamical infinity in either decoupling zone

$$\begin{bmatrix} F_e(r_\infty) \\ \\ F'_e(r_\infty) \end{bmatrix} = \begin{bmatrix} G^e_{\infty c} & G^e_{\infty c'} \\ \\ G^e_{\infty' c} & G^e_{\infty' c'} \end{bmatrix} \begin{bmatrix} F_e(r_c) \\ \\ F'_e(r_c) \end{bmatrix} \tag{15a}$$

$$\begin{bmatrix} F_n(R_\infty) \\ \\ F'_n(R_\infty) \end{bmatrix} = \begin{bmatrix} G^n_{\infty c} & G^n_{\infty c'} \\ \\ G^n_{\infty' c} & G^n_{\infty' c'} \end{bmatrix} \begin{bmatrix} F_n(R_c) \\ \\ F'_n(R_c) \end{bmatrix} \tag{15b}$$

The resultant linear relationship between the asymptotic electronic and nuclear channel functions

$$\begin{bmatrix} F_e(r_\infty) \\ \\ F'_e(r_\infty) \end{bmatrix} = G^e_{\infty c} \cdot G^c_{en} \cdot (G^n_{\infty c})^{-1} \begin{bmatrix} F_n(R_\infty) \\ \\ F'_n(R_\infty) \end{bmatrix} \tag{16}$$

can also be recast in $\mathcal{R}$-matrix form of a linear relationship between channel functions and their derivatives

$$\begin{bmatrix} F_e(r_\infty) \\ \\ F_n(R_\infty) \end{bmatrix} = \begin{bmatrix} \mathcal{R}^\infty_{ee} & \mathcal{R}^\infty_{en} \\ \\ \mathcal{R}^\infty_{ne} & \mathcal{R}^\infty_{nn} \end{bmatrix} \begin{bmatrix} F'_e(r_\infty) \\ \\ F'_n(R_\infty) \end{bmatrix} \tag{17}$$

and the global electronuclear K-matrix follows by asymptotic matching

$$\begin{bmatrix} F_{ee} & F_{en} \\ F_{ne} & F_{nn} \end{bmatrix} = \begin{bmatrix} J_e & 0 \\ 0 & J_n \end{bmatrix} - \begin{bmatrix} N_e & 0 \\ 0 & N_n \end{bmatrix} \begin{bmatrix} K_{ee} & K_{en} \\ K_{ne} & K_{nn} \end{bmatrix} . \qquad (18)$$

In this block matrix formulation, $J_{e,n}$ and $N_{e,n}$ refer to diagonal matrices of regular and irregular asymptotic solutions in the electronic and nuclear channels; and, in the full F and K matrices, the first and second indices specify a particular channel of a given solution.

It is clear from this short survey, that the R-matrix compound zone plays a central role in connecting asymptotic electronic and nuclear decoupling channels: pure electronic collisions, atom-atom collisions and hybrid processes ($e\text{-}AB \rightarrow A^- + B$) can then be treated in a uniform way.

III. THE ab initio R-MATRIX CALCULATION OF THE $^2\Pi_g$ RESONANCE IN VIBRATIONAL EXCITATION OF N_2

This resonance arises from the temporary capture of the colliding electron into the lowest unoccupied $1\pi_g$ valence orbital by tunelling through the ℓ=2 centrifugal barrier. Since direct vibrational excitation is negligible ($< 10^{-2}$ A^2 outside the resonance), the reaction is entirely determined by the formation and decay of the SR compound $N_2^-(^2\Pi_g)$. All the dynamics occur within a finite zone, whose extension is suggested by existing information (Fig. 5, from Krauss and Mies[16], later referred to as KM) on the potential energy curves of the $N_2(^1\Sigma_g^+)$ and $N_2^-(^2\Pi_g)$ states, and on the radial function of the $1\pi_g$ resonant orbital:

$$r_c = 8\ a_o \quad ; \quad R_c = 2.9\ a_o \quad . \qquad (19)$$

A. The Fixed Nuclei Electronic Dynamics

In order to establish the validity of the BO approximation for the SR compound, as well as the simplification brought by a local description of the electronic correlations, we have started from a fixed nuclei single configuration representation of the N_2^- R-matrix compound

$$N_2^-\ (1\sigma_g^2\ 1\sigma_u^2\ \ 2\sigma_g^2\ \ 2\sigma_u^2\ \ 1\pi_u^4\ \ 3\sigma_g^2\ \ 1\pi_g;\ ^2\Pi_g) \quad . \qquad (2o)$$

The orbitals are represented in prolate spheroidal coordinates, specially efficient to account for the nuclear singularities in homonuclear diatomics.

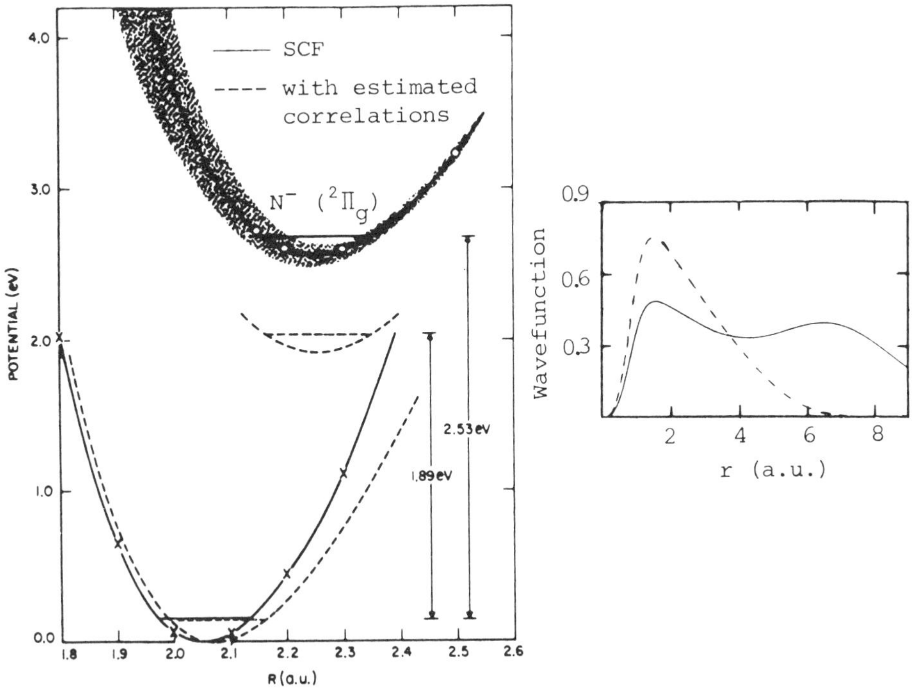

Fig. 5. Estimation of the R-matrix boundaries (r_c,R_c) using Krauss and Mies' calculations.
a. the N_2 and N_2^- potential curves from KM calculations.
b. the ℓ=2 components of the N_2^- ($1\pi_g$) orbital: the dashed curve corresponds to a bound type approximation and the solid curve to the unconstrained continuum function. Note the bimodal nature of the continuum function and the large amplitude of the inner peak at r = 1.2, very similar to its bound type approximation.

The fixed nuclei electronic RM Schrödinger equation (7a) reduces then to a set of SCF type monoelectronic radial equations, with a surface term for the $1\pi_g$ continuum orbital.

It is clear that, in the RM approach, the best single configuration description of the $N_2^-(^2\Pi_g)$ compound ($r<r_c$) is obtained by an SCF optimisation of the N+1 molecular orbitals at resonance (stabilization calculation within the R-matrix zone) followed by an R-matrix determination of the $1\pi_g$ continuum orbital in the vicinity of the resonance for the N_2 relaxed core. On the contrary, the standard static exchange approximation calculates the $1\pi_g$ continuum orbital in the N_2 frozen core, neglecting all the distortion of the core by the resonant electron.

The first four RM potential energy curves corresponding to the N_2 and N_2^- core approximation are reproduced on figures 6a and 6b (dashed-dot curves). A few qualitative features can already be pointed out. The two higher levels are practically parallel to the N_2 core potential energy curve (full line) and practically independent of the core representation: they obviously represent "fast" continuum states, weakly influenced by the molecular field but hardly influenced by the molecular vibration. On the contrary, the shape and position of the two lower levels are very sensitive to the core representation. In particular, the N_2^- core polarisation lowers the first RM state which becomes bound at $R>2.4\ a_o$. By continuity, one expects a nearly bound state (long lived shape resonance) associated with the first RM state for $R < 2.4\ a_o$.

The completion of the fixed nuclei scattering calculation, straightforward in the absence of long range couplings outside the RM zone, provides the electronic, characteristics of the resonance through the fitting of the eigenphase sum to a Breit Wigner formula. The resonance positions $E_r(R)$ are reported as dashed lines on Fig. 6a, 6b

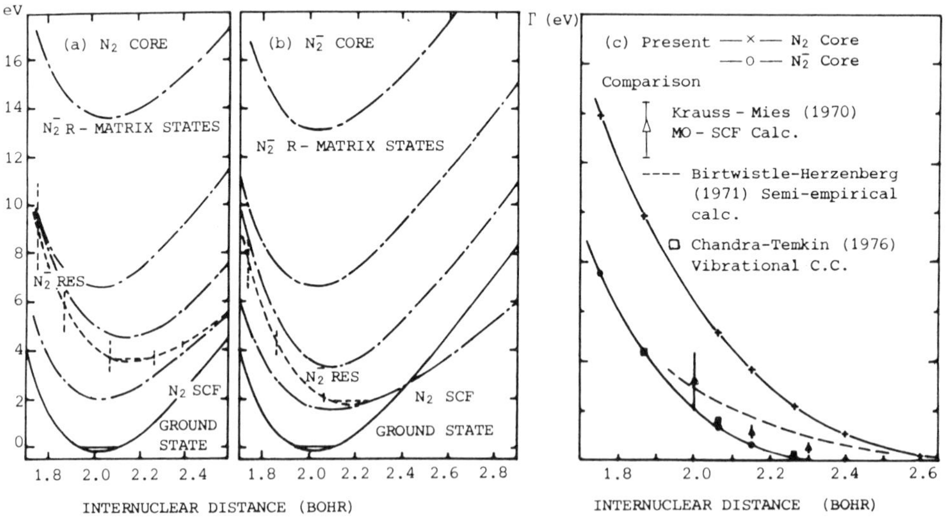

Fig. 6. The fixed nuclei electronic dynamics.
a. N_2 core approximation
b. N_2^- core approximation
c. Resonance width $\Gamma(P)$.

and the corresponding widths Γ(R) compared with previous results[16,17,5] on Fig. 6c. The efficiency of the N_2^- SCF representation of the short range compound state is illustrated by the close agreement with alternative predictions (Kraus Mies = KM, Dubé Herzenberg = DH[18]) for the resonance width, the relative position of the minimum of the N_2 and N_2^- resonant potential energy curves and their relative vibrational splittings.

Table 1. Characteristics of the N_2 and N_2^- potential energy curves

		Our N_2^- core	KM estimate[16]	DH semi-empirical[18]
$\Delta R = R_o^- - R_o$	a_o	0.1631	0.1796	0.1560
$\Delta E = E(R_o^-)-E(R_o)$	eV	1.9600	1.9250	1.9120
ω_-	eV	0.2652	0.2440	0.2440
$\omega - \omega_-$	eV	0.0493	0.0490	0.0490

Most of the qualitative features of the electronic shape resonance emerge directly from our calculation. Only the $\ell = 2$ partial wave is important asymptotically, but at SR, the additional electron, trapped inside the centrifugal barrier, correlates strongly to form a multicenter molecular compound electronic state. These strong correlations stabilize the N_2^- resonant curve by more than 2eV. However, the N_2^- resonant curve (dashed line on Fig. 6a, 6b) has its minimum shifted to larger internuclear distance and is flatter than the N_2 ground state potential. These results confirm the simple interpretation of the resonance as due to the capture of the incident electron into a π_g antibonding orbital.

A last feature which is worth some comment is the fact that the resonant $N_2^-(^2\Pi_g)$ potential curve appears mainly as a superposition of the first two R-matrix Born Oppenheimer compound states. When the N_2^- state becomes bound, at large internuclear distances, it coincides with the first most compact RM compound state which verifies the "inner region" BO separation of the nuclear and (N+1) electron compound motions. On the contrary, at very small internuclear distances, the width of the resonant state increases and the corresponding potential energy curve tends to coincide with the second RM state: the trapping of the colliding electron becoming very unlikely, the potential curve tends to be parallel to that of the N_2 ground state (direct scattering-type RM state). The nuclear motion is hardly affected by the colliding electron and the "outer region" BO separation of the nuclear and the N electron motions tend to prevail. At intermediate internuclear distances, the finite width is proportional to the coupling with the continuum, mainly described here by the mixing with the second RM state.

B. The Nuclear Dynamics

The second step of the vibrational RM calculation is the determination of the vibrational eigenstates of each adiabatic electronic RM curve. Eq. (7b) was solved using an expansion of the vibrational functions on an analytical basis of Gaussians. The comparison of the resonance width (Table 1) with the spacing of the vibrational levels $E_{k\nu}$, suggests already that there should be considerable structure in the excitation cross sections.

C. The Vibrational R-matrix and Cross Sections

The final step of the calculation relies upon the connection with the outer region

$$\Psi_E^{II}(N+1,R) = \phi_{N_2}({}^1\Sigma_g^+)(N,R) \sum_{vl} \chi_v^N(R) Y_{lm}(\hat{r}_{N+1}) \frac{F_{rl}(r_{N+1})}{r_{N+1}} \tag{21}$$

using the BO vibrational R-matrix

$$R^{BO}_{vlv'l'}(E,r_c) = \frac{1}{2} |<\phi_{N_2}| \phi_{core}>|^2$$

$$\sum_{k\nu} \frac{<\chi_v^N(R)|\chi_{k\nu}^{N+1}(R) F_{kl}(r_c, R)><F_{kl'}(r_c, R)\chi_{k\nu}^{N+1}(R)|\chi_{v'}^N(R)>}{E_{k\nu} - E} \tag{22}$$

$$+ <\chi_v^N(R)|R^{FN}_{ll'}(R)|\chi_{v'}^N(R)>$$

The first term includes explicitly the contribution of the vibrational sublevels $E_{k\nu}$ associated with the first four RM states which obviously describe the resonance. The second term accounts for all the higher levels (Buttle correction included) in the fixed nuclei (FN) approximation.

As in the FN electronic calculation, the residual long range couplings are neglected. The vibrational K-matrix and cross sections are therefore obtained by a simple matching at the core boundary, which allows the inclusion of all the relevant vibrational channels in the matching step (up to 2o vibrational sublevels are energetically allowed in the energy range of interest).

D. Discussion of the Results

Figure (7) which compares our two approximations (N_2 and N_2^- core potentials in dashed and full lines respectively), on the $^2\Pi_g$ contribution to the 0 → 0 and 0 → 1 transitions, illustrates the influence of the resonant state lifetime, since no vibrational structure appears in the N_2 core as opposed to the N_2^- core results. From now onwards, only the best N_2^- approximation will be considered.

The comparison of our 0→v' vibrational excitation cross sections with experiment (Fig. 8a) shows a close agreement with the shape of Ehrhardt and Willmann (1967)[19] relative measurements and with the magnitude of Wong's latest absolute measurements.[20]

As to the comparison with alternative calculations (Fig. 8b), our vibrational structure agrees at least as closely with experiment as the semiempirical Boomerang model of Dubé and Herzenberg (DH 1979) specially for large v'. A slight discrepancy is seen in the energy scale of the oscillatory structure which is related to the inaccuracy of the vibrational levels of the SCF N_2 target:

$$\omega_{HF} = 0.335 \text{ eV} \quad ; \quad \omega_{exp} = 0.293 \text{ eV}$$

while our predicted vibrational frequency difference ($\omega - \omega_-$, Table 1) agrees closely with DH estimates.

Our predictions are definitely more accurate than those of Chandra and Temkin (CT)[5] hybrid calculation. In fact, the electron-nuclei coupling is so strong in resonant scattering that the simple frame transformation to the Born-Oppenheimer representation of the short range compound state is closer to the exact result than an extensive dynamical coupling between the Born-Oppenheimer states of the target.

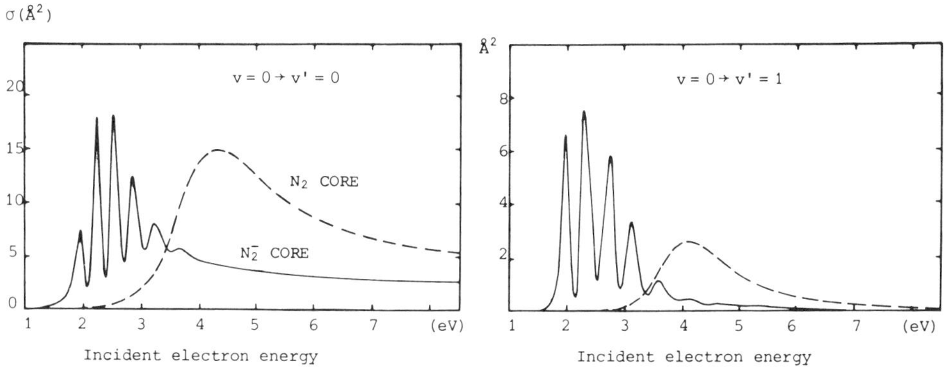

Fig. 7. The $^2\Pi_g$ contribution to the 0 → 0 and 0 → 1 cross sections for the N_2 and N_2^- representations.

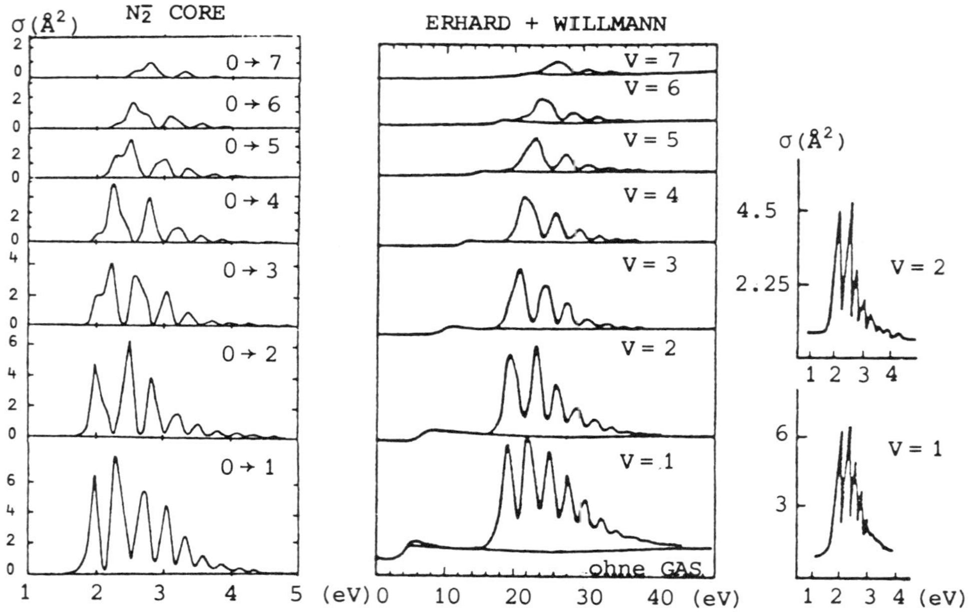

Fig. 8a. Comparison of vibrational excitation cross sections $\sigma_{0\to v}$ ($Å^2$) in the N_2^- core approximation with experiments.

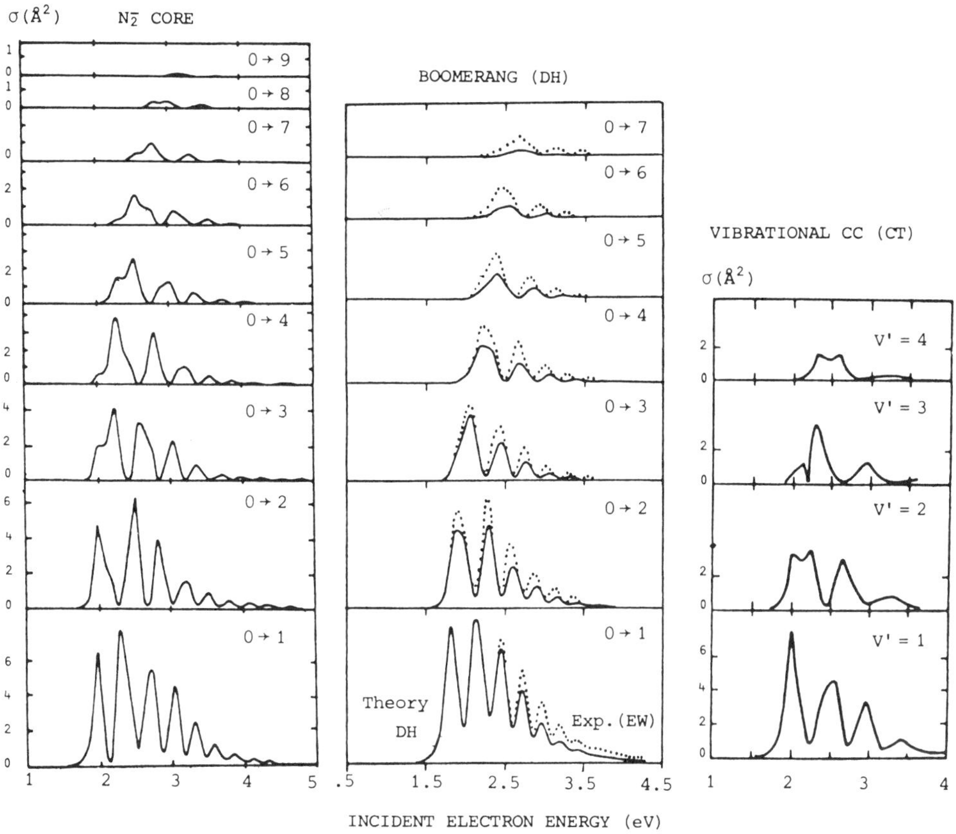

Fig. 8b. Comparison of vibrational excitation cross sections $\sigma_{0\to v}$(Å) in the N_2^- core approximation with previous calculations.

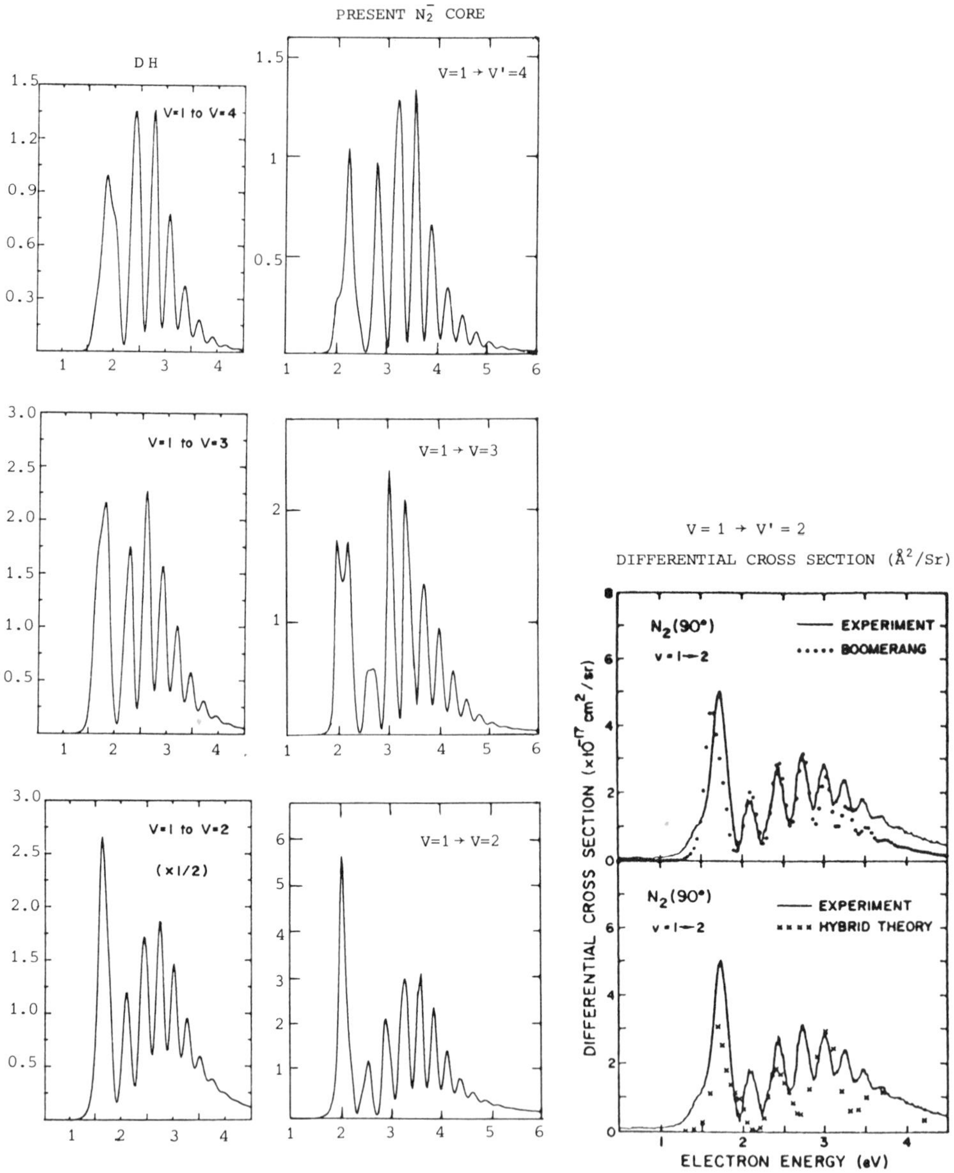

Fig. 9. Vibrational excitation cross sections $\sigma_{1\to v}$ (Å^2)
a. Dubé Herzenberg (DH) Boomerang results
b. N_2^- core approximation present results.
c. Comparison of Wong's experimental results with Dubé-Herzenberg and Chandra-Temkin theoretical results.

On figure (9), our results for the $1 \rightarrow v'$ vibrational excitation cross sections are compared with Wong's experiment. His differential cross section is just proportional to the integrated one, since only the $\ell=2$ partial wave contributes significantly, as well as with the DH semiempirical Boomerang results (Fig. 9a) and CT vibrational CC (crosses on Fig. 9c). In spite of the scale differences, a close agreement is suggested between the experiment, DH and the present results (Fig. 9b). It is interesting to remark that the minimum in the envelope of the $1 \rightarrow v'$ cross sections, coincide practically in energy with the maximum of the envelope of the $0 \rightarrow v$ cross sections, and can faithfully be related to the node in the initial $v = 1$ vibrational state at the maximum of the $v = 0$ state.

The $^2\Pi_g$ contribution to the $0 \rightarrow 0$ and $1 \rightarrow 1$ elastic transitions (Fig. 1o) shows similar trends of the envelope. However, the main difference with DH Boomerang results (inserts) concerns the energy dependence of the non resonant background. As suggested by the following comparison with experiment for the total cross section from the ground state (v=0), our behaviour (slow increase from threshold, more important value at large energy) is probably more satisfactory. As to our more pronounced substructure, it is consistent with our smaller resonant width at large R, which may be questioned (our SCF N_2 potential curve, leading to a wrong dissociation limit, being too stiff at large R).

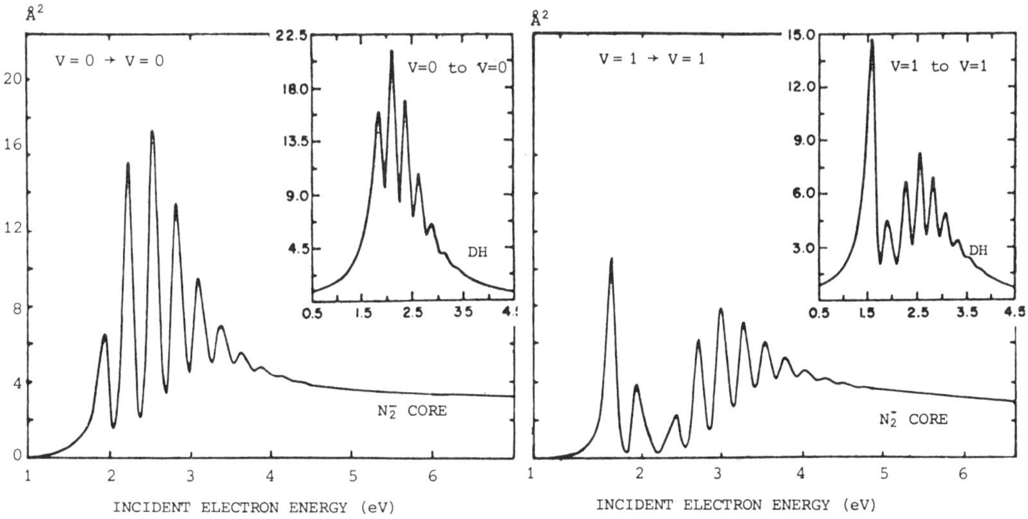

Fig. 10. The $^2\Pi_g$ contribution to $e+N_2(^1\Sigma_g^+)$ elastic scattering. Insets contain the Dubé-Herzenberg results.

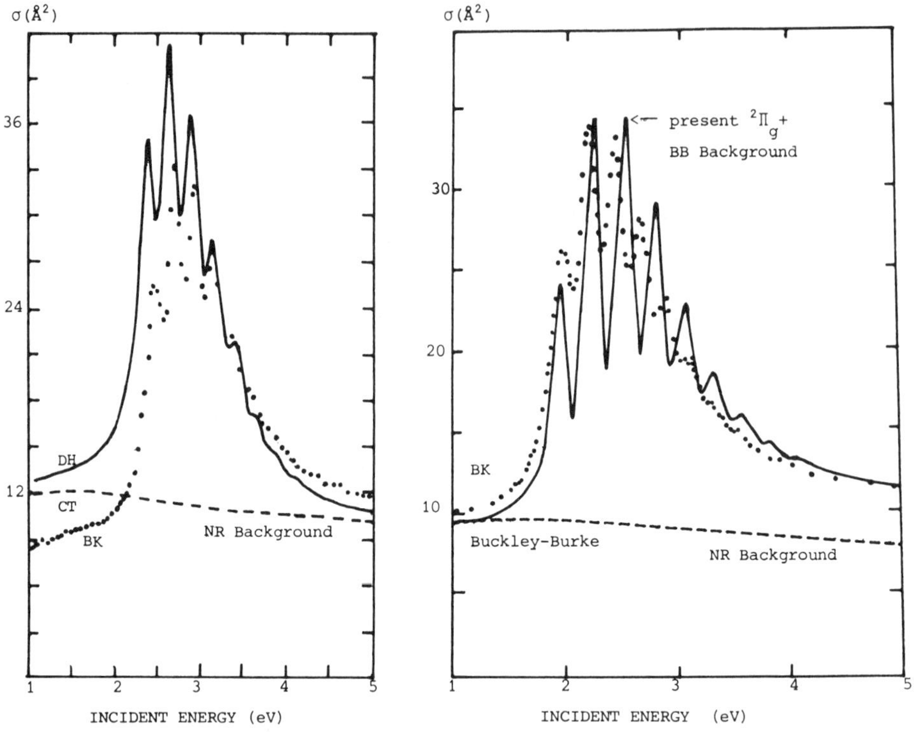

Fig. 11. The total scattering cross section (Å^2) for e+N_2 (v=0)

a. Bonham and Kennerly (·) compared with the theoretical sum (——) of Dubé-Herzenberg $^2\Pi_g$ resonant results and Chandra-Temkin non resonant background (-----).

b. Bonham and Kennerly experimental results (·) compared with the theoretical sum (——) of the present N_2^- core $^2\Pi_g$ resonant results and Buckley-Burke non resonant background (-----).

The total cross section on Fig. 11 provides the last important comparison with the absolute measurements of Golden[21] and Bonham and Kennerly (BK, Fig. 11a, 11b)[22]. Our $^2\Pi_g$ resonant contribution, augmented with the static exchange + polarisation results of Buckley and Burke[23] for the first three non resonant partial waves $^2\Sigma_g$, $^2\Sigma_u$, $^2\Pi_u$, agree closely in magnitude with the absolute experiments over all the energy range. On the contrary, the overestimation of DH results at low energy is directly related to the overestimation of the elastic resonant contribution (Fig. 10).

As shown recently by Hazi[24], this weakness is a direct consequence of the limited flexibility of the Boomerang model, as a simplified and parametrized form of the resonant scattering theory.

IV COMPARISON OF THE R-MATRIX WITH RELATED THEORIES

The N_2 example provides a convenient background for a comparison of the RM with the well established Boomerang model and the more general resonant scattering theory (§A). The full generality of the RM will emerge and additional comments will follow (§B). A short connection with the recent MQDT (Multichannel Quantum Defect Theory) application to dissociative process will close the discussion (§C).

A. Comparison with the Boomerang Model and the General Resonant Scattering Theory.

The formal similarity of the resonant scattering theory and the RM is the most fundamental point.

The resonant scattering theory[25] is based upon a non adiabatic configuration interaction[26] between diabatic representations of the resonant state (N+1 electron BO complex) and the non resonant continuum (N electron BO target + continuum electron,

$$\Psi = \phi_r^{BO}(NH)\zeta_d(R) + \sum\!\!\!\!\!\!\int dE'\phi_i^{BO}(N)\chi_v(R)\varphi_l(\vec{r}_{N+1},R)b_{il}(E')$$

Autoionisation is indeed the main non adiabatic mechanism.

The R-matrix method could be developed in a very similar form, starting from a diabatic electronic basis. Indeed, the Siegert resonant state could be selected by the complex Bloch operator

$$L_{el}^{res} = L_{el}^{b} \quad \text{with } b = ik_r \quad ; \quad k_r = (2W_r(R))^{1/2} \tag{24a}$$

$$W_r(R) = E_r(R) - i\Gamma(R)/2 \tag{24b}$$

and its configuration interaction (CI) with the orthogonal nonresonant continuum which is discretized by a real Bloch operator L_{el}^{b} could be performed.

More radically, the resonant scattering theory can be considered as a special case of the R-matrix formalism[10] using an orthonormal complex basis of Kapur Peierls (KP) outgoing waves

$$L_{el}^{KP} = L_{el}^{b} \quad \text{with } b = ik \quad ; \quad k \text{ arbitrary real} \tag{25}$$

The surface terms of the Schrödinger equation (9) are then proportional to the ingoing wave, namely to the scattering incoming wave $[\phi^N \chi_v^N]e^{-ikr}$ which is not an N+1 electron BO state.

By contrast, in the adiabatic R-matrix formulation successfully applied to N_2^-, the SR dynamics are described in their own, and the N+1 electron BO adiabatic approximation is shown to be valid. The non adiabatic effects of autoionization, which leads to an asymptotic N electron BO situation, are fully described by the Frame Transformation step.

This rapid survey allows us to suggest the relative advantage of both approaches:

The resonant scattering theory is physically more transparent in principle. However, the approximations ($\Gamma(R)$ form of the Boomerang model [18,24,25]; $\Gamma(E)$ form[26]) which lead to practicable semiempirical or ab initio parametrizations, reduce its flexibility and limit the validity of its predictive power.

The R-matrix is easier mathematically (real equations) and practicable in the general case. Moreover it allows a more symmetrical treatment of electronic and nuclear continua, as well as that of direct and indirect mechanisms.

B. Physical Insight Brought by the R-Matrix Formalism

It might be important at this stage to put to light the specific physical transparence of the R-matrix approach of resonant processes. Indeed, a careful analysis of the BO vibrational RM expression

$$R^{BO}_{vlv'l'}(E) = \frac{1}{2}\sum_{k\nu} \frac{\langle \phi_o^N \chi_v^N \psi_1^\lambda | \phi_{core}^N F_k^\lambda(r_c,R)\chi_{k\nu}^{N+1}\rangle\langle\phi_{core}^N F_k^\lambda(r_c,R)\chi_{k\nu}^{N+1} | \phi_o^N \chi_{v'}^N Y_{1'}^\lambda\rangle}{E_{k\nu} - E} \tag{26}$$

provides all the characteristics of the various types of resonances.

A very long lived resonant state ($\Gamma \ll \omega$), which corresponds to a compact electronic state, can be identified in practice with a single BO R-matrix state (as the N_2^- resonant state for $R > 2.2a_o$). Its electronic function is dominated by a single bound configuration and varies slowly with R. The varying part of the RM

$$R^{BO}_{vlv'l'} \simeq \frac{1}{2} \sum_{\nu} \frac{<\chi^{N}_{v}|\chi^{N+1}_{k\nu}><\chi^{N+1}_{k\nu}|\chi^{N}_{v'}>F_{kl}(r_c,\bar{R}_{av})F_{kl'}(r_c,\bar{R}_{av})}{E_{k\nu} - E} \qquad (27)$$

gives resonant peaks at the negative ion vibrational sublevels $E_{k\nu}$ with relative intensities related to the Franck Condon factors.

For $\Gamma \sim \omega$, as in our N_2^- calculation, the decay of the resonant state is described by the mixing of two or more BO R-matrix states. Then, the R dependence of the radial functions modifies the Franck Condon arguments, while the non negligible contributions of continuum RM states (with significant surface amplitudes) leads to important interference effects, modifying the position of the oscillation extrema.

For $\Gamma >> \omega$, at the other extreme, we have a large continuum component represented by the mixing of several RM states, with large surface amplitudes. These high levels having potential energy curves nearly parallel to that of the target, off-diagonal vibrational elements are due mainly to the R variation of the fixed nuclei electronic amplitude. For the few vibrational sublevels of each high lying RM electronic state, one can make the following approximation:

$$\sum_{\nu} \frac{\chi^{N+1}_{k\nu}(R)\chi^{N+1}_{k\nu}(R')}{E_{k\nu} - E} \sim \sum_{\nu} \frac{\chi^{N+1}_{k\nu}(R)\chi^{N+1}_{k\nu}(R')}{E_{k}(R) - E} \sim \frac{\delta(R-R')}{E_{k}(R)-E} \qquad (28)$$

and the vibrational RM reduces to the FN form.

C. Connection with the Eigenchannel + MQDT Approach of Dissociative Processes

The electronuclear R-matrix method can be considered as a formal generalisation and practical implementation of the eigen-channel + MQDT approach of dissociative recombinaison (ionic cores) proposed recently[27]. The main ingredient of this theory are the eigenchannel parameters, easily deduced from the RM calculation.

The SR eigenphases $\delta_{\alpha}(R)$ and eigenvectors $U_{i\alpha}(R)$, are obtained by mere diagonalisation of the SR fixed nuclei K-matrix, deduced from the R-matrix by matching eq. (18) on the SR hypersphere

$$K_{ij}U_{j\alpha} = tg\delta_{\alpha}(R,E)U_{j\alpha} \Longleftrightarrow \psi_{\alpha E} = \sum_{e}|e>F_{e\alpha} + \sum_{n}|n>F_{n\alpha} \qquad (29)$$

The SR solutions $\Psi_{\alpha E}$ and observables (dipole matrix elements in photoabsorption) are then readily obtained by R-matrix expansion

$$\Psi_{\alpha E} = \sum_k \frac{|\Psi_\alpha><\Psi_k|L|\Psi_{\alpha E}>}{E_k - E} \tag{3o}$$

The success of MQDT applications to photoabsorption (H_2[29], N_2[3o]) and model dissociative recombinaison of CH^+[28] already gives some confidence about the power of further applications of the ab initio R-matrix technology to a wide variety of molecular processes (Fig. 12).

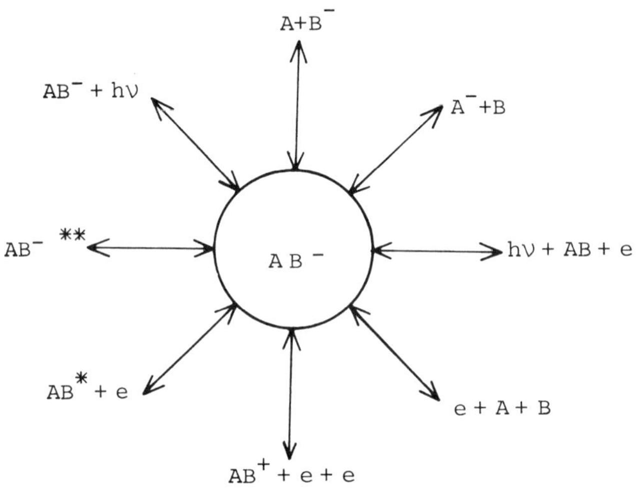

Fig. 12. The different dissociation modes of an unstable molecular compound.

V. CONCLUSION

The main conclusion of these discussions is that the BO electro-nuclear R-matrix approach provides a suitable method to describe SR interactions in electron-molecule processes. Specific tool for solving the Schrödinger equation independently in different regions of the configurational space, and for extending bound state techniques to the SR interactions in continuum processes, the present

RM approach suggests that the BO separation of the total electronic and the nuclear motion remains a rule for a penetrating scattered electron and that the compound SCF single electronic configuration provides a suitable description of the electronic state.

REFERENCES

1. D.M. Chase, Phys. Rev. 1o4, 838 (1956).
2. F.A. Gianturco and N.K. Rahman, Chem. Phys. Lett., 38o (1977); J. Phys. B1o, L 219 (1977); J. Phys. B11, 727 (1977). F.A. Gianturco and D.G. Thompson, J. Phys. B13, 613 (198o).
3. M.A. Morrison and L.A. Collins, Phys. Rev. A17, 918 (1978). L.A. Collins, W.D. Robb and M.A. Morrison, Phys. Rev. A21, 488 198o. L.A. Collins, R.J.W. Henry and D.W. Norcross, J. Phys. B13, 2299 (198o). W.D. Robb and L.A. Collins, Phys. Rev. A22, 2474 (198o). M.A. Morrison and L.A. Collins, Phys. Rev. A23, 127 (1981).
4. A.M. Arthurs and A. Dalgarno, Proc. Roy. Soc. A256, 334 (196o).
5. N. Chandra and A. Temkin, Phys. Rev. A13, 188 (1976). A. Temkin in "Electron and Photon Molecule Collisions", ed. by T.N. Rescigno, V. McKoy and B.I. Schneider (Plenum Press, New York, 1979) p. 173.
6. H.S. Taylor, E. Goldstein and G.A. Segal, J. Phys. B1o, 2253 (1977). E. Goldstein, G.A. Segal and R.W. Wetmore, J. Chem. Phys.68, 74 (1978).
7. R.K. Nesbet, J. Phys. B1o, L739 (1977). A. Herzenberg, Symposium on Electron Molecule Sollisions, ed. by I. Shimamura and M. Matsuzawa (University of Tokyo 1979) p.77.
8. M.A. Morrison and N.F. Lane, Phys. Rev. A12, 2361 (1975). C.W. McCurdy Jr., T.N. Rescigno and V. McKoy, J. Phys. B9, 691 (1976).
9. B.I. Schneider, M. Le Dourneuf and P.G. Burke, J. Phys. B12, L365 (1979).
1o. C. Bloch, Nuclear Phys. 4, 5o3 (1957).
11. E.S. Chang and U. Fano, Phys. Rev A6, 173 (1972).
12. P.G. Burke, I. Mackey and I. Shimamura, J. Phys. B1o, 2497 (1977). B.D. Buckley and P.G. Burke in "Electron and Photon Molecule Collisions", ed. by T.N. Rescigno, V. McKoy and B.I. Schneider (Plenum Press, New York 1979) p. 133. B.D. Buckley, P.G. Burke and Vo Ky Lan, CPC17, 175 (1979).
13. B.I. Schneider, CHem. Phys. Lett., 31, 237 (1975); Phys. Rev. A11, 1957 (1975); in "Electronic and Atomic Collisions", ed. by G. Watel (North Holland Publishing Co. 1978) p.257; in "Electron and Photon Molecule Collisions", ed. by T.N. Rescigno, V. McKoy and B.I. Schneider (Plenum Press, New York 1979) p.77.
14. T.F. O'Malley, Adv.At.Mol.Phys. 7, 223 (1969).
15. B.I. Schneider, M. Le Dourneuf and Vo Ky Lan, Phys. Rev. Lett., 43, 1927 (1979).

16. M. Krauss and F.H. Mies, Phys. Rev. A1, 1592 (197o).
17. D.T. Birtwistle and A. Herzenberg, J. Phys. B4, 53 (1971).
18. L. Dubé and A. Herzenberg, Phys. Rev. A2o, 194 (1979).
19. H. Ehrhardt and K. Willmann, Z. Phys., 2o4, 462 (1967).
2o. S.F. Wong, unpublished measurements reproduced in ref. 18.
21. D. Golden, Phys. Rev. Lett. 17, 847 (1966).
22. R.E. Kennerly and R.A. VOnham, Phys. Rev. A17, 1844 (1978). R.E. Kennerly, Phys. Rev. A21, 1876 (198o).
23. B.D. Buckley and P.G. Burke, J. Phys. B1o, 725 (1977).
24. A. Hazi, this issue and submitted to Phys. Rev. A (198o).
25. A. Herzenberg and F. Mandl, Proc. Roy. Soc., A27o, 48 (1962). A. Herzenberg, K.L. Kwok and F. Mandl, Proc. Phys. Soc., 84, 477 (1964). J.N. Bardsley, A. Herzenberg and F. Mandl, Proc. Phys. Soc., 89, 321 (1966). J.N. Bardsley, J. Phys. B1, 349 (1968); in "Electron and Photon Molecule Collisions", ed. by T.N. Rescigno... (Plenum Press) p. 267. T.F. O'Malley, Phys. Rev. 15o, 14 (1966); 156, 23o (1967).
26. F. Fiquet Fayard, Vacuum 24, 533 (1974).
27. C.M. Lee, Phys. Rev. A16, 1o9 (1977).
28. A. Giusti-Suzor, J. Phys. B13, (198o).
29. C. Jungen and O. Atabek, J. Chme. Phys. 66, 5584 (1977).
3o. H. Lefebvre-Brion and A. Giusti-Suzor, this issue.

RECENT DEVELOPMENTS OF THE FRAME TRANSFORMATION THEORY

OF ELECTRON MOLECULE PROCESSES

II. NEW SINGLE CENTER DESCRIPTION OF ELECTRONIC AND NUCLEAR CORRELATIONS IN ELASTIC SCATTERING AND IN THRESHOLD NUCLEAR EXCITATION

Vo Ky Lan, M. Le Dourneuf and J.M. Launay

Observatoire de Paris
92190 Meudon
France

A computationally efficient technique for an ab initio description of electron-molecule processes is investigated within the framework of Frame Transformations (FT). Complementary problems of the previous paper are examined. A new single center (SC) approach, replacing the standard partial wave expansion by an adiabatic angular expansion, provides an efficient way to account for the most strongly anisotropic electrostatic field of polar targets, still problematic in the present R-matrix (RM) codes. An optimal package, combining numerical and analytical techniques, allows systematic calculations of electronically elastic scattering by a wide variety of typical diatomics. The general FT between the short range (SR) representation (Born Oppenheimer (BO) N+1 electron compound) and the long range (LR) decoupling (BO N electron target + scattered electron), necessary to account for the dynamical decoupling from the nuclear field, which affects strongly threshold rotational or vibrational excitations of polar molecules, can at last be implemented satisfactorily. Preliminary results are presented for threshold rotational excitation of HF, HCl, the best candidates for "tomorrow-experimental-check".

I. INTRODUCTION

Paper I has illustrated the efficiency of a piecewise description of the complex electron molecule interaction, by locally optimal tools, on the important but relatively simple case of resonant vibrational excitation of N_2. The R-matrix (RM) plays a unique role in separating completely the dynamical description of the SR compound zone and the LR electronic and nuclear dissociation zones: its formal and practical advantages in rearrangement processes cannot be questioned. Because of the difficulties which arise due to the complexity of the electron molecule interaction, only a few were considered (strong SR electronuclear correlations), and only in a simple case: the formation of a "resonant compound" in a "gentle" homonuclear diatomic.

Let us first consider the electronic scattering problem at fixed nuclei (FN). The R-matrix SCF description of the N+1 electron compound is a simple way of accounting for exchange and the average effects of the strong SR and nearly stationary correlations which dominate the dynamics of a long lived resonant state. Direct scattering is to some extend more difficult to handle. The LR permanent multipoles of the target may cause a non trivial problem (strong permanent dipole near thresholds, specially). The induced polarisation, which leads to a LR $1/r^4$ interactions, is the dynamical resultant of virtual electronic excitation of the target. Its description by polarised pseudo states and its important effects at low energy are well known in electron-atom studies[1,2,3] but the practical extension of the technology to electron-molecule remains at its very beginnings[4,5].

The first objective of this paper is to report recent unpublished progresses in the FN electron-molecule scattering problem. Our distant aim is obviously to generalize the very sophisticated treatment developed with systematic success in electron-atom processes

hybrid close coupling (CC) expansion with

additional collisional channels, built from polarized pseudo states of the target, to incorporate rigourously, but in a compact form, the LR induced polarisation

bound "capture" terms or more general "correlation functions", built from multicenter molecular orbitals to describe SR correlations

hybrid resolution of the corresponding Schrödinger equation

R-matrix in the SR exchange region

CC solution in the LR potential region, by the variable phase method (VPM) or related techniques

The principal obstacle is probably the inefficiency of analytical basis used in present molecular RM codes (prolate spheriodal Gaussian[6] or Slater orbitals[7]) to produce converged static exchange (SE) results for polar targets. Due to the high density of strong anisotropic RM states that can accumulate into the deep anisotropic well of these "ionic structures" (see Fig. 3,4 below), the problems are to define the "smallest RM zone" compatible with the "non local exchange" and to find a systematic way of building a convergent continuum basis. This last problem has been solved from the start in atomic RM applications by defining a discrete continuum basis, numerical solution of a realistic model potential[8]. A step to overcome these difficulties is therefore to come back to traditional single center (SC) close coupling, to define a realistic model problem, to solve it most efficiently and to extract all possible physical information.

The new SC description of FN electron molecule scattering goes far beyond its prospective use in the development of RM technology. Indeed, it provides an efficient and physically transparent methodology for a unified description of the whole electrostatic interaction.

Going beyond traditional CC techniques, some of which (iterative techniques)[9] have recently reemerged as the most efficient way to get SE results for polar molecules, the adiabatic angular expansion provides a competitive alternative to the multicenter RM to handle the strong anisotropy of the interaction. Moreover, we will show that it gives a unique visualization of the dynamics, particularly useful in the interpretation and even the prediction of shape resonances.

The dynamical effect of the nuclei constitutes the second aspect of the electron molecule problem. The second objective of this paper is to consider the general implementation of the FT between the optimal SR description (Born Oppenheimer N+1 electron compound) and the LR decoupled situation (Born Oppenheimer (BO) N electron target + colliding electron). Resonant vibrational excitation, considered in paper I, was an exceptionally simple example since the electrostatic interaction in the outer region was negligible and the dynamical decoupling had therefore no observable effects. The easy implementation of the compound BO approximation within the RM scheme, as well as the relatively large vibrational spacings (about 1oo times larger than the rotational splittings) made the problem tractable without further "tricks". On the contrary, in threshold nuclear excitation of polar molecules, the LR competition between electrostatic interaction and dynamical decouplings from the nuclear motion is important, both effects being most significant for rotational excitation, which has been selected as the most challenging test of the FT general implementation.

The unified description of the whole electrostatic interaction by the SC approach, its convenience for a physical interpretation of exchange of angular momentum between the colliding electron and the molecular rotation, its ability to describe an angular N+1 electron BO compound as well as the R-matrix (mere coupling of angular momentum), and the performance of collisional technology used, have allowed us to examine the problem in full generality. A model case e + CO, having all important theoretical characteristics built in, while remaining sufficiently simple to allow an exact numerical solution, has first been considered[10] and two important extensions of the original rotational FT[11] (Chang and Fano, 1972)[12], have put us in a position to examine realistic cases (e + HF, HCl).

The remainder of the paper is organized as follows. In section II, the basic equations of the Fixed Nuclei Molecular Frame (MF) equations in the static + model exchange + model polarization are recalled and justified. Their extension to the exact rotational CC in the MF and Laboratory Frame (LF) is also mentioned. These equations provide a convenient background for section III which describes our single center package, built by selecting optimal numerical and analytical tools for the inner and outer regions, characterized respectively by strong non local potentials and weak local multipoles. The main and latest novelty of our new SC package is the introduction of adiabatic angular basis, and of diabatic FT between locally adiabatic representations, which reduce greatly the number of effective channels and their mutual couplings. In §IV, the physical insight and numerical simplifications brought by the new SC approach are illustrated for FN electron scattering. In §V, the concept of locally optimum angular basis is combined with the analytical techniques of MQDT in an efficient rotational FT package, whose pilot applications to threshold excitation of real polar molecules (HCl, HF) is sketched. The main results and perspectives of papers I, II are summarized (§VI).

II. BASIC EQUATIONS

A. Limitations

We will not speculate on the general CC equations for electron molecule. The interested reader can refer to the extensive discussion of the general electronic CC in the formally equivalent electron atom case[13], and to the detailed derivation of the rotational and vibrational CC equations in the molecular and laboratory frames (MF, LF)[14,12,15].

The bottleneck problem of the present electron molecule calculations is the accurate description of a single continuum HF orbital associated with a single configuration (HF) representation of the unperturbed, usually $^1\Sigma^+$ ground state of a closed shell molecular target

$$\Psi(N+1) = A[\phi_{^1\Sigma}(N)\ F(\vec{r}_{N+1})\chi(G_{N+1})] \tag{1}$$

In the FN approximation, the singularities of the electrostatic interaction lead to an extremely slow convergence of partial wave of the continuum orbital near the uncentered nuclei (up to ℓ=4o in the single center expansion with respect to the center of gravity of the molecule) and an extremely slow convergence of the radial integration due to the exceptionally violent and LR field of polar species (from a few hundreds to thousands bohr radii). As long as the Static Exchange (SE) approximation of the electrostatic interaction is not solved fully and simply, is it feasible to consider the dynamical effects of the nuclear motion in a fully converged (ℓ,j,v) CC, with the perspective of additional difficulties associated with strongly coupled nearly degenerate channels?

These considerations explain our selection of the simplest physically realistic approximation of the FN electronic interaction (FN equations, §B) which could be extended to rotational excitation (§C).

B. MF Equations for FN Elastic Scattering

The Static + OHFEG model exchange + model polarization approximation

Within the single configuration, unperturbed target, subspace eq. (1), the resolution of the Schrödinger equation leads to the exact static exchange (ESE) equation for the continuum orbital F

$$(2)\quad [-\frac{1}{2}\nabla_r^2 + V_{st.}(\vec{r}) - \frac{1}{2}k^2]\,F(\vec{r})=\sum_\alpha[(\varepsilon_\alpha-\frac{1}{2}k^2)\langle\phi_\alpha|F\rangle+\langle\phi_\alpha|\frac{1}{|\vec{r}-\vec{r}'|}|F\rangle]\phi_\alpha(\vec{r})$$

in which we have isolated the exchange kernel on the right hand side.

In the following, we will restrict our effort to the orthogonalisation + Hara Free Electron Gas model exchange (OHFEGE) approximation, since in the long term, the R-matrix is expected to give the most satisfactory solution, and in the mean time, iterative techniques, which reduce the exact solution of the SE integrodifferential equations (2) to the iterative solution of differential approximate equations, have provided valuable benchmarks[16] for approximate treatments of exchange and proved the OHFEGE to be the most accurate[17,18].

Let us recall briefly the basis of the OHFEGE approximation.

The explicit orthogonalisation of the continuum orbital to the bound orbitals is exact for a closed shell target and eliminates exactly the one electron exchange term. Indeed, if the a priori antisymmetrization eq. (1), insures the exclusion from closed shells whatever the components $\langle\phi_i|F\rangle$ on the bound orbitals, the explicit orthogonalisation suppresses this undesirable indetermination which may lead to numerical instability in the solution of the SE eq. (2). In addition to accounting easily for part of the Pauli principle, it has the further advantage of injecting in the unknown continuum orbital F, the precise SR nodal structure of the first bound states, that have usually been calculated accurately (HF for the target). Both advantages have made the success of the "Lagrange orthogonalisation procedure" or "pseudopotential method" in electron atom[8] or molecule[21] collisions.

The Hara Free Gas approximation provides a local approximation of the two electron exchange kernel, extending to the continuum[22] the Slater average exchange approximation for bound states.[23] The total wave function is approximated by a determinant of the plane waves, with the momentum of the bound electrons determined by the exclusion principle up to the local Fermi level

$$k_F(\vec{r}) = [\,3\pi^2 p(\vec{r})\,]^{1/3} \tag{3}$$

and the local impulsion of the colliding electron determined by the same potential as the bound electrons

$$E_{inc} = 1/2\ \kappa^2(\vec{r}) + V(\vec{r}) \quad ; \quad E_{inc} = 1/2\ k^2 \tag{4a}$$

$$E_{ion} = 1/2\ k_F^2(\vec{r}) + V(\vec{r}) \quad ; \quad E_{ion} < 0 \tag{4b}$$

$$\Rightarrow \quad 1/2\kappa^2(\vec{r}) = E_{inc} + 1/2\ k_F^2 - E_{ion} \tag{4c}$$

Its substitution in the calculation of the two electron exchange kernel leads to the HFEG local exchange potential dependent on the position and energy of the colliding electron

$$W_{ex}^{HFEG}(\vec{r},E_{inc}) = \frac{2}{\pi}\, k_F(\vec{r})\,F(\eta(\vec{r},E_{inc})) \tag{5a}$$

with

$$F(\eta) = \frac{1}{2} + \frac{1-\eta^2}{4\eta} \log\left|\frac{1+\eta}{1-\eta}\right| \tag{5b}$$

$$\eta(\vec{r},E_{inc}) = \frac{\kappa(\vec{r})}{k_F(\vec{r})}$$

We will not comment on variants of the FEG potentials which have been considered in the literature: asymptotic adjustments, tunnelling of the ionisation energy are typical ad hoc procedure which can always neutralize locally and globally all the inaccuracies of the approximations, whenever the exact answer is known!

As a conclusion of this presentation of the OHFEG approximation of exchange, let us stress once more the complementarity of its two ingredients. The explicit orthogonalisation to closed shells remains non local. It has the two important functions of insuring exactly the exclusion from closed shells and of making zero the one electron exchange terms, which are otherwise repulsive and therefore play a role equivalent to the orthogonalisation mentioned above (exclusion from the core). The HFEGE term is local and introduces approximately the attractive correction of the bielectronic repulsion term due to the Pauli exclusion principle.

The balance between these two terms is important to avoid spurious effects, as we pointed out to Collins in Feb. 1979 during his short visit at Meudon: the Σ resonance predicted in his e +LiF scattering calculation[19] with the HFEG model exchange disappeared completely once orthogonalisation had been imposed (OHFEGE)![20]

Before turning to the radial equations, let us mention that, like other workers, whenever our aim is to estimate the most reliable predictions on processes sensitive to LR effects, we introduce a model polarization potential having the correct asymptotic form

$$V_{pol}(\vec{r}) = -\frac{\alpha_0 + \alpha_2 P_2(\cos\Theta')}{2r^4} C(r) \tag{6a}$$

the average and anisotropic parts of the static polarizability deduced by the best estimates in the literature (experimental or ab initio) and the cut off function, taken in the most common form

$$C(r) = \Lambda - \exp\left(-\frac{r}{r_{cut}}\right)^6 \tag{6b}$$

Our total potential will generally be:

$$V(\vec{r}) = V_{st}(\vec{r}) + W_{ex}^{HFEG}(\vec{r}, E_{inc}) + V_{pol}(\vec{r}) \tag{7}$$

and our basic equations read:

$$\left[-\frac{1}{2}\nabla_r^2 + V(\vec{r}) - \frac{1}{2}k^2\right]F(r) = \sum_\alpha c_\alpha \phi_\alpha(\vec{r}) \tag{8a}$$

$$\langle F | \phi_\alpha \rangle = 0 \tag{8b}$$

The c_α being the Lagrange parameters insuring the explicit orthogonalisation to the bound orbital of the same symmetry.

SC partial wave expansion

The single center (SC) fixed nuclei electronic CC is based upon a partial wave expansion of the orbitals in the MF of reference, with the origin at the center of gravity of the nuclei and the z' axis along the internuclear axis (Fig. 1)

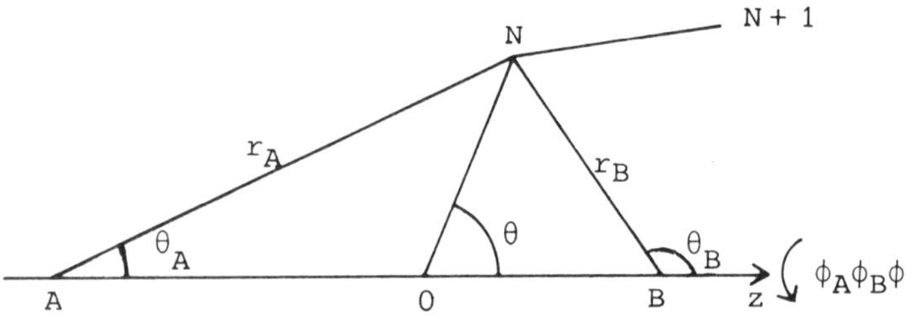

Fig. 1. Coordinate system for the single center expansion

$$F^{\lambda}(\vec{r}) = \sum_{l} Y_{l\lambda}(\hat{r}') \frac{F_l^{\lambda}(r)}{r} \tag{9a}$$

$$\phi_{\alpha}^{\lambda_\alpha}(\vec{r}) = \sum_{l} Y_{l\lambda_\alpha}(\hat{r}') \frac{\varphi_{l_\alpha}^{\lambda_\alpha}(r)}{r} \tag{9b}$$

where λ, λ_α are the projection of the orbital momenta on the internuclear axis.

Our calculations are based on published LCAO-MO-SCF bound states built from atomic Slater orbitals, whose expansion about the center of gravity (9b) is nalytical (Harris and Michels, 1965).[24] This allows an analytical Legendre expansion of the charge density of the molecule

$$\rho(r,B') = \sum_{\alpha=1}^{n_{occ}} N_\alpha |\phi_\alpha(\vec{r})|^2 = \sum_{\mu} a_\mu(r) P_\mu(\cos\Theta') \tag{10a}$$

$$\text{with } a_\mu(r) = \frac{2\mu+1}{2} \int_0^{\pi} \rho(r,\Theta') P_\mu(\cos\Theta') \sin\Theta' d\Theta' \tag{10b}$$

$$= \frac{1}{4\pi r^2} \sum_{\alpha=1}^{n_{occ}} N_\alpha \sum_{l=\lambda_\alpha}^{\infty} \sum_{l'=\lambda_\alpha}^{\infty} \varphi_{l_\alpha}^{\lambda_\alpha}(r) \varphi_{l'_\alpha}^{\lambda_\alpha}(r) [(2l+1)(2l'+1)]^{1/2}$$

$$(-1)^{\lambda_\alpha} \langle l o l' o | \mu o \rangle \langle l \lambda_\alpha l' - \lambda_\alpha | \mu o \rangle \tag{10c}$$

from which accurate multipolar expansions of the static[25] HFEG exchange potential eq. (3, 5a) and the total potential are easily deduced

$$V_{st.}(\vec{r}) = \int d^3\vec{r}'\rho(r,B')\frac{1}{|\vec{r}-\vec{r}'|} = \sum_\mu V_\mu(r)P_\mu(\cos\Theta') \quad (11a)$$

$$W_{ex}^{HFEG}(\vec{r}) = \sum_\mu W_\mu(r)P_\mu(\cos\Theta') \quad (11b)$$

$$V(\vec{r}) = \sum_\mu V_\mu(r)P_\mu(\cos\Theta') \quad (11c)$$

A useful check of the accuracy of the SC reduction is provided by the relation

$$N = 4\pi\int_o^\infty a_o(r)r^2dr \quad (12)$$

satisfied to about 8 decimal places in our quasi analytical treatment (the only approximation is the truncation of the partial wave expansion in 9b and 1oc).

Alternative numerical procedures may be less accurate. It is the case of the ALAM package[26], which evaluates $a_\mu(r)$ through the integral (1ob) with a 32 point Gauss-Legendre quadrature plus a fitting to an exponential form, leading to typical numerical mismatch of relation (12):

4.o15 for LiH instead of 4 (Morrison, ref. 26)
12.o17 for LiF instead of 12 (Collins, ref. 27)

In the FN approximation, the axial symmetry of the electrostatic interaction (11) leads to the decoupling of the radial equations in subspace of given projection λ of the electronic orbital momentum on the internuclear axis

$$[-\frac{d^2}{dr^2} + \frac{l(l+1)}{r^2} - k^2]F_l^\lambda(r) + 2\sum_{l'} V_{ll'}^\lambda(r)F_{l'}^\lambda(r) = \sum_\alpha \varphi_{l\alpha}^{\lambda_\alpha}\delta(\lambda,\lambda_\alpha)c_\alpha \quad (13a)$$

$$\sum_l F_l^\lambda|\varphi_{l\alpha}^{\lambda_\alpha}> = 0 \qquad \text{if } \lambda_\alpha=\lambda \quad (13b)$$

with $$V_{ll'}^\lambda(r) = [\frac{2l'+1}{2l+1}]^{1/2}\sum_\mu <l'o\mu o|lo><l'\lambda\mu o|l\lambda>V_\mu(r) \ . \quad (13c)$$

C. MF and LF equivalent formulations of the rotational CC

In order to have all the background relevant to the discussion of the numerical techniques, we recall here the main equations of the rotational CC, referring the reader to the original papers[12,14] for their derivation and detailed discussion[10-12].

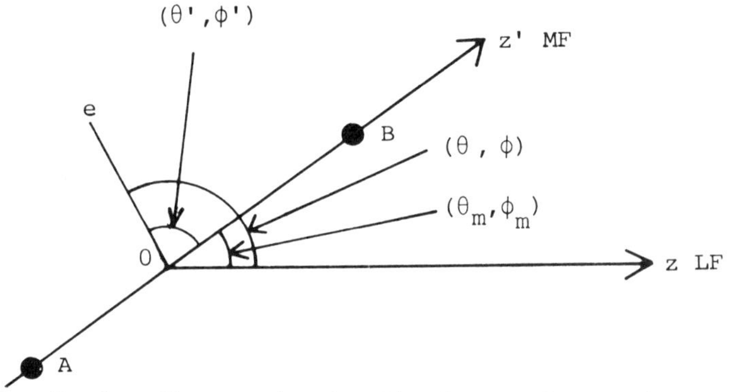

Fig. 2 Coordinates in the fixed LF and rotating MF.

The rotating MF can now be defined with the fixed LF by the instantaneous position of the internuclear axis (polar angles θ_m, φ_m on Fig. 2) and by the third arbitrary Euler angle. The internuclear distance being still assumed fixed and its parametric dependence on all dynamical variables will still be omitted.

$$H = \frac{1}{2}\left(\vec{P}_r^{\,2} + \frac{\vec{l}^2}{r^2}\right) + V(r,\theta') + B\vec{j}^2 \qquad (14)$$

where B is the rotational constant of the molecule and $\vec{j}$ its rotational angular momentum. Due to the introduction of the angular variables of the nuclei through $j(\theta_m, \varphi_m)$, the only good quantum numbers will be the total angular momentum $\vec{J}=\vec{l}+\vec{j}$, its projection M on a fixed axis, and the inversion parity $\Pi=(-1)^{l+j}$ of the whole system.

The exact formulation in the rotating MF uses the full angular basis

$$X_{l\lambda}^{JM\eta}(\vec{r},\hat{R}) = [Y_{l\lambda}(\hat{r}')D_{\lambda M}^{J*}(\hat{R})+\eta Y_{l-\lambda}(\hat{r}')D_{-\lambda M}^{J*}(\hat{R})]\left[\frac{2J+1}{8\pi(1+\delta_{\lambda o})}\right]^{1/2} \qquad (15a)$$

of the eigenvectors common to $\vec{l}^2$, $\vec{l}_z$, $\vec{J}^2$, $\vec{J}_z$ and Π as well as the reflexion wrt the x'Oz' plane with eigenvalue $\eta=\Pi(-1)^J$.

The exact formulation in the fixed LF uses the angular basis

$$Y_{lj}^{JM\eta}(\hat{r},\hat{R}) = \sum_m Y_{lm}(\hat{r})Y_{jM-m}(\hat{R})\langle lmjM-m|ljJM\rangle \quad (15b)$$

of the eigenvectors common to $\vec{l}^2,\vec{j}^2,\vec{J}^2,\vec{J}_z$ and $I=(-1)^{l+j}=\eta(-1)^J$.

The unitary transformation between these two angular representations

$$\Omega_{\lambda j}^{lJ\eta} = \langle X_{l\lambda}^{JM\eta}|Y_{lj}^{JM\eta}\rangle = \left[\frac{2j+1}{2J+1}\right]^{1/2}\langle l\lambda jo|ljJ\lambda\rangle\frac{1+\eta(-1)^{J-l-j}}{[2(1+\delta_{\eta o})]^{1/2}} \quad (15c)$$

describes the decoupling of the rotational quantum number

$$\vec{J} - \vec{l} = \vec{j} \quad (16)$$

Using the new expression of the Hamiltonian (14) and the properties of the two angular basis (15,16), one easily gets the exact radial equations:

In the MF, where the only modification is the addition of r independent rotational couplings

$$\left[-\frac{d^2}{dr^2}+\frac{l(l+1)}{r^2}-k^2\right]F_{l\lambda}^{J}(r)+2\left[\sum_{l'} V_{ll'}^{\lambda}(r)F_{l'\lambda}^{J\eta}(r)+\sum_{\lambda'} R_{\lambda\lambda'}^{lJ\eta}F_{l\lambda'}^{J\eta}(r)\right]$$

$$=\sum_\alpha \varphi_{l\alpha}^{\lambda_\alpha}(r)\delta(\lambda,\lambda_\alpha)c_\alpha \quad (17a)$$

$$\sum_l \langle F_{l\lambda}^{J\eta}|\varphi_{l\alpha}^{\lambda_\alpha}\rangle=0 \qquad \text{if } \lambda_\alpha=\lambda \quad (17b)$$

$$\text{with } R_{\lambda\lambda'}^{lJ\eta} = B\sum_{j=|J-l|}^{l+J}\Omega_{\lambda j}^{lJ\eta}\, j(j+1)[\Omega_{\lambda' j}^{lJ\eta}]^t \quad (18)$$

In the LF, where the rotational terms define the j dependent asymptotic channel energies

$$\left[-\frac{d^2}{dr^2}+\frac{l(l+1)}{r^2}-k_j^2\right]F_{lj}^{J\eta}(r) + 2\sum_{l'j'} V_{lj,l'j'}^{J\eta}(r)\cdot F_{l'j'}^{J\eta}(r)$$

$$= \sum_\alpha [\Omega_{\lambda_\alpha j}^{lJ\eta}]^t\varphi_{l\alpha}^{\lambda_\alpha}(r)c_\alpha \quad (19a)$$

$$\sum_{l}[\Omega^{lJ\eta}_{\lambda_\alpha j}]^t \langle F^{J\eta}_{lj}|\varphi^{\lambda_\alpha}_{l\alpha}\rangle = o \qquad \forall\, \lambda_\alpha \tag{19b}$$

with $\frac{1}{2}k_j^2 = \frac{1}{2}k^2 - Bj(j+1)$ (2oa)

$$V^{J\eta}_{lj,l'j'}(r)=[(2j'+1)(2l'+1)]^{1/2}\sum_{M}(-1)^{J+\mu}\langle l'o\mu o|lo\rangle\langle j'o\mu o|jo\rangle$$
$$\times\, W(ljl'j';J\mu)V_\mu(r) \tag{2ob}$$

Current approximations of the MF and LF exact rotational CC formulations will be discussed in section V on the general implementation of the rotational FT.

III. OUR NEW SC PACKAGE

Three SC Packages had already been developed and widely used to solve electron-molecule SE equations.

The first one, initially written by Sinfailam[28], was designed to solve the exact SE equations by replacing each radial term of the integrodifferential (ID) two electron kernel by an extra radial equation...[29] In spite of recent improvements by Raseev[3o], the scope of the program remains limited due to the rapidly growing incompressible number of differential equations added by the exchange terms; and its numerical accuracy becomes questionable if converged calculations are attempted (no stabilization procedure, to avoid the numerical dependence of the solutions, which result from the mixing of the exponentially growing closed channels, arising from the large $l>lo$ - even less - components locally closed by the centrifugal potential $l(l+1)/r^2$).

The second one, written by Chandra,[31] was designed to solve coupled inhomogeneous second order differential equations of our general type eq. (13,17,19) and has been applied to a wide variety of homonuclear (N_2)[32] and moderately polar molecules (CO[33], HF, HCl...[34]). The differential equations are solved directly by the De Vogelaere's method,[35] but a rather cumbersome package resulted from the implementation of the "Lagrange orthogonalisation" combined with a simple stabilisation procedure. The obtention of a particular solution of the inhomogeneous equation associated to each of the NB bound orbitals involved in the orthogonalisation, as well as a complete set of NC independent solutions of the homogeneous equations is necessary in order to satisfy the orthogonality conditions. Moreover, in addition to the usual division of the integration range in inner and outer region ($r>r_c$ where the inhomogeneous bound terms vanish), the inner region is itself divided in two zones ($r \gtrless r_o$) of outwards and inwards integrations, in order to avoid an excessive exponential growth of the high l channels, closed everywhere except asymptotically and in the neighbourhood of the nuclear wells. Much skill in the choice of r_o, computer time and computer storage were required to extract satisfactory NC squared K-matrix at the end of a calculation involving the integration of (3NC+2NB) independent solutions !

The third package,[36] extensively used by Collins, Morrison, Norcross and Robb in all their model exchange[17-20] and iterative ESE[9,16] calculations is designed to solve the same inhomogeneous coupled second order system as Chandra. Now, the coupled differential equations are converted to coupled integral equation and a simple trapezoidal quadrature allows a purely outwards propagation of the solutions up to the asymptotic region where an appropriate matching insures the orthogonality conditions and provides the final K-matrix. A reliable stabilization procedure is introduced: a set of linearly independent solutions being explicitly reconstructed by standard techniques[37] at every few integration points.

The common point to these three methods is that they rely heavily on numerical techniques, often driven to their limits, and provide little insight on the intermediate dynamics. Our new SC package intends to overcome these weaknesses. For convenience, we will describe successively numerical methods mainly designed for the inner region (strong couplings, inhomogeneous equation, integral orthogonality conditions) and semi-analytical methods optimal for the outer region (weak but LR couplings).

A. Inner Region

Inside the exchange zone ($r<r_c$), we must solve a general inhomogeneous second order differential system

$$\underbrace{\left[-\frac{d^2}{dr^2}+\frac{l(l+1)}{r^2}+V(r)-k^2\right]}_{\underset{\approx}{L}}\underset{\approx}{F}=\underset{\approx}{\varphi}\cdot\underset{\approx}{c} \qquad (21a)$$

subject to the integral orthogonality conditions

$$\int_0^{r_c} dr\,\underset{\approx}{\varphi}^t\cdot\underset{\approx}{F}=0. \qquad (21b)$$

For a partial wave expansion limited to NC collisional channels with NB orthogonality conditions, the matrix notation ($\underset{\approx}{A}$ for non diagonal matrix; $\underset{\sim}{A}$ for diagonal one) assumes that:

$\underset{\approx}{L}$ is the NCxNC Hamiltonian matrix

$\underset{\approx}{\varphi}$ is the NCxNB matrix of channel components of the bound orbitals involved in the orthogonality constraints

$\underset{\approx}{c}$ is a row matrix of NB Lagrange multipliers for each independent solution $\underset{\approx}{F}$.

Let us now give a few comments about interesting amendments (§1) we have brought to the usual integration and stabilization procedures involved in the solution of (21), and insist more on the radical improvements brought by the reformulation of the equations in the adiabatic angular basis (§2).

1) Improved numerical solution of (21) by stabilized outwards integration

For symmetries which require no explicit orthogonalisation, the most efficient way to obtain a scattering boundary condition on the surface of the inner region, is to follow Johnson's[38] algorithm for a direct, stable, outward propagation of the NCxNC logarithmic derivative matrix selected at the origin by the regularity condition.

For the other symmetries, explicit solutions are required in order to impose the orthogonality condition. Our approach follows Chandra's general philosophy but improves its implementation. Like him, we have selected the De Vogelaere's algorithm, more accurate than most standard finite difference formulae like Runge Kutta[39] and specially convenient for an accurate calculation of the overlaps by Simpson's rule (solutions calculated at each half mesh point).

The solution of the inhomogeneous system (21) is defined as a suitable linear combination (orthogonality conditions) of the NC regular solutions of the homogeneous system

$$\underset{\approx}{L} \cdot \underset{\approx}{F}_o = 0 \Rightarrow \underset{\approx}{F}_o \quad (NCxNC) \tag{22a}$$

and a particular solution of each inhomogeneous equation

$$\underset{\approx}{L} \cdot \underset{\approx}{F}_\alpha = \underset{\approx}{\varphi}_\alpha \Rightarrow \underset{\approx}{F}_\alpha \quad (NCxNB) \tag{22b}$$

Its determination proceeds in two steps:

A set of NS=NC+NB solutions $(\underset{\approx}{F}_o, \underset{\approx}{F}_\alpha)$ is propagated outwards, from the initial condition

$$\underset{\sim}{F}_o(r_1) = r_1^{\underset{\sim}{l}+1}\,\underset{\sim}{1} \qquad \underset{\sim}{F}'_o(r_1) = (\underset{\sim}{l}+1)r_1^{\underset{\sim}{l}}\,\underset{\sim}{1} \tag{23a}$$

$$\underset{\sim}{F}_\alpha(r_1) = 0 \qquad \underset{\sim}{F}'_\alpha(r_1) = 0 \tag{23b}$$

using De Vegelaere's algorithm with variable stepsize and, incrementation of the corresponding overlap matrices

$$\underset{\approx}{I}_{o,\alpha}(r_{n+1}) = \underset{\approx}{I}_{o,\alpha}(r_n) + \int_{r_n}^{r_{n+1}} \underset{\approx}{\varphi}_{\alpha}^{t} \cdot \underset{\approx}{F}_{o,\alpha}\, dr \tag{24}$$

A particular set of NC regular solutions, satisfying the orthogonality constraints (21b) is selected by imposing "R-matrix" (or "K-matrix" type boundary conditions) at the surface of the exchange zone r_c

$$\underset{\approx}{F}_o(r_c)\underset{\approx}{c}_o + \underset{\approx}{F}_\alpha(r_c)\underset{\approx}{c}_\alpha = \underset{\approx}{F}_{R \text{ or } K}(r_c) \tag{25a}$$

$$\underset{\approx}{F}'_o(r_c)\underset{\approx}{c}_o + \underset{\approx}{F}'_\alpha(r_c)\underset{\approx}{c}_\alpha = \underset{\approx}{F}'_{R \text{ or } K}(r_c) \tag{25b}$$

$$\underset{\approx}{I}_o(r_c)\underset{\approx}{c}_o + \underset{\approx}{I}_\alpha(r_c)\underset{\approx}{c}_\alpha = 0 \tag{25c}$$

where

$$\underset{\approx}{F}_R(r_c) = \underset{\sim}{1} \Rightarrow \underset{\approx}{F}'_R(r_c) = \underset{\approx}{R}^{-1}(rc) \tag{26a}$$

or

$$\underset{\approx}{F}_K(r_c) = \underset{\sim}{J} - \underset{\sim}{N}\,\underset{\approx}{K}(r_c) \ ; \quad \underset{\approx}{F}'_K(r_c) = \underset{\sim}{J}' - \underset{\sim}{N}'\,\underset{\approx}{K}(r_c) \tag{26b}$$

$\underset{\sim}{J}$, $\underset{\sim}{N}$ being the diagonal matrices of Bessel or Coulomb functions solutions. The matching equations reduce easily to a Kramer system for (2NC+NB)xNC unknown $\underset{\approx}{c}_o$, $\underset{\approx}{R}$, $\underset{\approx}{c}_\alpha$

$$\begin{array}{c} \\ \text{NC} \\ \text{NC} \\ \text{NB} \end{array}
\begin{array}{c} \begin{array}{ccc} \text{NC} & \text{NC} & \text{NB} \end{array} \\
\begin{bmatrix} \underset{\approx}{F}_o(r_c) & 0 & \underset{\approx}{F}_\alpha(r_c) \\ \underset{\approx}{F}'_o(r_c) & -\underset{\sim}{1} & \underset{\approx}{F}'_\alpha(r_c) \\ \underset{\approx}{I}_o(r_c) & 0 & \underset{\approx}{I}_\alpha(r_c) \end{bmatrix} \end{array}
\begin{array}{c} \text{NC} \\ \begin{bmatrix} \underset{\approx}{c}_o \\ \underset{\approx}{R} \\ \underset{\approx}{c}_\alpha \end{bmatrix} \end{array}
=
\begin{array}{c} \text{NC} \\ \begin{bmatrix} \underset{\sim}{1} \\ 0 \\ 0 \end{bmatrix} \end{array} \tag{27}$$

Eq. (27) shows clearly that the accuracy of the matching depends closely on the "numerical linear independence" of the propagated solutions up ro r_c. Since numerical dependence arises from contamination of all solutions by coupling to the exponentially growing solutions of high locally closed channels, stabilization is performed regularly to reselect a linearly independent set of NC homogeneous solutions and convenient particular inhomogeneous solutions. One can see immediately that stabilized (S) solutions at r_{s_i}:

$$\underset{\approx}{F}_o^S(r_{s_i}) = \underset{\sim}{1} \qquad \underset{\approx}{F}_\alpha^S(r_{s_i}) = 0 \tag{28}$$

and the corresponding overlaps are simply deduced from the matrix of unstatilized (US) solutions and overlaps by right matrix multiplication

$$\begin{bmatrix} \underset{\approx}{F}_o^S(r) & \underset{\approx}{F}_\alpha^S(r) \\ \underset{\approx}{F}'^S_o(r) & \underset{\approx}{F}'^S(r) \\ \underset{\approx}{I}_o^S(r) & \underset{\approx}{I}_\alpha^S(r) \end{bmatrix} = \begin{bmatrix} \underset{\approx}{F}_o^{US}(r) & \underset{\approx}{F}_\alpha^{US}(r) \\ \underset{\approx}{F}'^{US}_o(r) & \underset{\approx}{F}'^{US}(r) \\ \underset{\approx}{I}_o^{US}(r) & \underset{\approx}{I}_\alpha^{US}(r) \end{bmatrix} \begin{bmatrix} \underset{\approx}{F}_o^{US}(r_{s_i})^{-1} & -\underset{\approx}{F}_o^{US}(r_{s_i})^{-1} \cdot \underset{\approx}{F}_\alpha^{US}(r_{s_i}) \\ 0 & \underset{\sim}{1} \end{bmatrix} \tag{29}$$

In practice, stabilization needs to be more frequent near the nuclei where the coupling to highly closed channels is strong. But one stabilisation every 2o integration step is usually enough to get less than $1o^{-4}$ asymmetry in the matched R (or K)-matrix.

2) <u>New formulation in the angular adiabatic basis.</u>

The intrinsic weakness of the traditional SC approach is that many partial waves are necessary to represent the cusp of the electronic wavefunction near the nuclei, producing high and locally closed channels which interact very strongly with the low lying channels.

A solution to circumvent this drawback is to use a more appropriate angular basis: the adiabatic basis which diagonalizes locally the effective potential

$$\sum_{l'} [V^\lambda_{ll'}(r) + \frac{l(l+1)}{r^2} \delta_{ll'}] A^\lambda_{l'p(r)} = E^\lambda_{p(r)} A^\lambda_{lp(r)} \tag{3oa}$$

$$\underset{\approx}{A}^\lambda \text{ orthogonal} \tag{3ob}$$

The smooth radial parametric dependence of the adiabatic angular eigenstates

$$Z^\lambda_{p(r)}(\hat{r}) = \sum_l Y_{l\lambda}(\hat{r}) A^\lambda_{lp(r)} \tag{31}$$

combined with the wide splitting of the adiabatic potential curves at small r

$$A^\lambda_{lp(r)} \underset{r\to o}{\sim} \delta_{lp} \qquad E^\lambda_{p(r)} \underset{r\to o}{\sim} \frac{l(l+1)}{r^2} \tag{32}$$

and usually still near the nuclei, suggest that the dynamics will depend only on a few low lying adiabatic channels p, if the spherical harmonics expansion of eq. (9a) is replaced by the adiabatic angular expansion

$$F^\lambda(\vec{r}) = \sum_p Z^\lambda_{p(r)}(\hat{r}') \frac{F^\lambda_{p(r)}(r)}{r} \tag{33}$$

Considerable simplifications brought by similar adiabatic expansions have indeed been established in hyperspherical description of electronic correlations[40], colinear reactive scattering[41,42] rotational excitation in molecule collisions[43]. It has also already been introduced in electron molecule sollisions by Clark et al (1980)[44] but Clark's suggested application to rotational excitation of polar molecules has not yet been fully investigated (Clark, 1979).[45]

We will depart from the most previous works in the practical use of the adiabatic basis. In order to avoid numerically inconvenient non adiabatic couplings, we chose a "diabatic" approach. The integration range is divided into sectors $\{r_o^i, r_f^i\}$ within which the continuum orbital is expanded with respect to the midpoint r_m^i adiabatic basis eq. (33)

$$F^{\lambda}_{(\vec{r})} = \sum_p Z^{\lambda}_{p(r_m^i)}(\hat{r}') \frac{F^{\lambda}_{p(r_m^i)}(r)}{r} \qquad (34a)$$

$$\Updownarrow$$

$$F^{\lambda}_{p(r_m^i)}(\vec{r}) = \langle p(r_m^i)|F^{\lambda}(\vec{r})\rangle = \sum_l [A^{\lambda}_{lp(r_m^i)}]^t F^{\lambda}_l(r) \qquad (34b)$$

which we will call the "quasi adiabatic basis". The weakly coupled quasi adiabatic equations are then solved exactly:

$$\sum_{p'} [-\frac{d^2}{dr^2}\delta_{pp'} + V^{\lambda}_{pp'(r_m^i)}(r) - k^2] F^{\lambda}_{p'(r_m^i)}(r) = \varphi^{\lambda}_{p(r_m^i)\alpha} c_{\alpha} \qquad (35a)$$

with

$$V^{\lambda}_{pp'(r_m^i)}(r) = \sum_{ll'} [A^{\lambda}_{lp(r_m^i)}]^t [\frac{l(l+1)}{r^2}\delta_{ll'} + V_{ll'}(r)] A^{\lambda}_{l'p'(r_m^i)} \qquad (35b)$$

(diagonal at r_m^i by definition)

$$\varphi^{\lambda}_{p(r_m^i)\alpha}(r) = \langle p(r_m^i)|\varphi^{\lambda}_{\alpha}(r)\rangle = \sum_l [A^{\lambda}_{lp(r_m^i)}]^t \varphi^{\lambda}_{l\alpha}(r). \qquad (35c)$$

At the boundary of each sector, the adiabatic channel functions and their derivatives are transformed to the next sector by left matrix multiplication

$$F^{\lambda}_{p(r_m^i)}(r) = \sum_{p'} \underbrace{\langle p(r_m^{i+1})|p'(r_m^i)\rangle}_{T^{\lambda}_{p(r_m^{i+1})p'(r_m^i)}} F^{\lambda}_{p'(r_m^i)}(r) \qquad (36a)$$

with

$$\underset{\approx}{T}^{\lambda}(r_m^{i+1}, r_m^{i}) = [\underset{\approx}{A}^{\lambda}(r_m^{i+1})]^t \cdot \underset{\approx}{A}^{\lambda}(r_m^{i}) \ . \tag{36b}$$

As to the overlaps, they are expressed in terms of the adiabatic radial functions exactly in the same way as in the original spherical representation

$$\int_{r_o}^{r_f} d^3r \varphi_\alpha^\lambda(\vec{r}) \cdot F\,(r) = \int_{r_o}^{r_f} dr \sum_p \varphi^\lambda_{p(r_m^i)\alpha}(r) F^\lambda_{p(r_m^i)}(r) \ . \tag{36c}$$

Let us now summarize the main advantages of this diabatic formulation on the quasi-adiabatic angular representation:

a) The number of partial waves (l) to include in the calculation can be estimated from the diagonalization of the effective potential (3oa) in each radial region, before any scattering calculation: therefore, the l convergence is straightforward.

b) This number of partial (l) may be optimized in each radial range, using simple criteria on the expansion coefficients of the adiabatic basis. Full convergence near the nuclear singularities and reasonable l truncation outside their range are easily estimated, by contrast with the rather arbitrary l truncation in standard approaches (Collins et al...[19,36]).

c) A much smaller number of adiabatic channels (p) gives practically the same results as the primitive l basis, due to the weakness of the residual couplings.

d) The smallness of the number of adiabatic channels which contribute significantly to the dynamics allows usually a simple physical interpretation of the dynamics in terms of one or a few adiabatic potential curves (see § IV, specially for the interpretation of shape resonances)

e) Numerical instabilities disappear and the stabilization becomes practically unnecessary, since the high adiabatic channels are omitted from the start !

f) The adiabatic basis is calculated once, so that calculations at subsequent energies are very fast. This feature stresses the close analogy of this method with the R-matrix propagator of Light and Walker.[46] Although the HFEG exchange potential is energy dependent, its angular energy dependence is sufficiently weak so that the adiabatic basis at one average energy suffices to give weakly coupled representations over a wide energy range.

B. Outer Region

Outside the exchange zone, we are left, in all cases, with a set of differential homogeneous equations coupled by a small number of LR multipoles, with all channel energy degenerate in the FN approximation, and nearly degenerate in the exact LF formulation of the rotational CC. It becomes convenient to isolate the residual LR couplings on the RHS of the equations:

$$\left[\frac{d^2}{dr^2} - \frac{l(l+1)}{r^2} + \frac{2z}{r} + k_i^2\right] F_i^\lambda(r) = 2 \sum_{i'} V_{ii'}^\lambda(r) F_{i'}(r) \qquad (37a)$$

$$z = \text{residual charge of the molecular target} \qquad (37b)$$

$$V_{ii'}^\lambda(r) \underset{r \geqslant r_c}{\sim} \sum_{\mu=\mu_{min} \geqslant 1}^{\mu_{max}} \frac{a_{ii'}^\mu}{r^{\mu+1}} \qquad (37c)$$

Although the De Vogelaere numerical integration in quasi adiabatic representations, is an efficient way of solving the outer equations and providing physical insight about the dynamical ranges of the various multipoles, it is desirable to replace numerical techniques by analytical ones. A good example of this philosophy is provided by the analytical treatment of the FN dipole field, which simplifies partly the numerical work in electron polar molecule scattering.[44]

By including in the adiabatic angular basis most of the potential anisotropy, we have considerably simplified the angular problem. Similarly, one can simplify the radial problem of the outer region by replacing the determination of the oscillatory channel functions by that of their slowly varying expansion on an analytical radial basis accounting for most of the radial and eventually energy dependence of the equations (37). These are the key ideas of the variable phase method (VPM)[47], Gailitis analytical asymptotic expansion (GAE)[48] and the generalized multichannel quantum defect (MQDT)[49], whose combination allows a fast and reliable solution in the most critical cases (rotational excitation of strongly polar molecules near threshold). We will mention only the main advantages of each module, referring the reader to a recent presentation of these methods.[50]

1) The variable phase method

By expanding the channel functions on the basis of two independent solutions J(r),N(r) of the asymptotic LHS equation (37a)

$$\underset{\approx}{F}(r) = \underset{\sim}{J}(r)\, \underset{\approx}{A}(r) - \underset{\sim}{N}(r)\, \underset{\approx}{B}(r) \qquad (38a)$$

with the additional relationship

$$\underset{\approx}{F}'(r) = \underset{\sim}{J}'(r)\, \underset{\approx}{A}(r) - \underset{\sim}{N}'(r)\, \underset{\approx}{B}(r) \qquad (38b)$$

the VPM reduces the scattering problem to the determination of the slowly verying amplitude ratio $B(r).A^{-1}(r)$; the generalized K(r) matrix which satisfies a decoupled first order non linear system:

$$\underset{\approx}{K}'(r) = -\frac{2}{\omega}[\underset{\sim}{J}(r)-\underset{\approx}{K}(r)\underset{\sim}{N}(r)]\underset{\approx}{V}(r)[\underset{\sim}{J}(r)-\underset{\sim}{N}(r)\underset{\approx}{K}(r)] \qquad (39)$$

where $\omega = \underset{\sim}{J}\underset{\sim}{N}'-\underset{\sim}{J}'\underset{\sim}{N}$ is the constant Wronskian of the two basis functions.

The most natural choice of the {J,N} basis in terms of convenient asymptotic forms (sine, cosine for open channels, increasing and decreasing exponential for closed channels) leads to a generalized K(r) matrix which corresponds asymptotically to the usual definition (only the open portion is considered) and represents in each point the cumulated effect of the potential up to that point. The evaluation of K(r) (directly related to the scattering observable) appears clearly in (39) to result from the competition between the couplings and dynamical factors. Furthermore, by choosing a radial basis {J,N} including part of the LR potentials (dipole[44], induced polarisation[51]), the numerical integration is faster and the physical analysis of the dynamics more detailed.

2) The Gailitis asymptotic expansion

The efficiency of the VPM can still be increased and its convergence rigourously insured by replacing the tail of the numerical integration by an analytical asymptotic expansion. This is particularly important near threshold and in the presence of strongly coupled nearly degenerate channels, in which case the asymptotic form is very slowly reached. In close analogy to the VPM, the Gailitis expansion looks for the distortion of asymptotically decoupled exponential-type exact solution $\mathcal{E}$ of the LHS of eq. (37), as an analytical power series:

$$\underset{\approx}{G}(r) = [\sum_{p=0}^{\infty} \underset{\approx}{c}^{p} r^{-p}]\, \underset{\approx}{\mathcal{E}}(r) \qquad (4o)$$

The convergence of the traditional Burke and Schey[52] expansion on a purely exponential asymptotic basis is greatly improved by incorporating the effect of the Coulombic and centrifugal terms, and even larger portions of the potential (dipole....), the convergence radius depending monotonically on the size of the residual couplings (eq. (15) of ref. 5o).

3) The generalized MQDT

The basic paper on the generalized MQDT[49] and recent extension[51] provide the analytical expression of the J,N,$\mathcal{E}$ basis for a wide variety of potentials.

It also suggests the possibility of formulating the VPM in terms of an energy-analytical basis $\{J^o, N^o\}$, the connection between analytical quantities $\{J^o, N^o, K^o\}$ and physically interesting quantities $\{J,N,K\}$ reducing to a "mere formula". The generalisation of the VPM into a generalized (r-dependent) MQDT is therefore straightforward, but of considerable practical interest since it reduces the numerical calculation to a loose energy mesh, typical of SR interactions.[53] This feature allows an essential simplification of the FT implementation near nuclear thresholds (§V).

IV. FIXED NUCLEI ELASTIC SCATTERING RESULTS

Our new SC approach on the quasi adiabatic basis has been tested and applied systematically to a variety of typical diatomics homonuclear (H_2,N_2) and heteronuclear of increasing polarity (CO, HCl, HF, LiH, LiF). Illustrative preliminary results are discussed here, while final extensive studies will be published shortly.[54]

A. Visualisation of the Scattering Interaction

The intrinsic advantage of the collisional formulation is to isolate the interaction seen by the projectile. The graphical representation of its verious components[54] helps physical intuition. As an illustration, we present on figures 3 and 4 the static potential of the typical molecules we have chosen to study. Their main characteristics are summarized in table 1 (reference to the molecular

Table 1. Characteristics of our molecular targets

	H_2	N_2	CO	HF	HCl	LiH	LiF
molecular orbitals	Ransil Ref 55	Ransil Ref 56	McLean Ref 57	McLean Ref 57	McLean Ref 57	Ransil Ref 56	McLean Ref 57
$R_{eq}(a_o)$	1.402	2.068	2.132	1.733	2.409	3.015	2.987
D (ea_o)			-0.105	-0.736	-0.458	-2.330	-2.613
D_{exp}			+0.044	-0.719	-0.436	-2.31	-2.49
Q (ea_o^2)	0.480	-1.015	-1.547	-1.728	-2.790	-3.158	+4.566
Q_{exp}	0.474± 0.3	-1.04±.2	-1.859				
$-E_{ion}$(a.u)	0.564	0.545	0.554	0.650	0.476	0.304	0.472
B (a.u) 10^{-5}	27.694	0.916	0.976	9.577	4.824	3.422	0.596

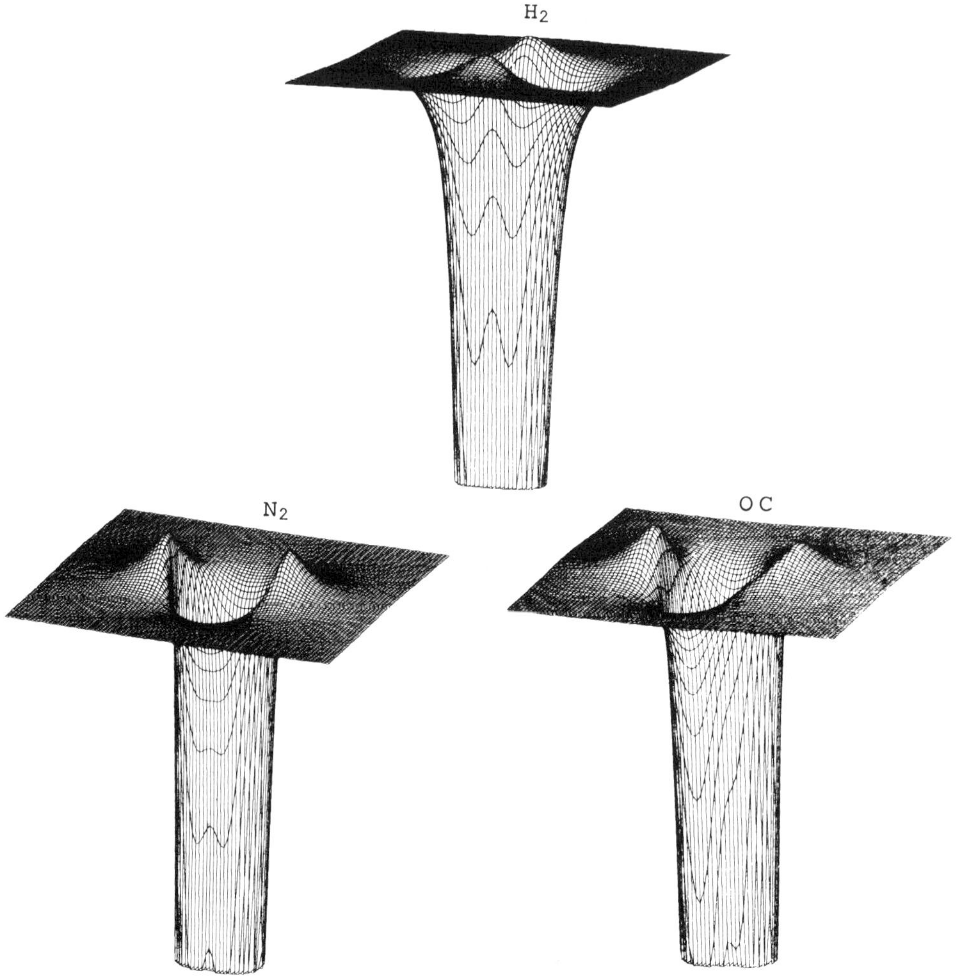

Fig. 3. Static interaction potential for e + H_2, N_2, CO

LCAO-MO-STO basis used, equilibrium internuclear distance, multipoles ......). Incidentally, this table should help the reader to clarify some sign inconsistancies in table 1 of re. 18b: we have found it convenient to refer the algebraic value of the dipole to an implicit orientation of the internuclear axis from the left to the right atom.

Fig. 3 shows three examples of "mild" static potential: indeed, the potential scale, similar for these three species, has been adapted to the weakness of the structures (-.1 eV).

H_2 (fig. 3a) is not much more than a large atom, with its small equilibrium internuclear distance (1.4 a_o), the small anisotropy of its charge density,[54] the only deep splitting of the two nuclear wells at -24.4 eV, not visible at the scale of the figure. Notice also the tiny repulsive quadrupolar bumps:

$\simeq$ .15eV, at 2.85a_o perpendicular to the internuclear axis,

which results from the concentration of the two 1sσg bonding electrons between the nuclei.

N_2 (fig. 3b) is somewhat more anisotropic, with its equilibrium internuclear distance of 2.o68a_o, and the larger bumps:

$\simeq$ .66eV, at 3.4a_o along the internuclear axis,

of the negative quadrupole which tends to repel the colliding electron from the diffuse $1\pi_u^4$ target electrons, stretched along the internuclear axis.

For CO (fig. 3c, table 1), the similarities with the isoelectronic N_2 molecule are more striking than its polarity. The dissymmetric quadrupole bumps

.53eV at 3.7au from the center of gravity on the oxygen side
.6oeV at 4.18au from the center of gravity on the carbon side

are quite comparable to those in N_2, while the tiny polarity competes only weakly with the quadrupole on the limited radial range presented on the graph. Note that the predicted polarity ($O^- - C^+$, attracting slightly the incident electron on the carbon side) is opposite to the observed one (table 1 and ref. 58), so that an imperical adjustment of the tail of the dipole potential

$$V_1(r) = V_1^{th}(r) + \frac{D_{exp} - D_{th}}{r^2} C(r) \qquad (41)$$

C(r) = cut off function similar to eq. 6b,

rather than the adjustment proposed by Chandra (eqs. 4.9,4.1o of ref. 33), may improve slightly the quality of detailed calculations.

Fig.4 displays results of some real polar molecules since three out of four have a supercritical dipole (D_c=.639ea_o). Note that the potential scale had been reduced (1/2 for HF, HCl; 1/1o for LiH, LiF) in order to adapt to the eV scale of their structure.

HF, HCl (fig. 4a,4b) cases can be considered as intermediate between purely covalent targets (H_2,N_2,CO) and highly ionic one (LiH,LiF). The static potential well remains deep between the two nuclei ($\gtrsim$4oeV for HF; $\gtrsim$25eV for HCl which is less tight). The large

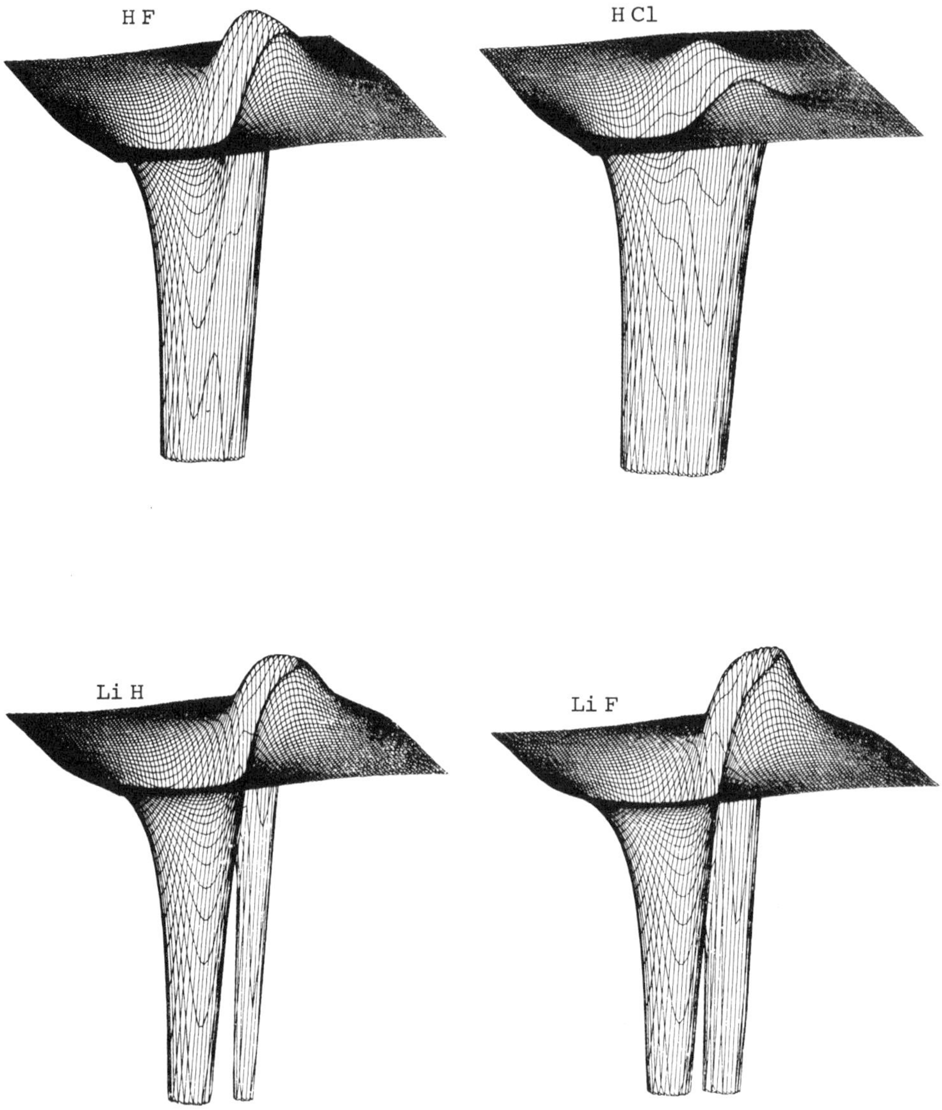

Fig. 4. Static interaction potential for e + HF, HCl, LiH, LiF

electron affinities of F and Cl explain the electronic accumulation on their side and the corresponding repulsive barriers

1.13eV for HF ; .427eV for HCl

for an incident electron arriving on the halogen side. On the contrary, the H side provides a wide well in which a s wave can accumulate at low energy, as suggested by Herzenberg to account for the threshold peaks obwerved in vibrational excitation. The considerable variation of the ratio $\frac{D}{Q}$ - .43 for HF and .166 for HCl explains the remaining qualitative differences between the two pictures.

LiH and LiF (fig. 4c, 4d) deserve their qualification as highly polar molecules. The two potential wells are nearly separate (V_{max} - 3.eV between the two nuclei) while on negative sides (H,F,) appears a huge repulsive barrier of

3.12eV for LiH ; 4.46eV for LiF.

This huge anisotropy of the potential, which will force the incident electron to penetrate on the Li side, requires a detailed angular description.

B. Compact visualisation of the shape resonance in the adiabatic angular approach

Most of the qualitative features of the low energy eigenphases are better understood in view of the spatial representation of the interaction.[54]

A complementary visualisation of the dynamics is given by the adiabatic potential curves (eq. 3o). We propose, now, to show how these potential curves $\varepsilon_p^\lambda(r)$ provide the expected one dimension representation of molecular shape resonances. Two examples of increasing

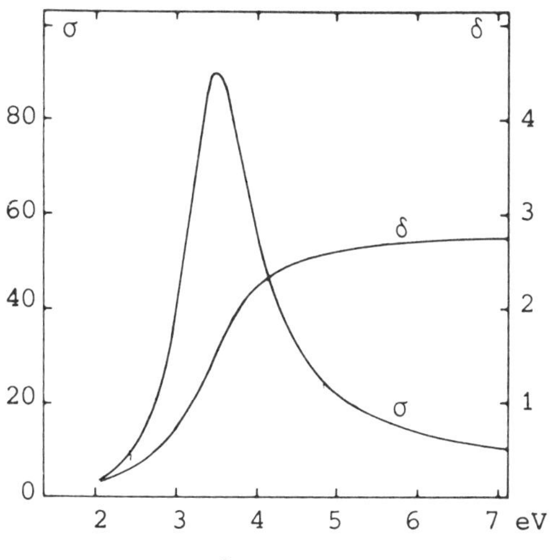

Fig. 5. $e+N_2(^2\Pi_g)$ eigenphase sum δ and cross section $\sigma(a_0^2)$

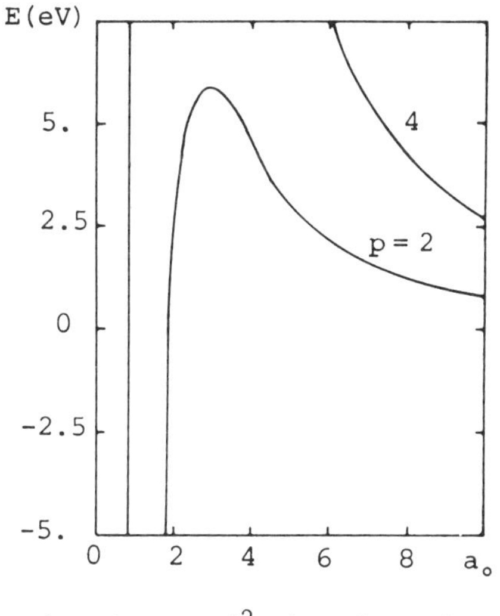

Fig. 6. $e+N_2(^2\Pi_g)$ adiabatic energy curves

complexity will be considered

1) The e + $N_2(^2\Pi_g)$ partial wave

Figure 5 shows the e + $N_2(^2\Pi_g)$ eigenphase sum and partial elastic cross section in a calculation based upon Ransil best limited LCAO-MO-STO representation of the N_2 core, the HFEG energy dependent model exchange potential and the ad hoc polarisation potential and the ad hoc polarisation potential suggested in ref. 17.

The fixed nuclei (FN) $^2\Pi_g$ resonance appears with its wellknown qualitative benaviour. The resonance energy ($E_r \simeq 3.9$eV) is smaller than the SE Ransil results ($E_r = 4.47$eV)[16a] as expected (polarization partly included), but still higher than the experimental position (polarization cut off not yet optimized). Let us forget about this weakness or its trivial remedy which has no influence on the important point we want to make

These results were first obtained by the traditional single center (SC) approach including 1o partial waves (1=2....2o, with $\mu_{el} \leq 26$, $\mu_{nuc} \leq 4o$ accounted in the potential). Very recently, numerically identical results have been obtained with only 4 quasi adiabatic channels. Although at the time of the present report we have not had the time to check it, qualitative exact results are even expected with one quasi adiabatic channel.

This is suggested by the adiabatic diagram presented on fig.6. The adiabatic potential curves are indeed well separated and only the first one penetrates deeply into the molecular core. Inspection of the adiabatic eigenvectors shows that the angular mixing is weak except near the nuclei, and even then it varies smoothly (atypical non-adiabatic behaviour). In particular, the first adiabatic mode, which tends to 1 = 2 at small and large distances, has an appreciable 1 = 4 amplitude ($\gtrsim$.1) cnly between .7 and 2.1a_o (nuclei at 1.o32a_o). A closer analysis of this eigenvector shows, in addition, a change of sign for this 1 = 4 amplitude at 3.7a_o: at small distances the colliding electron has a maximal amplitude along the internuclear axis (nuclear well) while at large distances the maximal amplitude is in the plane perpendicular to the internuclear axis (weak influence of the negative quadrupole of N_2, visualized on fig. 3a).

Since, in the particular case of the $^2\Pi_g$ resonance in N_2, the local HFEG is expected to give a good approximation of exchange effect (there is no occupied π_g orbital, therefore no "non local" orthogonality to impose), the first adiabatic eigenvalue of fig. 6 is expected to give a quantitative one dimension graphical representation of the $^2\Pi_g$ shape resonance. Calculations are underway to assess this statement.

2) The e + HCL($^2\Sigma$) partial wave

Figure 7 compares our results (.) with Collins et al (198o)[18a] recent calculations (———).

In fig. 7a our static + OHFEG model exchange results agree systematically with Collins et al OHFEGE, the tiny mismatch being attributed to the difference in the target representation (Collins et al using the slightly less accurate wavefunction of Cade and Huo[59]). It is interesting to note that the explicit orthogonalisation has negligible effect in the case of HCl. Moreover, both approximations are close to the exact ESE results (+), illustrating the reliability of the OHFEGE approximation now established in a wide variety of molecular systems.[18b]

In fig. 7b, our static + OHFEG model exchange + ad hoc polarization (.) shows a similar systematic agreement with Collins et al static local exchange + polarization (SLEP ———). Although we choose the same ad hoc polarization (optimized for this resonance), the two calculations differ now in two points: the target representation mentioned above and the dipole description. We have adjusted dipole tail to experiment on the following eq. 41, while their theoretical value (.4714) is substantially overestimated (see table 1).

These results, which were first obtained in the traditional CC approach including 1o partial waves (1=0...9) and all multipoles compatible with the triangular rule, have been quantitatively reproduced by the new SC package, including only 5 quasi adiabatic channels in the exchange zone.

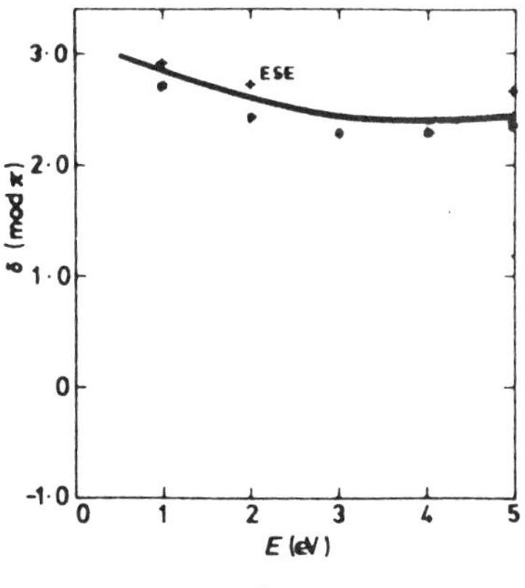

Fig. 7a. e+HCl($^2\Sigma$) eigenphase sum without polarization

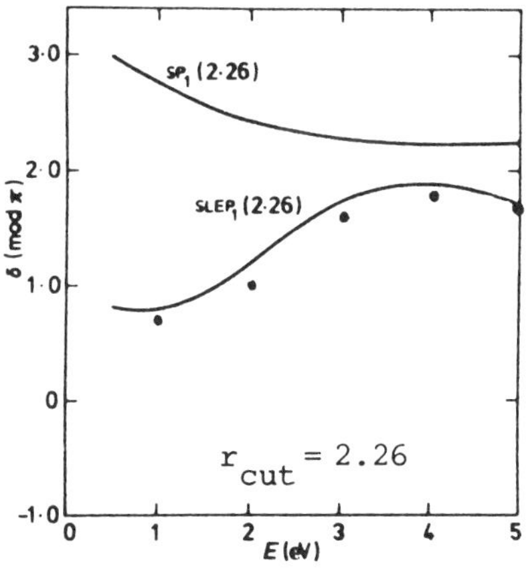

Fig. 7b. with polarization

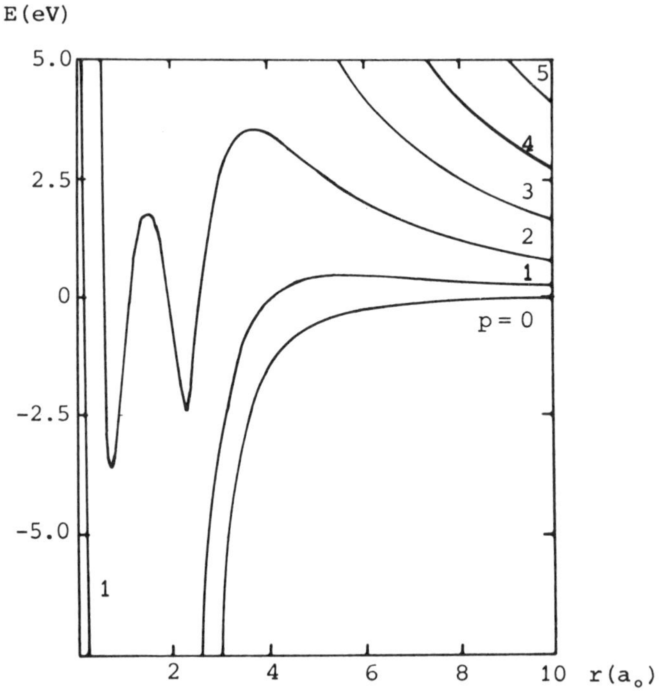

Fig. 8 e + HCl ($^2\Sigma$) adiabatic energy curves

Figure 8 shows the e + HCl ($^2\Sigma$) adiabatic diagram at the same scale as the previous fig. 6 for e + N_2 ($^2\Pi_g$). The asymmetry of the HCl molecule and the violence of the e + HCl interaction lead to a much higher density of adiabatic levels (6 below 5eV at $10a_0$ instead of 2 for N_2). At large distances ($r > 8a_0$) the adiabatic channels become close together but get a nearly pure dipole character and the scattering problem decouples completely. At smaller r, the situation is more complex, but the higher ($p > 4$) adiabatic channels, which are dominated by the centrifugal barrier, decouple completely. In this intricate short range region, only 3 adiabatic channels penetrate into the molecular core: being locally open in the energy range of our calculation ($0 \rightarrow 5$eV), they are expected to participate in the dynamics. The two cusps which appear on the third curve correspond obviously to the nuclear singularities and similar features appear on the two lower ones but at strongly negative energies.

With its large bump of 3.5eV at $3.6a_0$,

the third adiabatic level gives a qualitative interpretation of the calculated and observed shape resonance around 3eV. The l=2 centrifugal barrier provides a simple trapping mechanism. At smaller distance, the strong angular mixing within each adiabatic state and their strong r variation forbids any definitive conclusion without further study. However, there is some indication that the electron could be trapped inside the walls of the Cl quadrupole (fig. 4b).

The second adiabatic level, which has a large bump with

a maximum of .5eV at $5.5a_0$,

has analogous characteristics, with an l=1 centrifugal barrier.

The first adiabatic level is quite different: without barrier, it seems to describe the s wave piling up on the H side suggested by Herzenberg.

Although much more analysis is required to understand the full dynamics of e + HCl ($^2\Sigma$), the adiabatic formulation may be the right tool to clarify the situation.

V. THRESHOLD NUCLEAR EXCITATION OF POLAR MOLECULES

In paper I, we have selected a resonant process, dominated by SR interactions. We had focussed on their description, pointing out the tremendous simplification brought by the restoration of the BO approximation for the N+1 electron compound (factorization of the electronic and nuclear problem). But, we had paid little attention to the FT implementation since the outer region made no significant contribution to the dynamics. In this section, we will focuss on the complementary situation, occuring mainly near thresholds, where the LR competition between the electrostatic interaction and the dynamical decoupling from the nuclear motion becomes important.

A typical example is the rotational excitation of polar molecules by thermal or subthermal electrons.[60] The impossibility of neglecting the rotational decoupling was made evident by the theoretical divergence of the forward and therefore also the total cross sections of the fixed dipole field.[61] In fact, this divergence is not a practical problem since the large partial waves can be properly described by the analytical first order Born approximation (FBA) for a rotating point dipole (Clark, 1977).[62] However, the large angle scattering and the momentum transfer cross sections - which are the important quantities for plasma applications - depend strongly on the first partial waves which experience the whole interaction zone.

A. Limitations of the previous approaches

The exact formulation in the fixed laboratory frame (LF) which retains the rotational Hamiltonian and couples explicitly the rotational sublevels of the target in the rotational CC expansion (eq. 19) is the rigourous approach in principle; but in practice, it cannot account for the (jl) couplings up to complete convergence in the whole configuration space because of the high anisotropy and strength of the SR interaction, and because of the huge number of rotational levels which should be included (large number of asymptotically open rotational states, huge number of locally open states near the nuclei).

The rotating molecular frame (MF), which takes care of the axial symmetry of the molecular field (eq. 17) is more appropriate to the description of the electrostatic effects, but the centrifugal effects associated to the molecular rotation lead to small but infinite range couplings.

It is precisely to solve this problem that Fano[63] introduced first the concept of FT, in order to take advantage of the qualitatively different features of the interaction between the electron and the molecule as their mutual distance decreases. At large distance, the LF rotational CC is essential to describe the dynamical decoupling to the asymptotical dissociation channels. But at short distance, the accelerated colliding electron is rigidly coupled to the axial field and partakes to the rotation: the N+1 electron BO approximation allows to solve the simpler FN equations (13) in the MF. These two locally optimal formulations should be combined: the solution of the FN-MF equations, suitable at SR, being expressed to the LF angular scheme by a mere orthogonal transformation, and providing the starting point of the exact LF equations, suitable at LR.

However, in the original example - 1 uncoupling the autoionizing photoabsorption spectrum of H_2 - the LR electrostatic interaction reduces to the isotropic Coulomb field and the implementation of the FT is straightforward using MQDT analytical techniques. On the contrary, for neutral targets, specially for polar molecules, the tail of the molecular potential is weaker and anisotropic. The dynamical decoupling is expected at shorter distances r_t, but the solution of the rotational CC-LF equations is a real numerical problem. None of the three previous attempts to implement the FT for polar molecules hat solved the problem completely.

Only Chandra (1977)[33] examined the near threshold problem on e+CO scattering. But, his implementation of the FT was not very efficient nor conclusive. Extracting the K-matrix only at the end of the calculation (matching of the inner and outer solutions), he could not isolate SR and LR effects and the FT point r_{rot} appeared to result from a numerical compromise rather than a physical choice. Furthermore, a suspicious instability of the results with increasing r_t, put some doubt on the practical reliability of the FT.

Clark (1979)[44] added to this pessimism by setting a lower limit

E_{min} to the collision energy where the FT could be practically applied. But, his E_{min} was so high that, for the species considered, the adiabatic nuclei approximation ($r_t=\infty$) gave the same result. The implicit assumption of his discussion was that non perturbative treatments of the LF-LR couplings would be computationally prohibitive.

Collins and Norcross (1978, referred later to as CN)[19] encountered no practical difficulties in studying highly polar molecules (LiF,CsF,KI) below Clark's energy limit. In fact, they proposed an alternative FT. Instead of selecting a FT point r_t, they selected a partial wave l_t such that the penetrating partial waves $l<l_t$, mainly influenced by the SR interactions, are fully described in the FN-MF approximation, while the distant partial waves $l>l_t$, which feel the LR tail of the potential outside the centrifugal barrier, are fully described in the LF, either by exact rotational CC solutions in a model potential which accurately represent only the LR interactions, or by the FBA for the very large l. Their FT on l relies upon the following argumentation:"For small l, the wavefunction for the scattered particle, will penetrate deeply into the molecular charge cloud, and its properties will be determined primarily by the strong SR interaction potential. The additional accumulation of phase owing to the rotational Hamiltonian will be small by comparison and may be neglected, despite the fact that at some large radii the rotational and kinetic contributions may be comparable". This simplified FT on l would fail near threshold. In a private communication,[63] Norcross mentioned that this limitation could easily be removed by extracting the dipole interaction in the MF and putting it back in the LF, in the FBA approximation. Norcross noticed that his improved l-FT verifies nicely Chandra's result on CO, but this seems to us not conclusive since the dipole interaction is very weak in this case, as opposite to hydrogen - and alkali - halides, where it is well known that the dipole interaction cannot be trated perturbatively.[64]

B. Our implementation of the FT

We have brought three contributions to the traditional "r-FT" of Chang and Fano.

1) The VPM implementation of the traditional FT.

In previous papers[10,11,50] we have shown that the traditional FT do apply to the rotational excitation of polar molecules near threshold and is most simply and efficiently handled by generalizing the concepts of effective range theory to the VPM. For this development step, we chose the same e+CO model problem[65] as Chandra.[33] The relative weakness of the potential (fig. 3c) compared to strongly polar species (LiH,LiF on fig. 4c,4d) allows an exact solution of the rotational CC. Although the dipole of CO is very small, the ratio B/D is rather large (table 1), so that the effects of rotational couplings appear in full generality at sufficiently low energy.

The main conclusion of this study is that the choice of the FT point r_t is less crucial than the choice of the FT observable Θ on which the orthogonal angular transformation (eq. 15) must be performed

$$\underset{\approx}{\Theta}^{LF} = \underset{\approx}{\Omega}^{t}\,\underset{\approx}{\Theta}^{MF}\,\underset{\approx}{\Omega} \qquad (42)$$

This choice would have no importance for two equivalent formulations (solutions of the exact MF equations transformed to solve the exact LF equations), but becomes crucial when approcimate equations are used (FN-MF af $r \leq r_t$). This is illustrated by fig. 9a, showing the results obtained for a typical case (J=2 partial cross section $j=2 \rightarrow j'=2$), when the FT is performed on three different observables ($\underset{\approx}{R}$-matrix, $\underset{\approx}{K}$-matrix, analytical $\underset{\approx}{K}^{o}$-matrix) as a function of the FT point r_t (horizontal axis).

The transformation on the $\underset{\approx}{R}$-matrix, adopted by Chandra,[33] is satisfactory at small r_t, but diverges completely at large r_t. This systematic aggravation results from the sensitivity of the logarithmic derivative ($\sim$ cotg kr) to the mismarch of the exact and approximate argument $(k_j - k)r_t$ which increases linearly with r_t. The corresponding poles in the RM get closer and closer as the radius r_t increases.

The transformation on the $\underset{\approx}{K}$-matrix seems to follow directly from Chang and Fano suggestion of matching the variable amplitude A and B assuming the basis functions to satisfy

$$J_{1k}(r_t) \simeq J_{1k_j}(r_t)$$

$$N_{1k}(r_t) \simeq N_{1k'_j}(r_t). \qquad (43)$$

The adiabatic nuclei approximation is in fact a particular case of this approach when the FT point goes beyond the range of the interaction. The K-matrix results of fig. 9a follow this prediction, but do not converge to the exact results when the FT point tends to the origin (exact LF in whole space !). The aggravation when the incident energy decreases, suggests that the divergence is related to the non analytical behaviour of the K-matrix, or more simply to the rapid energy dependence of the Bessel function amplitude.[43] The pessimistic conclusions of Clark[44] are related to this problem.

The $\underset{\approx}{K}^{o}$ matrix, obtained by removing the non analytical kinematic factor

$$\underset{\approx}{K}^{oMF} = \underset{\sim}{k}^{-(1+\,1/2)}\,\underset{\approx}{K}^{MF}\,\underset{\sim}{k}^{-(1+\,1/2)} \qquad (44)$$

or its inverse, the M matrix of effective range theory, provides finally the right observable, combining a smooth dependence in E and in r.

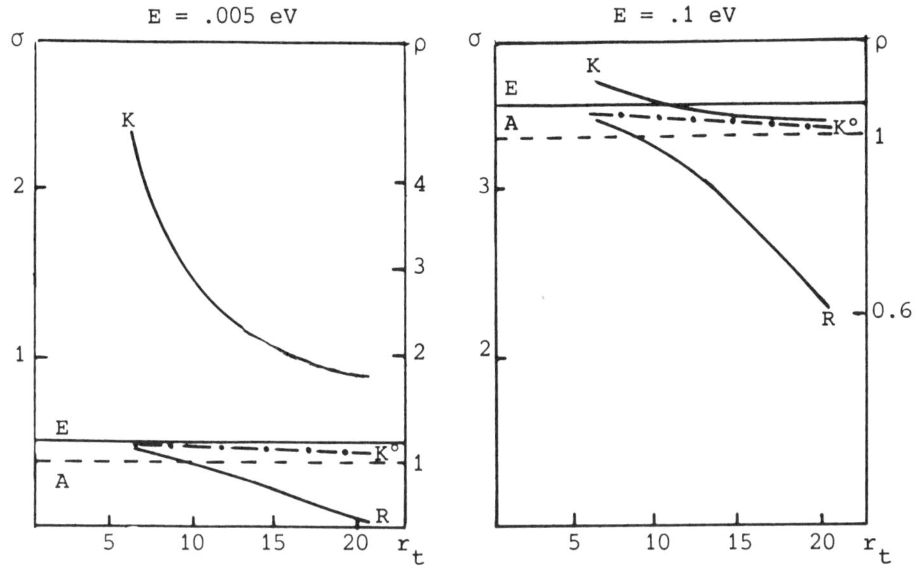

Fig. 9. e+CO scattering. J=2 partial cross section for j=2→j'=2
a. Frame transformation on R, K, K^{o} using the FN approximation in the MF. ρ is the ratio $\sigma_{calc}/\sigma_{ex}$. E:exact results; A:adiabatic nuclei results.

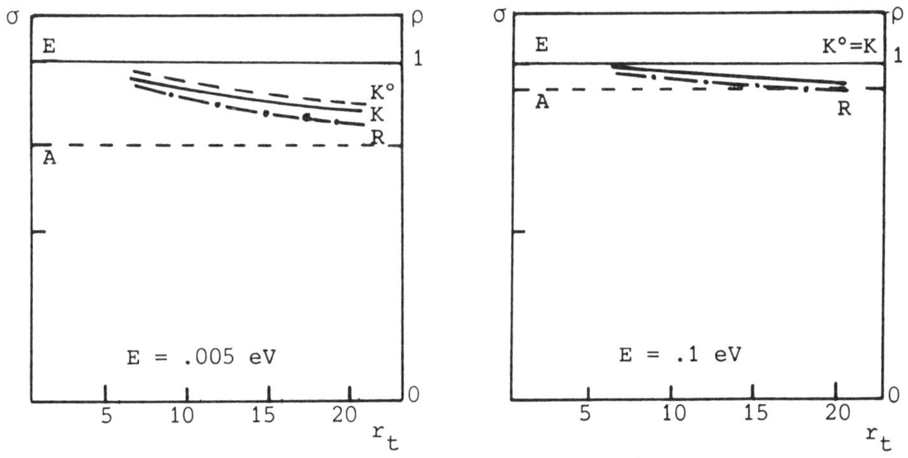

b. Frame transformation on R, K, K^{o} using the l_z conserving approximation in MF.

Let us point out two advantages of the FT on $\underset{\approx}{K}^o$

a) Formally, it makes a smooth junction between the threshold behaviour and the adiabatic nuclei approximation at sufficiently high energy.[10] By integrating the VPM equations in energy normalized basis (eq. 39 with $\underset{\sim}{J},\underset{\sim}{N},\underset{\approx}{K}$ formally replaced by $\underset{\sim}{J}^o,\underset{\sim}{N}^o,\underset{\approx}{K}^o$) the asymptotic K matrix given by

$$\underset{\approx}{K} = \underset{\sim}{k}_j^{l+1/2}\, \underset{\approx}{K}^{oLF}(r_\infty)\, \underset{\sim}{k}_j^{l+1/2} \tag{45}$$

will have the correct threshold dependence. It is also clear from eq. (44,42,45) that far from threshold ($k_j \simeq k$), this procedure will lead to the adiabatic results

b) Practically, the VPM is superior to Chandra's original techniques since it allows

straightforward convergence studies, since it integrates outwards the $\underset{\approx}{K}^o$ matrix describing the cumulated effect of the potential to that point,

straightforward determination of r_t, since the VPM equation expresses directly the evolution of the $\underset{\approx}{K}$ matrix in terms of each type of couplings.

2) The adiabatic rotation approximation

An even more refined way to implement the FT is to make only the l_z conserving approximation in the MF equations. This approximation neglects only the off diagonal λ coupling centrifugal terms, but accounts for the average change of the kinetic energy of the electron relative motion in the rotating molecular frame

$$\frac{1}{2}\, k^2_{l\lambda J\eta} = \frac{1}{2}\, k^2 - R^{lJ\eta}_{\lambda\lambda} \tag{46}$$

as it follows adiabatically the MF rotation. This simple correction restores the consistency among the FT on K -, R - and K^o-matrices, (fig. 9b) suggesting that the physical picture of a SR - N+1 electron BO compound, already established in vibrational excitation, extends to the rotation. Although not necessary in the present case, this refined approach is expected to be useful whenever the rotation cannot be neglected during the SR collision (high rotational quantum number, long lived resonance). Its interest is strengthened by the fact that the extra amount of work introduced by the J dependence in the MF equations becomes negligible when later implemented within the R-matrix framework, using the full adiabatic MF formulation for bound states.[65]

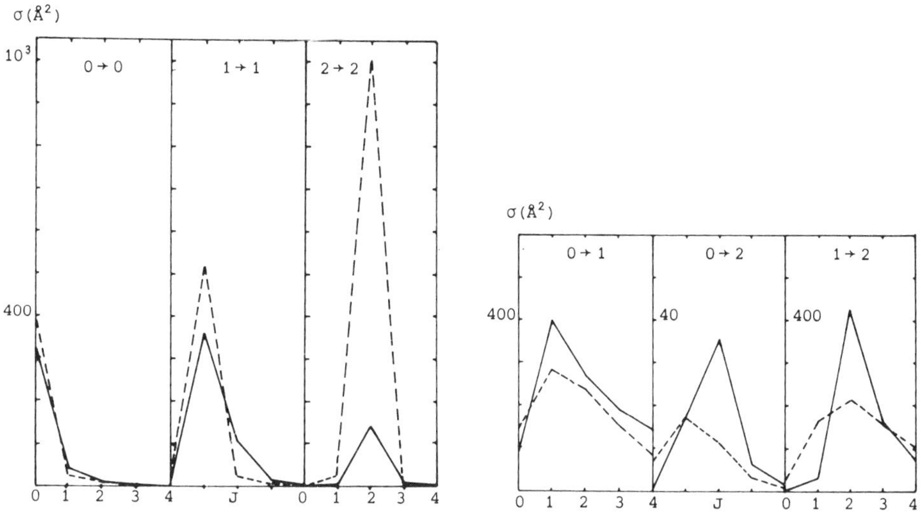

Fig. 10. Rotational excitation partial cross sections for e+HF as a function of J, at E = o.ool a.u.: the frame transformation results (———)and adiabatic nuclei approximation (---).

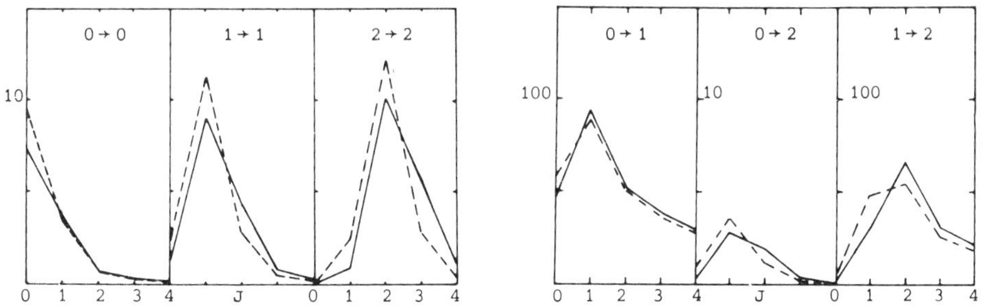

Fig. 11. Rotational excitation partial cross sections as a function of J for e+HCl, at o.o5 eV. Full lines and dashed lines as in Fig. 1o.

3) The adiabatic angular basis

In the previous section, we have seen that the use of an electronic angular adiabatic basis, diagonalizing

$$\frac{l(l+1)}{r^2}\,\delta_{ll'} + V_{ll'}(r)$$

could reduce considerably the number of channels and provide valuable insight about the dynamics.

Its extension to the rotational CC, by diagonalizing

$$\frac{l(l+1)}{r^2}\,\delta_{ll'} + V_{lj,lj'}(r) + Bj(j+1)\,\delta_{jj'} \tag{47}$$

introduced by Clark[44] provides a gradual FT from the λ coupling regime at SR to the (jl) decoupling at infinity. It provides an estimate of the effective range of dynamical coupling before any scattering calculation and shows the range of strong electrorotational couplings due to avoided crossings.

B. PRELIMINARY RESULTS ON REAL POLAR MOLECULES

Figure 1o compares the J=0,...4 contributions to elastic (fig. 1oa) and inelastic (fig. 1ob) rotational cross sections in e + HF at .001au, in the static + OHFEGE approximation and figure 11 gives similar results for e + HCl at .o5eV. The partial cross sections differ significantly, suggesting that the FT is necessary to predict

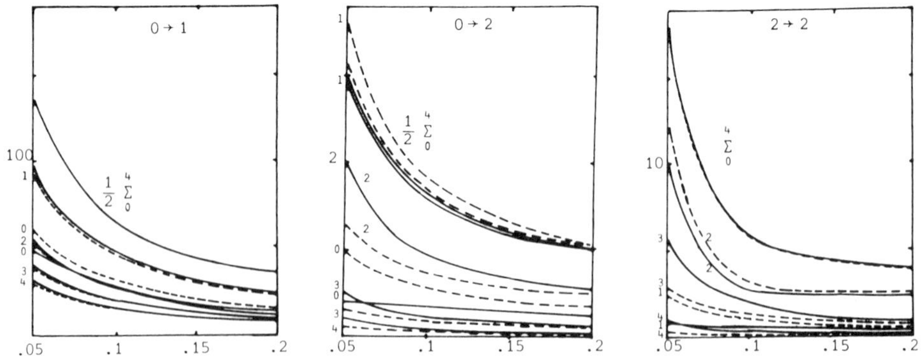

Fig. 12. Rotational excitation partial cross sections for e+HCl as a function of energy E between o.o5 and o.2o eV. Full lines and dashed lines as in Fig. 1o.

differential and momentum transfer cross sections. Amongst the systematic trends of the FT effects, one should remark specially the enhancement of the partial wave J which produces an l=0 exit channel, showing the drastic influence of adiabatic barriers (see adiabatic diagram for e + HCl, on fig. 8), particularly effective at such low energies. Figure 12 presents the energy dependence of the e + HCl results, suggesting that the adiabatic nuclei approximation becomes valid around .2eV in this case.

More extensive studies, including the important polarization effects, are underway and will be published shortly.[66]

VI. CONCLUSION

The main conclusions of these papers are the confirmation of the physical picture which underlies the PT theory of electron-molecule collisions and the broad applications of the FT concept in promoting locally optimal descriptions. We have shown how it could be implemented in the description of electronic and nuclear correlations which arise in resonant vibrational excitation and threshold rotational excitation.

The R-Matrix is the optimal tool to isolate the SR (N+1) electron compound and extend bound state techniques to the description of the SR electronuclear complex.

The adiabatic angular SC approach provides an alternative efficient tool to describe angular electronuclear correlations; by selecting dynamically important channels, it reaches physical insight and numerical efficiency simultaneously. Its implementation in the LR decoupling region defines a gradual FT and isolates the regions of strong electronuclear coupling (avoided crossings).

The variable phase method is a powerful tool for analysis by the influence of the LR potentials: it facilitates calculations up to very large radii without much computational efforts, by extracting an observable which varies slowly as a function of r and E.

Our conclusions derived from the success of two complementary simple studies: the resonant vibrational excitation of N_2, using the generalized R-Matrix to describe the SR vibrating compound and the threshold rotation excitation of polar molecules using the adiabatic angular single center approach and the VPM to describe the rotational FT and LR decoupling.

The conjunction of these progresses permits to attack the majority of diatomic molecules without prohibitive computational effort while preserving a clear physical insight in the collision process.

REFERENCES

1. L. Castillejo, I.C. Percival and M.J. Seaton, Proc.Roy.Soc.Lond. A254, 259 (1959)
2. R. Damburg and E. Karule, Proc.Phys.Soc.90, 637 (1967)
   P.G. Burke, D.F. Gallaher and S. Geltman, J.Phys.B2, 1142 (1969)
3. M. Le Dourneuf, Thesis, University of Paris, CNRS Archives A012658
   M. Le Dourneuf, Progress Report at Xth ICPEAC, Paris 1977, in "Electronic and Atomic Collisions", ed. by G.Watel (North Holland Publishing Co.1978)p.143; M. Le Dourneuf, Vo Ky Lan and P.G.Burke Comm.At.Mol.Phys.7, 1 (1977).
4. B.I. Schneider, Chem.Phys.Lett.51, 578 (1977)
5. A. Klonover and U. Kaldor, J.Phys.B12, 323 (1979); and in "Electron and Photon Molecule Collisions", ed. by T.N. Rescigno, V. McKoy and B.I. Schneider (Plenum Press, New York 1979)p.123
6. B.I. Schneider, Chem.Phys.Lett.31, 237 (1975); Phys.Rev.A11, 1957 (1975); Xth ICPEAC Paris 1977, in "Electronic and Atomic Collisions", ed. by G.Watel (North Holland Publishing Co.1978)p.257; and in "Electron and Photon Molecule Collisions", ed. by T.N. Rescigno, V. McKoy and B.I. Schneider (Plenum Press, New York 1979) p.77
7. P.G. Burke, I. Mackey and I. Shimamura, J.Phys.B10, 2497 (1977)
   B.D. Buckley and P.G. Burke in "Electron and Photon Molecule Collisions", ed. by T.N. Rescigno....(Plenum Press 1979)p.133
   B.D. Buckley, P.G. Burke and Vo Ky Lan, CPC.17, 175 (1979)
8. P.G. Burke, A. Hibbert and W.D. Robb, J.Phys.B4, 153 (1971)
   K.A. Berrington, P.G. Burke, M. Le Dourneuf, W.D. Robb, K.T. Taylor and Vo Ky Lan, CPC, 14, 367 (1978)
9. L.A. Collins, W.D. Robb and M.A. Morrison, Phys.Rev.A21, 488(1980)
10. M. Le Dourneuf and Vo Ky Lan in "Electron and Photon Molecule Collisions", ed. by T.N. Rescigno...(Plenum Press 1979)p.51
11. M. Le Dourneuf, Vo Ky Lan and B.I. Schneider, Symposium on Electron Molecule Collisions, ed. by I. Shimamura and M. Matsuzawa (University of Tokyo 1979) addenda as post deadline invited paper
12. E.S. Chang and U. Fano, Phys.Rev.A6, 173 (1972)
13. P.G. Burke and M.J. Seaton, Meth.in Comp.Phys.10, 1 (1971)
14. A.M. Arthurs and A. Dalgarno, Proc.Roy.Soc.A256, 334 (1960)
15. B.H. Choi and R.T. Poe, Phys.Rev.A16, 1821 and 1831 (1977)
16. a/ L.A. Collins, W.D. Robb and M.A. Morrison, J.Phys.B11L777(1978) and Phys.Rev.A21, 488 (1980)
    b/ W.D. Robb and L.A. Collins, Phys.Rev.A22, 2474 (1980)
17. M.A. Morrison and L.A. Collins, Phys.Rev.A17, 918 (1978)
18. a/ L.A. Collins, R.J.W. Henry and D.W. Norcross, J.Phys.B13, 2299 (1980)
    b/ M.A. Morrison and L.A. Collins, Phys.Rev.A23, 127 (1981)
19. L.A. Collins and D.W. Norcross, Phys.Rev.A18, 467 (1978)
20. L.A. Collins, W.D. Robb and D.W. Norcross, Phys.Rev.A20, 1838(1979)
21. P.G. Burke and N. Chandra, J.Phys.B5, 1696 (1972)
22. S. Hara, J.Phys.Soc.Japan, 22, 710 (1967)

23. J.C. Slater,Quantum Theory of Atomic Structure,Vols 1 and 2, (McGraw Hill, New York 1960)
24. F.E. Harris and H.H. Michels,J.Chem.Phys.43,165 (1965)
25. F.H.M. Faisal,J.Phys.B3,636 (1970)
F.H.M. Faisal and A.L.V. Tench,CPC,2,261 (1971)
F.A. Gianturco,CPC,11,237 (1976)
26. M.A. Morrison,CPC 21,63 (1980)
27. L.A. Collins, private communication during his visit at Meudon (Fev. 1979)
28. P.G. Burke and A.L. Sinfailam,J.Phys.B3,641 (1970)
A.L. Sinfailam,CPC 1,445 (1970)
29. R. Mariott,Proc.Phys.Soc.72,121 (1958)
30. G. Raseev,CPC 20,267 (1980) and this conference.
31. N. Chandra CPC5,417 (1973)
32. N. Chandra and A. Temkin,Phys.Rev.A13,188 (1976)
A. Temkin in "Electron and Photon Molecule Collisions",ed.by T.N. Rescigno....(Plenum Press 1979)p.173
33. N. Chandra,Phys.Rev.A16,80 (1977)
34. F.A. Gianturco and D.G. Thompson,J.Phys.B10,L21 (1977)
35. R. De Vogelaere,J.Res.Natl.Bur.Std.,54,119 (1955)
W.A. Lester Jr.,J.Comp.Phys.3,322 (1968)
36. W.N. Sams,D.J. Kouri,J.Chem.Phys.51,4809 (1969)
E.R. Smith,R.J.W. Henry,Phys.Rev.A7,1585 (1973)
M.A. Morrison,N.F. Lane and L.A. Collins,Phys.Rev.A15,2186(1977)
M.A. Morrison in "Electron and Photon Molecule Collisions",ed. by T.N. Rescigno....(Plenum Press 1979)p.15
37. R.A. White and E.F. Hayes,J.Chem.Phys.57,2985 (1972)
38. B.R. Johnson,J.Comp.Phys.13,445 (1973)
39. A.C. Allison,J.Comp.Phys.6,378 (1970)
40. J.H. Macek,Phys.Rev.160,170 (1967)
C.D. Lin,Phys.Rev.A10,1986 (1974);A14,30 (1976)
U. Fano and C.D. Lin,AT.Phys.4 (New York Plenum)p.47(1975)
H. Klar and U. Fano,Phys.Rev.Lett37,1134 (1976)
H. Klar and M. Klar,J.Phys.B13,1057 (1980)
C.H. Greene,J.Phys.B13,L39 (1980)
C.H. Greene, submitted to Phys.Rev A (1980)
41. A. Kuppermann,J.A. Kaye and J.P. Dwyer,J.Chem.Phys.Lett.74,257 (1980)
J. Römelt,Chem.Phys.Lett.74,263 (1980)
42. J.M. Launay...to be published
43. K. Takayanagi,J.Phys.Soc.Japan 45,976 (1978)
K. Sakimoto,Institute of Space and Aeronautical Science Research Note (1980) to be published
44. C.W. Clark and J.Siegel,J.Phys.B13,L31 (1980)
45. C.W. Clark,Phys.Rev.A20,1875 (1979)
46. J.C. Light and R.B. Walker,J.Chem.Phys.65,4272 (1976)
47. M. Le Dourneuf and Vo Ky Lan,J.Phys.B10,L35 (1977)
48. M. Gailitis,J.Phys.B9,843 (1976)
49. C.H. Greene,U. Fano and G. Strinati,Phys.Rev.A19,1485 (1979)

50. Vo Ky Lan and M. Le Dourneuf, XIth ICPEAC, Kyoto 1979, in "Electronic and Atomic Collisions", ed. by K.Takayanagi (North Holl. Co 1980)p.751
51. S. Watanabe and C.H. Greene, Phys.Rev.A22,158 (1980)
52. P.G. Burke and H.M. Schey, Phys.Rev.126,147 (1962)
53. U. Fano, J.Opt.Soc.Am.65,979 (1975)
54. M. Le Dourneuf and Vo Ky Lan, to be published
55. S. Fraga and B.J. Ransil, J.Chem.Phys.35,1967 (1961)
56. B.J. Ransil, Rev.Mod.Phys.32,245 (1968)
57. A.D. McLean and M. Yoshimine, IBM J.Res.Dev.Suppl.12,206 (1967)
58. A.D. McLean and M. Yoshimine, J.Chem.Phys.46,3862 (1967)
59. P.A. Cade and W.M. Huo, J.Chem.Phys.47,649 (1967)
60. K. Takayanagi, in "Electronic and Atomic Collisions", IXthICPEAC University of Washington Press, Seattle 1975, p.219
Y. Itikawa, Phys.Rep.46,117 (1978)
61. W.R. Garrett, Phys.Rev.A4,2229 (1972)
62. C.W. Clark, Phys.Rev.A16,1419 (1977)
63. U. Fano, Phys.Rev.A2,353 (1970)
64. I. Shimamura, Symposium on Electron Molecule Collisions, ed. by I. Shimamura and M. Matsuzawa (University of Tokyo 1979)p.13
65. R.T. Pack and J.O. Hirschfelder, J.Chem.Phys.49,4009 (1968)
66. Vo Ky Lan and M. Le Dourneuf, to be published

ON THE INFLUENCE OF EXCHANGE
ON ELECTRONIC CONTINUUM FUNCTION IN DIATOMIC MOLECULES

G. Raşeev

Institut de Chimie, Université de Liège

B-4000 Sart-Tilman par Liège 1, Belgium

ABSTRACT

A recent method we have developped allows for exchange at a level comparable to the static interaction in the potential term of the one centre continuum radial equations. Two different convergence behaviours are found in N_2 + e and LiH + e. These behaviours are viewed as two extreme cases and lead us to a general conclusion about the exchange interaction convergence.

INTRODUCTION

In the last few years two different approaches have mainly been used in ab-initio calculations of low-energy electron molecules collision cross sections. As detailed by Buckley and Burke [1] these approaches are either the L^2 type or the close coupling techniques. The single-centre close coupling approaches have been implemented as a noniterative solution of differential equations (Burke and Sinfailam [2], Buckley and Burke [3], Raseev et al. [4], Raseev [5]), a non iterative or iterative solution of integral equations (Sams and Kouri [6], Morrison [7], and a solution of algebraic equations (Burke et Seaton [8], Seaton and Wilson [9], Crees et al. [10], Crees and Moores [11]).

The main difficulty in the single-centre close coupling molecular methods is the convergence of different observables with the bound, continuum and potential expansions. But in general this convergence can be studied in a very systematic way. Only in the case of the static potential, this study has already been performed for $e + H_2$ and $e + N_2$ by Morrison and Collins [12]. Their convergence cutoffs of the static potential can be, at least in a first approximation, used in a static exchange calculation as a static calculation is more difficult to converge then a static exchange one.

The close coupling noniterative solution of differential equations method (Raseev [5]), is used in this paper to study the convergence of the exchange interaction in a given static potential. Exchange convergence studies have been performed with respect to eigenphase sum of the continuum function. We have chosen the eigenphase sum as it gives a convergence mesure regardless the type of calculation we are performing (e.g. elastic e-molecule scattering or photoionization). Generally the convergence of the exchange with respect to eigenphase have been judged from graphs but some numerical evaluations of the convergence (following Morrison [7]) are given in the text. In a resonant channel the exchange convergence calculations are made at fixed kinetic energy of the colliding electron. For a nonresonant channel two energies are selected. With the cutoffs in the bound, continuum and exchange interaction expansions corresponding to the converged exchange interaction we calculate the eigenphase sum as a function of the kinetic energy of the electron. For reasons explained here after we have considered the $e - N_2$ shape resonance collision in the $^2\Pi_g$ state and the $LiH + e$ collision in the $^2\Sigma^+$ state. We have attempted to introduce the polarization of the N_2 target by the use of molecular orbitals of N_2^- instead of N_2. When it is appropriate the orthogonalization between bound and continuum functions is also introduced with little effect on the eigenphase of the converged results.

Let us recall briefly our close coupling method. More detailed description can be found in Raseev [5] (here after refered to as I) or in Raseev et al. [4]. In the fixed nuclei approximation (Temkin and Vasavada [13], Hara [14]), the static exchange radial continuum equations are

$$\sum_{\ell'} \left[\left(\frac{d^2}{dr^2} - \frac{\ell(\ell+\ell)}{r^2} + k^2 \right) \delta_{\ell'\ell} - 2\, V^{ST}_{\ell'\ell}(r) \right] f_{\ell i}(r) = $$

$$= -2 \sum_{\ell' b} V^{Ex}_{\ell'\ell}(r)\, u_{\ell' b}(r) \qquad \ell \in [1, \ell_{max}] \quad (1)$$

The single centre expansion of bound and continuum functions corresponding to the radial terms of the equation (1) is

$$\phi_{bm_b}(\vec{r}) = \sum_{\ell \geq |m_b|}^{\infty} \frac{1}{r} u_{\ell b}(r) Y_{\ell, m_b}(\hat{r})$$

and

$$\phi_{i,\lambda}(k,\vec{r}) = \sum_{\ell \geq |\lambda|}^{\infty} f_{\ell i}(k,r) Y_{\ell,\lambda}(\hat{r})$$

where ℓ and b label the single centre expansion and bound orbitals. k, $\vec{r}$, $\hat{r}$, λ, m_b are the modulus of the wave vector, the position vector, the angular variables associates with r, the projection of the angular momentum on the internuclear axis for the continuum and bound electrons. $u_{\ell b}(r)$ and $f_{\ell i}(k,r)$ are the radial functions and $Y_{\ell,\lambda}$ is the spherical harmonics. The static potential $V^{ST}_{\ell\ell'}(r)$ is a sum of one electron-nucleus attractive and of two electron repulsive potentials fully integrated over the angular coordinates. The exchange potential $V^{Ex}_{\ell'\ell}(r)$ is a two electron exchange interaction integrated over the radial coordinate of the first electron and angular coordinates of the two electrons. Its expansion in terms of bound orbitals b and of m_γ and γ of the $\frac{1}{r_{12}}$ expansion (I) is

$$V^{Ex}_{\ell'\ell}(r) = \sum_{b} \sum_{m_\gamma} \sum_{\gamma} C_b^{Km_\gamma} C_b^{\gamma m_\gamma} (\ell', \ell | m_b, \lambda)$$

$$\mathcal{Y}^{\gamma m_\gamma}(\phi_{bm_b}, f_i, r) \qquad (2)$$

where $C_b^{Km_\gamma}$ is the electron occupation of the bound orbital, $C_b^{\gamma m_\gamma}(\ell', \ell | m_b, \lambda)$ is a coefficient from the angular integration over the second electron and ℓ and ℓ' come from spherical harmonics expansions of the bound and continuum functions of the second electron. The compact exchange function $\mathcal{Y}^{\gamma m_\gamma}$ (Raseev et al. [4])is given as

$$\mathcal{Y}^{\gamma m_\gamma}(\phi_{bm_b}, \phi_{i,\lambda}, r) = \sum_{\ell\ell'} C_E^{\gamma m_\gamma}(\ell,\ell'|m_b,\lambda)\, \mathcal{Y}_\gamma(u_{\ell b}, f_{\ell' i}, r) \tag{3}$$

where the standard exchange function (Sinfailam [15] , Hartree [16]) is

$$\mathcal{Y}_\gamma(u_{\ell b}, f_{\ell' i}, r) = r^{-\gamma-1}\int_0^r r'^{\gamma} u_{\ell b}(r')\, f_{\ell' i}(r')\, dr' + r^{\gamma}\int_r^\infty r'^{-\gamma-1} u_{\ell b}(r')\, f_{\ell' i}(r')\, dr' \; . \tag{4}$$

From the relation (3) it is clear that the number of the compact exchange functions is smaller than the number of standard exchange functions. The additional exchange equations in terms of compact exchange functions are

$$\frac{d^2}{dr^2}(r\,\mathcal{Y}^{\gamma m_\gamma}) = \frac{\gamma(\gamma+1)}{r^2}(r\,\mathcal{Y}^{\gamma m_\gamma}) - (2\gamma+1)\sum_{\ell\ell'} C_E^{\gamma m_\gamma}(\ell,\ell'|m_b,\lambda)\,\frac{u_{\ell b}(r)\, f_{\ell' i}(r)}{r} \; . \tag{5}$$

The equations (1) and (5) are coupled through the second term of the equation (5) which contains the radial continuum function.

The procedure developped for molecules by Burke and Sinfailam [2] solves simultaneously the equations (1) and (5) written for the standard exchange function (4). Working with the function (4) in the additional exchange equations (5) will quickly rise the number of total coupled equations (1) and (5), as we have one equation (5) for each orbital and ℓ, ℓ', γ selected values. Using equation (3) instead of (4) the number of equations (5) will only be a function of number of orbitals and γ.

The exact number of additional exchange equations (N_{ex}) in the last case is given by the following formula

$$N_{ex}^{\lambda} = \sum_{m_b}\sum_{M} p_{m_b\lambda M}\, n_{m_b} \tag{5}$$

In (6) n_{m_b} is the number of bound orbitals of the symmetry m_b and $P_{m_b\lambda M}$ is the number of terms in the $1/r_{12}$ Legendre expansion allowed by the Clebsh Gordon triangulat rule (see e.g. Rotenberg et al. [17]). λ and m_b are the continuum function and bound orbital projections on the internuclear axis and M is the charge density projection on the internuclear axis.

The number of exchange equations generated from the standard exchange function (4) is given by multiplying N^{λ}_{ex} by ℓ_b and ℓ_c of the Table 1 and by reducing this number following the Clebsh-Gordon triangular rule. In the Table 1, we list the number of standard and compact exchange functions for some of the calculation displayed in the Figures 1 and 2 for the $N_2 + e$ collision. As it can be seen the reduction in working with compact exchange functions is significant.

We are considering here a closed shall target and calculating the continuum function following (1). The orthogonalization is explicitely introduced by an inhomogeneous term. In this case the $V^{Ex}_{\ell\ell'}(r)$ term (full exchange) is an attractive potential and it will deepen the molecular potential well. From the standard collision theory (see e.g. Rodberg and Thaler [18]) given a kinetic energy of the electron more attractive is the potential higher is

TABLE 1 : The number of exchange functions or additional equations for $N_2^-(^2\Pi_g)$

γ*	ℓ_b*	ℓ_c*	NO OF EXCHANGE FUNCTIONS	
			STANDARD	COMPACT
9	8	8	155	31
	14	14	404	31
11	8	8	173	38
	14	14	516	38
13	8	8	182	45
	14	14	617	45

* γ, ℓ_b, ℓ_c are exchange potential, bound and continuum cutoff in the corresponding expansions.

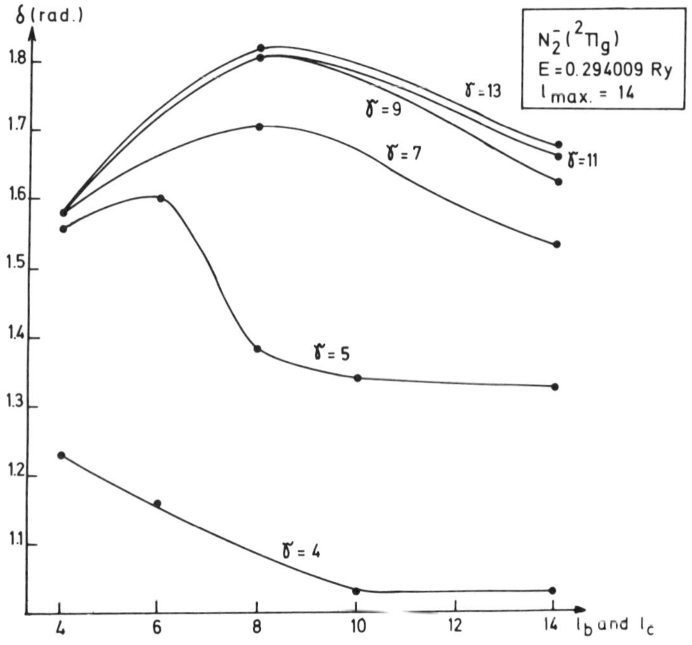

Fig. 1. Eigenphase sum of $N_2^-(^2\Pi_g)$ as function of the cutoffs in the bound and continuum expansions of exchange (Eq. (3)) for different values of γ of exchange (ℓ_b and ℓ_c are the cutoffs corresponding to ℓ, ℓ' in the equations). The Nesbet [25] basis set have been used.

the eigenphase. Calculations at fixed energy as the ones displayed in the Figures 1, 2 and 5 show exactly this behaviour. And in the Figure 3, the π_g resonance is moved to lower kinetic energies of the electron.

The systematic introduction of the exchange and the corresponding convergence studies are very tricky problems. We have studied them with respect of three expansions : the ℓ expansion of bound orbital, the ℓ' of the continuum function and the γ of the $1/r_{12}$ term. From the collision theory (Rodberg and Thaler [18]) we know that the high ℓ terms in the continuum function expansion do not penetrate in the internal region where the exchange is important. But from the Clebsh-Gordon triangular rule $\ell + \ell' < \gamma < | \ell - \ell' |$ the high ℓ, ℓ' terms in the bound and continuum expansion will contribute to the low terms in the expansion. Moreover in the case of polar molecules such at LiH the ℓ, ℓ' coupling is so strong that it propagates the interaction to high ℓ.

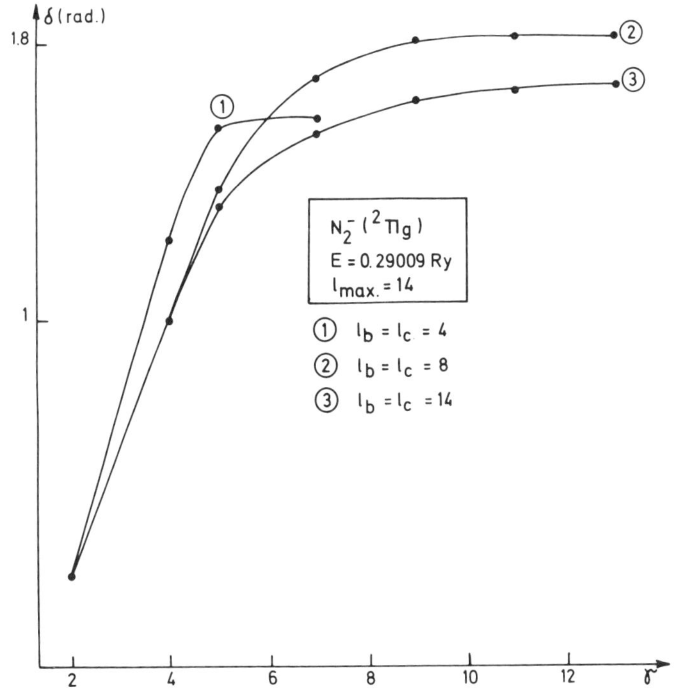

Fig. 2. Eigenphase sum of $N_2^-(^2\Pi_g)$ as function of γ of exchange (Eq.(3)) for different values of ℓ_b and ℓ_c. The Nesbet [25] basis set has been used.

Let us first consider the exchange interaction convergence behaviour in $N_2^-(^2\Pi_g)$. The πg resonance has been used as a test for a number of new methods in e-molecule collisions (Collins et al. [19] and references therein). We shall study the convergence in the middle of the resonance (E = 0.294009 Ry) as the eigenphase is very sensitive to small changes in the potential. Moreover we can examine it either in terms of ℓ_b and ℓ_c the cutoffs in bound and continuum functions (corresponding to ℓ' and ℓ in the equations) allowed in exchange (Fig. 1) or in terms of γ (Fig. 2). In this molecule the ℓ, ℓ' coupling is not very strong but the potential is non spherical and slowly converging in terms of γ. Consequently the relevant analysis of the exchange convergence will be in terms of γ of Figure 2. From this Figure we select the converged to 8% cutoff of $\gamma = 7$; $\ell_b = 14$; $\ell_c = 14$; as compared to $\gamma = 13$;

$\ell_b = 14$; $\ell_c = 14$*. A better choice converged to within 3% will be $\gamma = 9$; $\ell_b = 14$; $\ell_c - 14$ but the number of exchange equations is too high (see Table 1) and the calculation too expensive to allow for routine calculations. Therefore with the cutoff of $\gamma = 7$; $\ell_b = 14$; $\ell_c = 14$ (Figure 3) we calculate the energy dependence of the π_g resonance.

We obtain mainly the same results as Collins et al. [19] with the Nesbet molecular basis for the target. Their exchange cutoff is of $\gamma = 8$; $\ell_b = 4$; $\ell_c = 4$ as compared with our $\gamma = 7$;

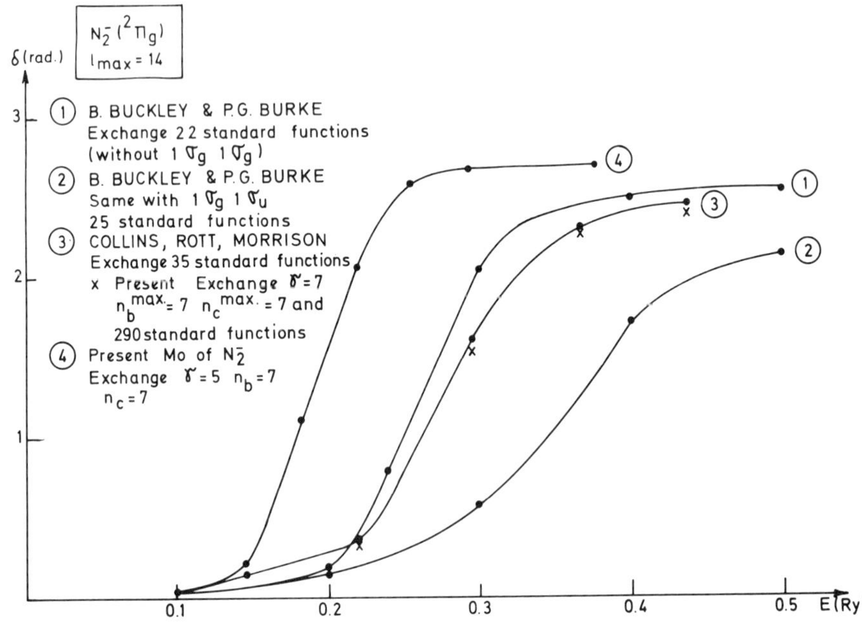

Fig. 3. The eigenphase sum of $N_2^-(^2\Pi_g)$ as function of the colliding electron energy.
1.- 23 standard exchange functions with all the bound orbitals introduced in exchange.
2.- Buckley and Burke [21] ; 18 standard exchange functions same as 1 (without $1\sigma_g$ and $1\sigma_u$ in exchange).
3.- Collins, Robb and Morrisson [9] ; 35 standard exchange functions. Dots are the present results 290 standard exchange functions $\gamma = 7$; $\ell_b = 14$; $\ell_c = 14$.
4.- Present work : MO of $N_2^-(^2\Pi_g)$ exchange $\gamma = 7$; $\ell_b = 14$; $\ell_c = 14$.

* We have performed a series of calculations with $\ell_{max} = 16$ in (1) and different cutoffs in the exchange interaction with results similar to that dispayed on the Figures 1 and 2. The exchange cutoff of $\gamma = 7$; $\ell_b = 14$; $\ell_c = 14$ is only 4% from our $\ell_{max} = 14$; $\gamma = 7$; $\ell_b = 14$; $\ell_c = 14$.

ℓ_b = 14; ℓ_c = 14 and at least for this basis their calculation is not converged. The irregular behaviour of the exchange convergence with the ℓ, ℓ' expansion displayed on the Figure 1 can explain the fact that our and their results are very close to each other.

Generally all the N_2^- π_g resonance calculations in the literature have been performed with the MO of N_2. Le Dourneuf et al. [2o] have used the N_2^- MO (the last electron is on $1\pi_g$ orbital) in the R matrix framework with very good results. We have reproduced with our method these calculations and found the resonance energy of 2.64 eV as compared with the value 2.39 eV calculated by Buckley and Burke [3] and Burke and Chandra [21] with an adjustable polarization potential. The corresponding Γ in our calculations is 0.6 eV as compared with the value of 0.399 eV of Burke and Chandra [21] and the value of 0.57 eV obtained by Dubé and Herzenberg [22]. To some extend using N_2^- orbitals is equivalent to polarizing the target. As compared to the N_2 SCF calculation the $1\pi_g$ unbound orbital is occupied in the N_2^- SCF calculation. If the continuum π_g ($k\pi_g$) and $1\pi_g$ orbital are similar close to the molecule then the N_2^- MO basis is an appropriate description of the behaviour of the continuum electron in the internal region. This is already the case as the $1\pi_g$ is mainly a d orbital and that the main contribution to the $k\pi_g$ function comes from the d term in the sperical harmonic expansion. In the external region the N_2^- MO approximation will be inappropriate as the continuum electron is far away from the molecule and the target has been relaxed from the N_2^- MO to the N_2 MO's. As the eigenphase sum (Taylor [23]) is a commulative contribution from all over the electronic space the correctness of the N_2^- MO approximation lies in the localization of the $k\pi_g$ electron in the internal region. This is generally the case for a shape resonance before the electron is escaping to infinity and particularly for the π_g N_2^- shape resonance we are studying.

Now let us turn to the LiH + e ($^2\Sigma^+$) case where the selected σ wave is the most penetrant and consequently the most sensitive to the exchange interaction. Moreover, as this state is non-resonant (Collins et al. [24]), we have first selected a relatively high energy (1.0 Ry) to increase the penetration of the high partial wave and to allow a change in the angular behaviour of the continuum electron. The results of our studies with respect to the convergence of the exchange term are displayed in Figures 4 and 5 (ℓ_b and ℓ_c are cutoffs corresponding to ℓ' and ℓ in the equations). As it can be seen, from these figures the only relevant picture of the exchange convergence in this case is in terms of ℓ_b, ℓ_c expansions. The step function displayed in Figure 4 is rapidly converging in terms of γ and slowly converging in terms of ℓ_b and ℓ_c. This behaviour is due to the nearly atomic character of LiH and its high dipole moment which couples the different ℓ. If we compare two of our

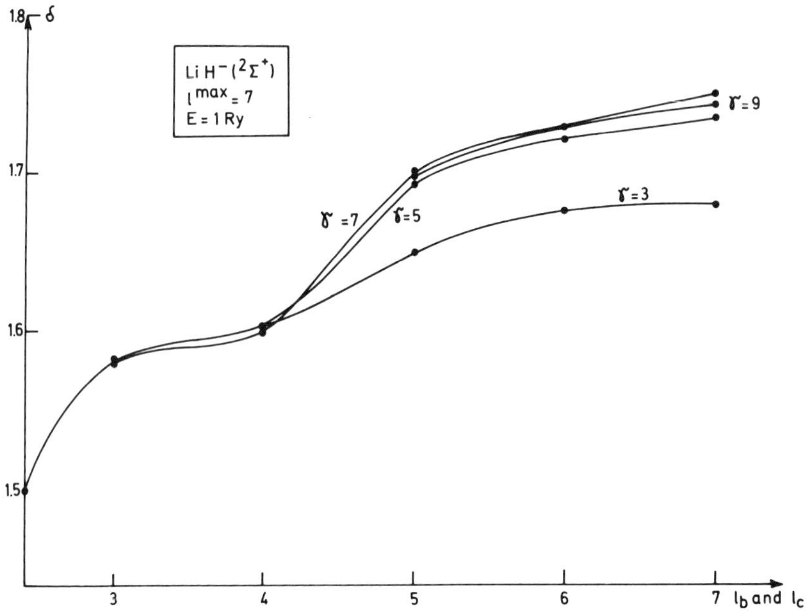

Fig. 4. The eigenphase sum of LiH^- $(^2\Sigma^+)$ as function of ℓ_b, ℓ_c bound and continuum expansions cutoffs (corresponding to ℓ, ℓ' in the equations) for different values of γ. The Cade and Huo [26] basis set have been used.

calculations in terms of convergence criteria as developped by Morrisson [7] , we find that the calculation with the exchange cutoff of $\ell_b = 3$; $\ell_c = 2$ and $\gamma = 5$ (exchange cutoffs recommended by Collins et al. [24] is converged to within 14 % as compared to our best result $\ell_b = 7$; $\ell_c = 7$ and $\gamma = 9$. For the last calculation the result is the same within two significant digits for the orthogonalized and nonorthogonalized calculations. We have repeated the exchange convergence studies for 0.1 Ry and calculated the energy dependence of Σ^+ state of LiH + e (exchange cutoffs of $\gamma = 7$; $\ell_b = 7$; $\ell_c = 7$ (Fig. 6)). The exchange convergence studies for 0.1 Ry gives mainly the same results as for 1 Ry but the exchange convergence is much faster. The Collins et al. [24] cutoff (see above) is converged to 3 % as compared to our converged result ($\ell_b = 7$; $\ell_c = 7$; $\gamma = 9$). Energy dependence

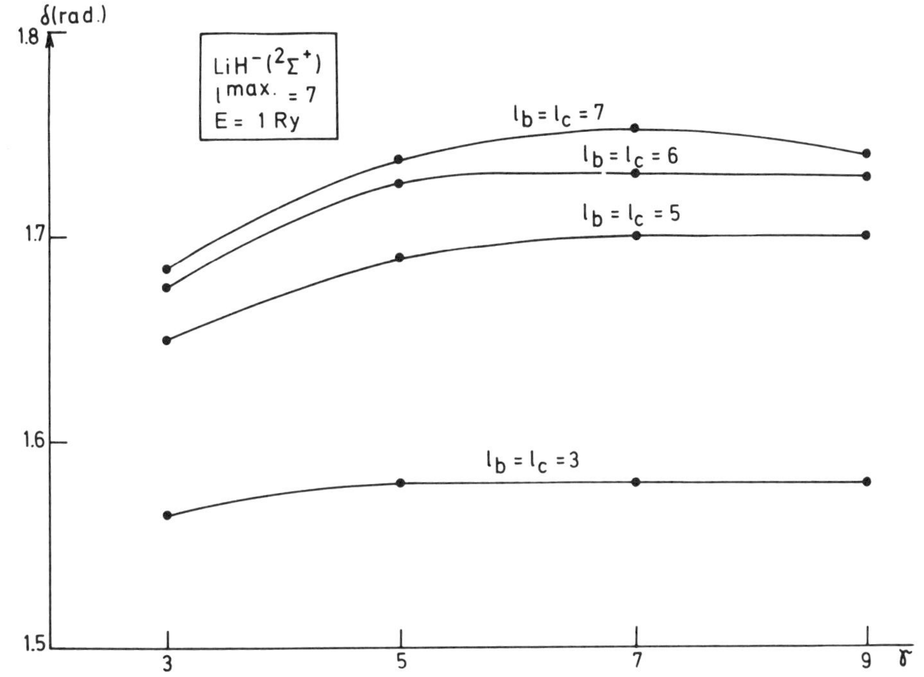

Fig. 5. The eigenphase sum of LiH ($^2\Sigma^+$) as function of γ exchange for different values of ℓ_b, ℓ_c bound and continuum expansion. The Cade and Huo [26] basis set have been used.

displayed on the Figure 6 shows the same behaviour as that of Collins et al. [24] but as explained above their calculation is not fully converged at high energies.

As a conclusion of the exchange convergence studies, we have considered here two very different molecules and found a particular behaviour in each case. Regardless the resonant and non resonant character of the states considered, it is believed that any molecule will have some intermediate behaviour between these two extreme cases. We therefore suggest that in the case of a resonance the convergence studies would be performed at least at one energy (in the middle of the resonance). In a non resonant channel the choice of the energy is less obvious but the arguments developped here (namely that the studies are to be performed at high energy) still stands. The problem is that in some channels

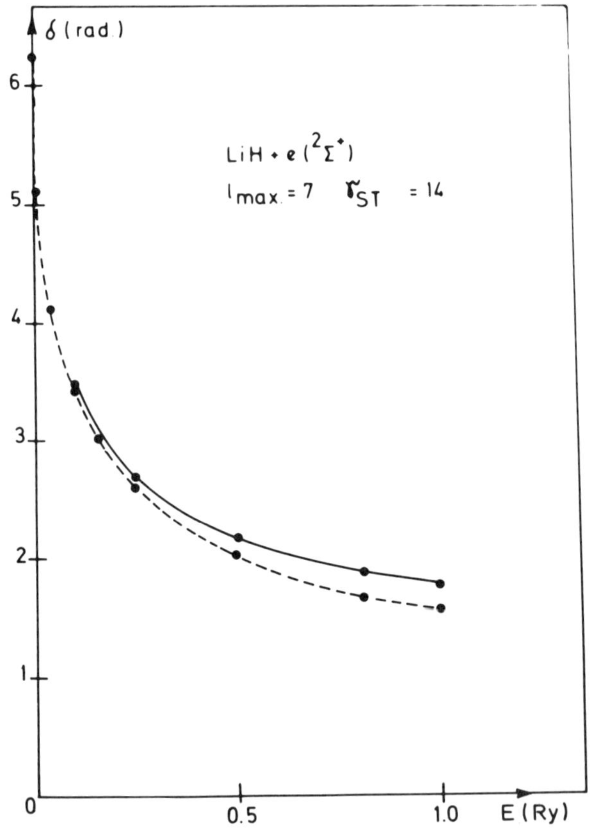

Fig. 6. The eigenphase sum of LiH^- ($^2\Sigma^+$) as function of the colliding electron energy.

------ Collins et al. [24] with exchange cutoff of $\gamma = 5$; $\ell_b = 3$; $\ell_c = 2$.

——— Present work with exchange cutoff of $\gamma = 7$; $\ell_b = 7$; $\ell_c = 7$.

the eigenphase is very small near the threshold and therefore difficult to converge. It follows that it is better for a non-resonant channel to perform convergence studies at more than one energy.

ACKNOWLEDGEMENTS

The author would like to acknowledge very useful discussions with H. Le Rouzo, H. Lefebvre-Brion and V. Mc Koy . He acknowledges the support from a NATO grant at early stage of this work and the computing facilities offered by V. Mc Koy during his stay at Caltech.

REFERENCES

1 B.D. Buckley and P.G. Burke, Symposium on Electron Molecule Collisions, Edited by Simamura I. and Hatsuzawa M., University of Tokyo, (1979), p. 37

2 P.G. Burke, A.L. Sinfailam, J. Phys. B3, 641 (1970).

3 B.D. Buckley and P.G. Burke, J. Phys. B10, 725 (1977).

4 G. Raseev, A. Giusti-Suzor and H. Lefebvre-Brion, J. Phys. B11, 2735 (1978).

5 G. Raseev, Compt. Phys. Comm. (1980), 20, 267, 275 (1980).

6 W.N. Sams, D.J. Koury, J. Chem. Phys. 51, 4809 (1969).

7 M.A. Morrisson : "The coupled channels integral equations method in the theory of low energy electron-molecule scattering", T.N. Rescigno, B.V. McKoy and B. Schneider, Editors (Plenum, N.Y. 1979), p. 15.

8 P.G. Burke and M.J. Seaton, Meth. in Comp. Phys. 10, 2 (1971).

9 M.J. Seaton, P.M.H. Wilson, J. Phys. B5, L5 (1972).

10 M.A. Crees, M.J. Seaton, P.M.H. Wilson, Comp. Phys. Comm. 15, 23 (1978).

11 M.A. Crees, D.L. Moores, J. Phys. B8, L195 (1975).

12 M.A. Morrisson, L.A. Collins, J. Phys. B10, L119 (1977).

13 A. Temkin, K.V. Vasavada, Phys. Rev. 160, 109 (1967).

14 S. Hara, J. Phys. Soc. Japan, 22, 710 (1967).

15 A.L. Sinfailam, Comp. Comm. 1, 445 (1971).

16 D.R. Hartree, The Calculation of Atomic Structure (New-York, Wiley, 1957).

17 M. Rotenberg, R. Biving, N. Metropolis, J.K. Wooten, The 3j and 6j symbols, The Technology Press, Massachussetts Institute of Technology (1959).

18 L.S. Rodberg, R.M. Thaler, Introduction to Quantum Scattering Theory, Academic Press, New-York (1967).

19 L.A. Collins, W.D. Robb and M.A. Morrisson, J. Phys. B11, L777 (1978).

20 M.Y. Ledourneuf, Vo Ky Lan, B.I. Schneider, XIth I.C.P.E.A.C., Kyoto (1979).

21 P.G. Burke, N. Chandra, J. Phys. B5, 1696 (1972).

22 L. Dube, A. Herzenberg, Phys. Rev. A20, 194 (1979).

23 J.R. Taylor, Scattering Theory, John Wiley and Sons Inc. (1972).

24 L.A. Collins, W.D. Robb and M.A. Morrisson, Phys. Rev. A21, 488 (1980).

25 R.K. Nesbet, J. Chem. Phys. 40, 3619 (1964).

26 P.E. Cade, W.M. Huo, J. Chem. Phys. 47, 614 (1967).

DECAY OF FESHBACH RESONANCES

IN DIATOMIC MOLECULES

H. Lefebvre-Brion and A. Giusti-Suzor

Laboratoire de Photophysique Moléculaire*
Bât. 213, Université Paris-Sud
91405 Orsay France

INTRODUCTION

The analogy between the decay modes of the Feshbach resonances in negative molecular ions and in neutral molecules is illustrated from recent experiments.

Two main processes of decay for the Feshbach resonances can be distinguished
i) Electronic decay where the kinetic energy of the ejected electron is borrowed from the electronic energy of the core. In a simple picture, this process is a two electron process.
ii) Vibrational decay where the vibrational excitation of the resonance is transformed into electron kinetic energy. This can be represented by an one-electron process.

We do not consider here the possible decay of these resonances by predissociation.

For the negative ions, the example of the Feshbach resonances in NO will be given. The variation of the electronic width with the energy of the ejected electron is studied theoretically and compared with experimental results.

For neutral molecules, correlation between vibrational and electronic autoionizations will be studied, using a multichannel quantum defect theory (MQDT). The autoionizing states in the

*Laboratoire associé à l'Université Paris-Sud

photoionization spectrum of N_2 will be taken as an example.

FESHBACH RESONANCES IN NEGATIVE IONS

Electronic decay

Feshbach resonances in NO^- have been observed for the first time by Sanche and Schulz [1]. Recently, these resonances have been again studied by Gresteau et al. [2]. Table 1 compares the experimental results with theoretical predictions made by one of us [3]. Using the stabilization method of Hazi and Taylor [4], the main configurations of the resonant states are well described by two Rydberg electrons in the field of the positive ion. The predicted nature of the states is not in contradiction with the experimental results, except for the d state. The configuration predicted for this state has been attributed by Gresteau et al. [2] to the d' state, a newly observed resonance.

Attention will be restricted here to the $b^3\Pi$ resonance, for which many results have been obtained. Four vibrational levels of this resonance have been observed and shown to decay into excited vibrational levels of the $X^2\Pi$ ground state of the NO neutral molecule up to $v_f = 18$, following the reaction (see Fig. 1)

$$\begin{aligned} &NO\ (X^2\Pi\ \ v=0) + e\ (E_r) \rightarrow NO^-\ (b^3\Pi\ \ v=v_r) \rightarrow \\ &\qquad NO\ (X^2\Pi\ \ v=v_f) + e\ (E_f) \end{aligned} \tag{1}$$

TABLE 1 : Energy location of observed resonances in NO^- (v=0) (eV)

Resonance	a	b	c	d	d'
Obs (Ref. 1)	5.04	5.41	5.46	6.44	
Obs (Ref. 2)	5.03	5.408	5.465	6.45	6.41
Calc (Ref. 3)	5.08	5.41	5.46		6.47
Predicted state	$^1\Sigma^+$	$^3\Pi$	$^3\Sigma^+$		$^3\Sigma^-$
Predicted Configuration	$(NO^+)(3s\sigma)^2$	$(3s\sigma)(3p\pi)$	$(3s\sigma)(3p\sigma)$		$(3p\pi)^2$

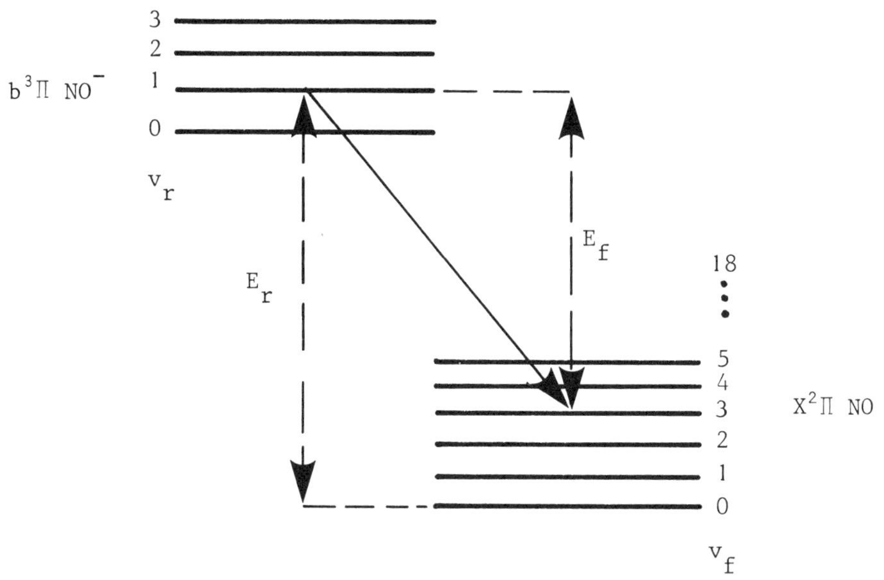

Fig. 1. Electronic decay of the $b^3\Pi$ Feshbach resonance in NO

The transition intensities are related to the strength of the electronic interaction, given by the electronic width $\Gamma_{el}(E)$, and to the Franck-Condon factors between the initial and final vibrational levels $<v_i|v_f>^2$. The excitation of the resonance is proportional to the product $\Gamma_{el}(E_r) \times <v_o|v_r>^2$. The decay of the resonance is proportional to the product $\Gamma_{el}(E_f) \times <v_r|v_f>^2$. Consequently the experimental intensity I_f^r of the transition (1) has been estimated as

$$I_f^r \propto \frac{\Gamma_{el}(E_r) \times <v_o|v_r>^2 \times \Gamma_{el}(E_f) \times <v_r|v_f>^2}{E_r \Gamma_r} \qquad (2)$$

where

$$\Gamma_r = \sum_f \Gamma_{el}(E_f) \times <v_r|v_f>^2 . \qquad (3)$$

The intensity has been found to be not proportional to the Franck Condon factors. To explain this fact, a dependence of the electronic width on the electron energy has been assumed, following the conclusions[5] of theoretical calculations. From (2), we obtain

$$\Gamma_{el}(E_f) \quad \propto \quad \frac{I_f^r \text{ (exp)}}{<v_o|v_r>^2 \text{ x } <v_r|v_f>^2} \tag{4}$$

where the proportionality factor depends only on the level v_r. The relative values of Γ_{el} deduced from experimental intensities and calculated Franck-Condon factors are shown in Fig. 2 for $v_r = 0$ to 3. They vary strongly with the electron energy E_f. The slope of the curve is the same for each v_r vibrational level and this justifies the hypothesis of a dependence of Γ_{el} only on the electron energy. No variation with R, the internuclear distance, seems to appear.

Ab-initio calculations of Γ_{el} have been made at R = 2.05 a.u., following the method developed in Ref. 5. The configurations of the $b^3\Pi$ NO^- Feshbach resonance

$$(NO^+)\ (3s\sigma)\ (3p\pi)$$

and of the continuum of the $X^2\Pi$ NO ground state of the same symmetry

$$(NO^+)\ (2\pi)\ (\chi\sigma_{k,\ell})$$

($\chi\sigma$ is the continuum wave function of the ejected electron), differ by two electrons. Consequently, the expression of Γ_{el} can be simply given in terms of two electron integrals

$$\Gamma_{el}(E_f) = \frac{k}{\pi} \sum_{\ell} \{ <3s\sigma\ 3p\pi\ |\frac{1}{r_{12}}|\ \chi\sigma_{k,\ell}\ 2\pi> - <3s\sigma\ 3p\pi\ |\frac{1}{r_{12}}|\ 2\pi\ \ \chi\sigma_{k,\ell}> \}^2 \tag{5}$$

where ℓ is the angular momentum of the electron and $k = \sqrt{2m\, E_f}$. With the convention

$$<b_1 b_2\ |\frac{1}{r_{12}}|\ b_3\ \chi_{k,\ell}> = <b_1(1)\ b_2(2)\ |\frac{1}{r_{12}}|\ b_3(1)\ \chi_{k,\ell}(2)>.$$

The continuum wave function has been assumed to be a spherical Bessel function

$$\chi\sigma_{k,\ell} = \frac{2\pi}{k}\ j_\ell\ (k_r)\ Y_{\ell,0}(\hat{r}) \ . \tag{6}$$

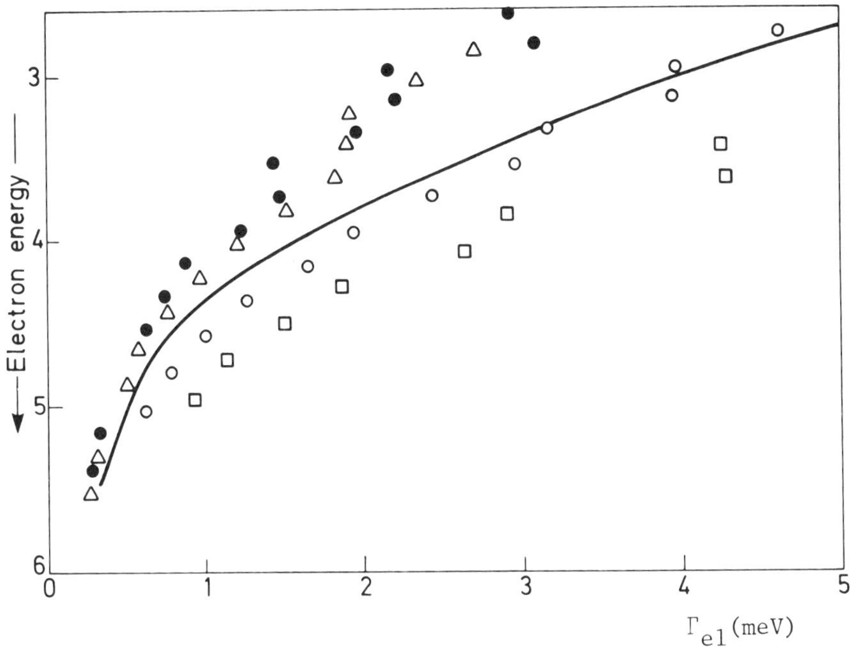

Fig. 2. Comparison of the experimental relative variation of Γ_{el} : (□ $v_r = 0$; o, $v_r = 1$; Δ $v_r = 2$; ● $v_r = 3$) (Ref. 2) with calculated absolute values (———) (meV).

The absolute values are given in Fig. 2 and the observed and calculated energy dependences are in excellent agreement. This seems to confirm that the variation with R of the resonance width can be neglected. The total width calculated for $v_f = 0$ using Eq. (3) is about 0.6 meV. Experimentally, one can only say that the width is less than 10 meV.

Further calculations using a static exchange approximation[6], instead of the Bessel approximation, to describe the continuum wave function in Eq. (5) are in progress.

Vibrational decay

The NO^- Feshbach resonances yield the first observations of vibrational decay in negative ions[7]. For example, the v = 1 level of the $b^3\Pi$ resonance decay in the v = 0 level of the $A^2\Sigma^+$ Rydberg state which has the configuration (NO^+) ($3s\sigma$) and can be considered as the parent of the resonance. In the transition

$$NO^{-}\ (b^3\Pi\ v = v_r) \rightarrow NO\ (A^2\Sigma^+\ v = v_r - 1) + e$$

the configuration of the $b^3\Pi$ resonance (NO^+) $(3s\sigma)$ $(3p\pi)$ differs from that of the continuum of the $A^2\Sigma^+$ state (NO^+) $(3s\sigma)$ $(\chi\pi_{k,\ell})$ by only one orbital and the strength of the decay is given by the following expression [8]

$$\Gamma = 2\pi \{ <\psi (b^3\Pi)\ |\frac{\partial}{\partial R}|H_{el}|\ \psi (A^2\Sigma^+ \cdot \chi\pi_k)_{\ell}> < v_r\ |R - R_e|\ v_f >\}^2 \quad (7)$$

The $\Delta v = 1$ propensity rule and the proportionality of the decay with v which follow from the harmonic approximation

$$< v\ |R - R_e|\ v - 1 > \ \propto\ v\ \delta_{v,\ v-1} \quad (8)$$

are experimentally verified for the $b^3\Pi$ resonance.

FESHBACH RESONANCES (AUTOIONIZED STATES) IN NEUTRAL MOLECULES

In molecules the main autoionization processes result also from electronic or vibrational interactions. Other possible processes are the rotational autoionization, which appears in light molecules as H_2, and the spin-orbit autoionization which could be observed in heavy molecules. In a molecule such as N_2, electronic autoionization is much more important than vibrational autoionization. This has been shown by Duzy and Berry [9], using the Golden rule formula separately for each process. We shall treat here a case where the vibrational autoionization is actually an indirect electronic autoionization. In the other example, we shall show how the branching ratio can be modified if the electronic coupling between continua is taken into account.

Indirect electronic autoionization

The total photoionization cross section of N_2, measured at very high resolution (0.016 Å) [10] shows two types of peaks in the region 784 - 785 Å (Fig. 3). At 783.2 Å, a very large and structured peak corresponds to the m = 4 member of the Worley's third Rydberg series [11] converging to the v = 0 level of the $A^2\Pi_u$ state of N_2^+. It is autoionized electronically by the continuum of the v = 0 level of the $X^2\Sigma_g^+$ ion ground state which is the only open channel at this energy. Just below this energy a set of small peaks appears which correspond to the highest members (m > 15) of the Worley Jenkins'series [11] converging to the v = 1 level of

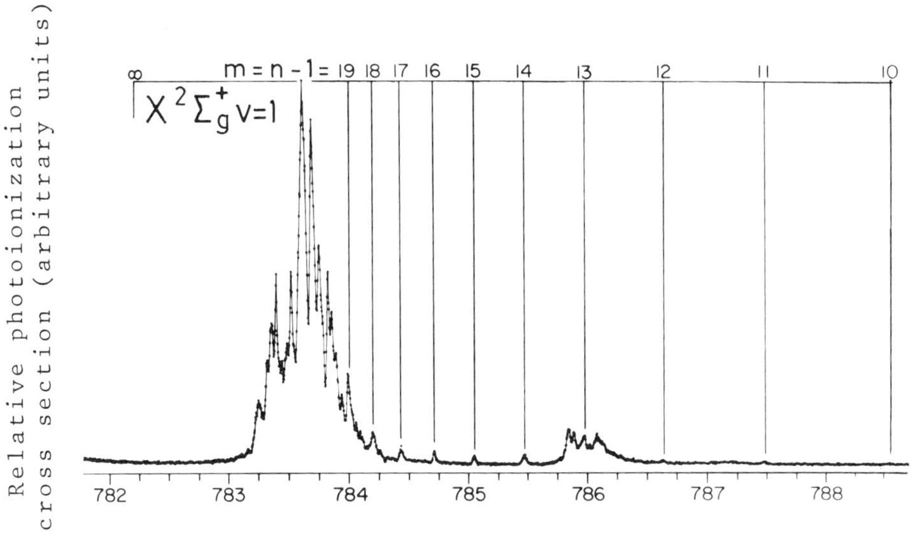

Fig. 3. Photoionization cross section of N_2 taken at 77°K with a resolution of 0.016 Å (from Ref. 10 reproduced with the courtesy of the authors).

the $X^2\Sigma_g^+$ state. The continuum of the v = 0 level of the ground state of N_2^+ is again responsible for these autoionized peaks but here through a vibrational coupling. The situation is illustrated in Fig. 4. For a Rydberg state with effective quantum number $n^* = n - \delta$ (δ is the quantum defect), both the coupling with the continuum due to the vibrational interaction[12] and the transition moment are inversely proportional to n^{*3}. It results that the intensity maxima of the members of a Rydberg series should be constant with n^*. Experimentally, in the mentioned region, the intensity maxima of the ($X^2\Sigma_g^+$) $np\pi u$ series increase as the peaks come nearer and nearer to the center of the electronic large resonance.

We have studied this problem using a theory based on the MQDT approach. The details of the theory are published elsewhere[13]. This theory needs several parameters which can be calculated, deduced from semi-empirical studies, or adjusted around these values. Here, the parameters which fix the relative energies of the levels (the quantum defects δ_i, the thresholds for the different continua E_v^i) are taken from experimental data[11].

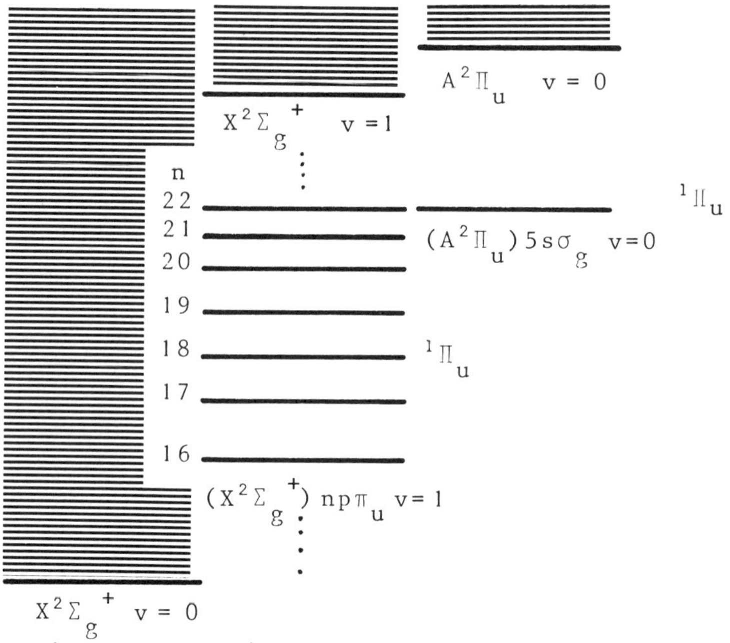

Fig. 4. Schematic representation of the Rydberg levels of Fig. 3 autoionized by the continuum of $X^2\Sigma_g^+$ (v = 0)

The electronic transition moments D_{el} have been taken from theoretical works[14,15] and the vibrational overlaps involved in the vibronic transition moments come from the review of Lofthus and Krupenie[16].

The variation with the internuclear distance R of the pπ wave phase shift in the continuum of the $X^2\Sigma_g^+$ state is taken from the results of static exchange calculations[14]. This phase shift is simply related to the quantum defect which is plotted in Fig. 5 as a function of R. The derivative $\frac{d\delta}{dR}$ is introduced in the MQDT treatment. The last quantity needed for the calculation is the electronic interaction between the different channels which have the same symmetry $^1\Pi_u$. The interaction V_{el} between the two continua $(X^2\Sigma_g^+)$ $\varepsilon p\pi u$ and $(A^2\Pi_u)$ $\varepsilon s\sigma g$ can be related to the interaction $V_{nn'}$ between corresponding discrete n and n' Rydberg states by the relation

$$V_{el} = n^{*3/2} \, n'^{*3/2} \, V_{nn'} \tag{9}$$

The electronic interaction between the two first members of Rydberg series, namely the $[(X^2\Sigma_g^+)\ 3p\pi]\ c^1\Pi_u$ state and the $[(A^2\Pi_u)\ 3s\sigma g]$ $o^1\Pi_u$ state converging to the $A^2\Pi_u$ state has been studied theore-

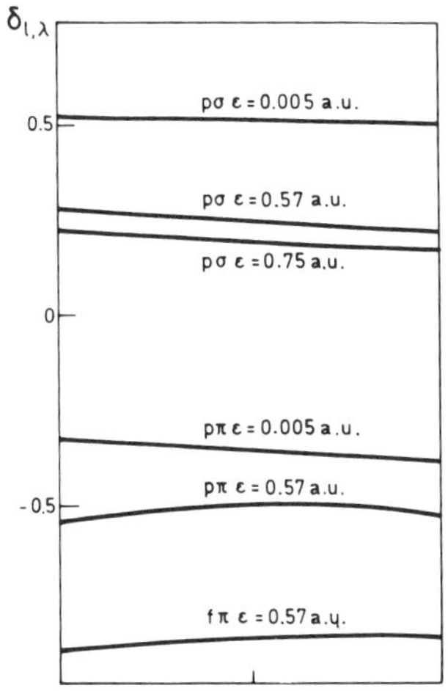

Fig. 5. Variation of the quantum defect with the internuclear distance of the wave in $X^2\Sigma_g^+$ core of N_2^+.

tically by Leoni[17]. Its value varies with R but it is less than 500 cm^{-1}. We have taken a slightly larger value in order to try to reproduce a width for the peak m = n − 1 = 4 of about 80 cm^{-1}. From the Golden rule formula, we have thus another estimation for V_{el} (a dimensionless quantity)

$$\Gamma_n(cm^{-1}) = 2\ \text{Ryd.}(cm^{-1})\ x\ 2\pi\ x\ |V_{el}|^2\ x\ \langle v\ |\ v'\rangle|^2\ x\ n^{*\ -3} \quad . \qquad (10)$$

Here, the vibrational overlap is taken between the level v = 0 of the $X^2\Sigma_g^+$ state and the level v' = 0 of the $A^2\Pi_u$ state. With this set of parameters whose numerical values are given in the caption of Fig. 6, a first calculation has been made, in the usual MQDT scheme[18], including only the vibrational autoionization in the Worley Jenkins' series. The cross section depicts narrow peaks (Γ (n = 16) = 0.04 cm^{-1}), whose width decreases when n increases, with a constant height as expected. Then we have included the electronic interaction between the two electronic channels.
We have obtained much larger peaks (Γ (n = 16) = 0.35 cm^{-1}) whose height increases with n in qualitative agreement with experiment. The phenomenon can be explained by a mechanism of indirect autoionization analogous to the mechanism of indirect predissociation: the (5sσg) resonance is strongly coupled with the continuum v = 0 of the $X^2\Sigma_g^+$ state. The resonances (m = 15 ...) are very weakly coupled by vibrational interaction with this continuum but they interact with the (5sσg) resonance by the same electronic factor

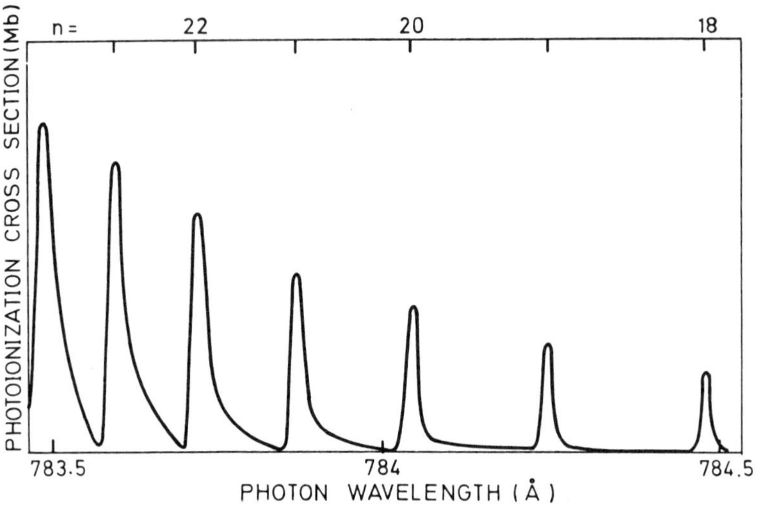

Fig. 6. Indirect vibrational autoionization of the Worley Jenkins series. On the left hand of the figure is the n = 5 member of the Worley's third Rydberg series. Chosen parameters : $\delta_1 = 0.65$ $\delta_2 = 1.055$ $d\delta_1/dR = -0.1\ a_o^{-1}$, $\langle v = o\,A \mid v = o\,X \rangle = 0.71$ $\langle v = o\,A \mid v = 1\,X \rangle = 0.61$, $D_{el}\ (X' - X) = 1.046$ $D_{el}\ (X' - A) = 2.02$, $\langle v = o\,X' \mid v = o\,X \rangle = 0.956$ $\langle v = o\,X' \mid v = o\,A \rangle = 0.495$, $\langle v = o\,X' \mid v = 1\,X \rangle = 0.283$, $V_{el} = 0.085$.

$|V_{el}|^2$ as in Eq.(10), now multiplied by the vibrational factor $|\langle v = 1\,X \mid v = o\,A \rangle|^2$. By this mixing, these states borrow some width to the large $5s\sigma_g$ resonance and the intensity maxima increase as the states come closer to the resonance.

Note that even when the vibrational coupling is neglected, these apparent vibrationally autoionized resonances appear in the calculated cross section.

A quantitative agreement with experiment has not been searched out because of the simplicity of our model. Actually, two electronic resonances are probably superimposed at 127674 cm^{-1} = 783.24 Å.

On the other hand, the members of the Worley-Jenkins' series are not pure $^1\Pi_u$ states but are a mixing of $^1\Pi_u$ and $^1\Sigma_u^+$ states in case (d). Then the rotational structure of the states has to be included in a complete calculation.

Electronic branching ratio

Experimental partial photoionization cross sections[19] for the $A^2\Pi_u$ and $B^2\Sigma_u^+$ of N_2^+ are given in Fig. 7. The n = 9, 10 members of the Hopfield' series converging to the v = 1 level of the $B^2\Sigma_u^+$ state show small but distinct peaks in the channel v = 0 of the $B^2\Sigma_u^+$ state. These peaks, in principle, are due to vibrational autoionization and should be so narrow that it would be impossible to detect them with the resolution of the apparatus. Here again, the observed peaks may be explained by a contamination of the vibrational width by the electronic coupling between the two electronic channels.

We have tried to reproduce the situation which is schematically represented in Fig. 8. Using for the derivative of the quantum defect the same value as in the first example, the width for the level n=9 due to the vibrational interaction is small only 0,07 cm^{-1}. Now the Rydberg states are coupled electroni-

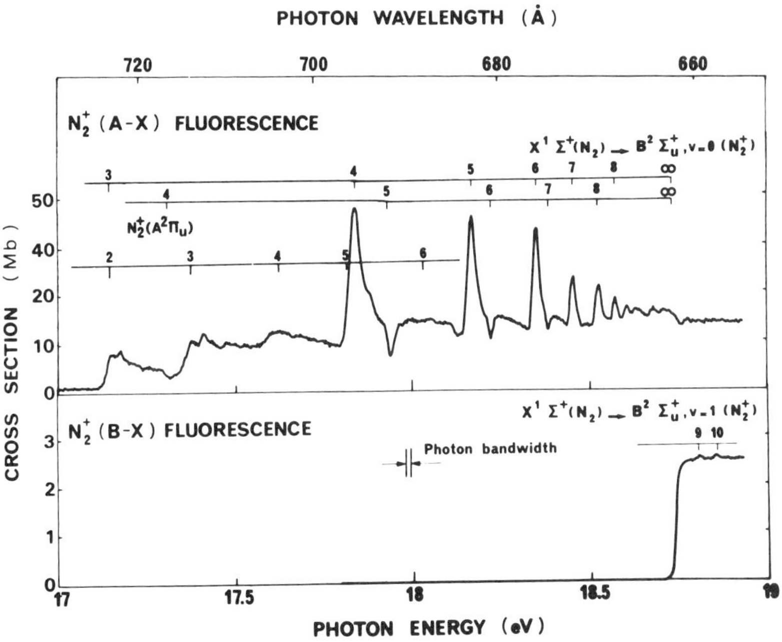

Fig. 7. Partial cross section for N_2^+ (A-X) and (B-X) fluorescence in the 17-19 eV excitation range (from Ref. 19 reproduced with the courtesy of the authors).

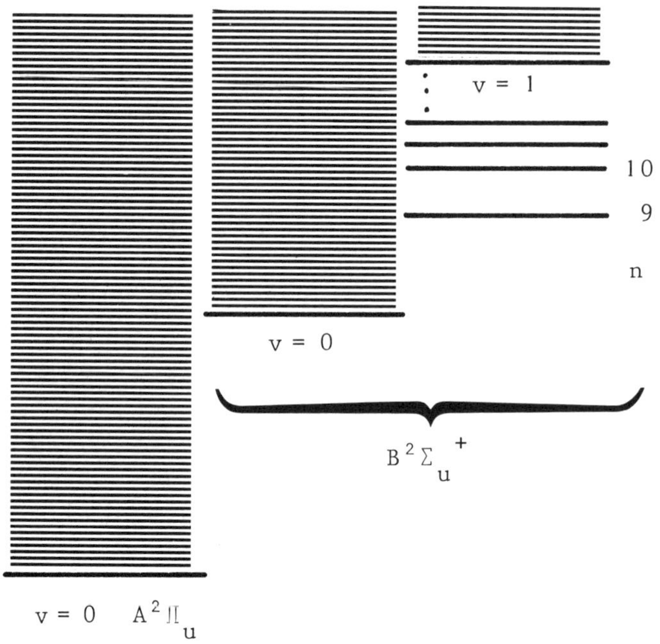

Fig. 8. Schematic representation of the Rydberg levels of N_2 autoionized by two continua (see Fig. 7).

cally with the continuum of the $A^2\Pi_u$ state. This electronic coupling is probably strong since it is responsible for the very well known strong autoionization of the same Hopfield' series converging to v = 0 of the $B^2\Sigma_u^+$ state. The ratio of the vibronic couplings is equal to the ratio of the vibrational overlaps between the level v = 0 of the $A^2\Pi_u$ state and the levels v = 0 and 1 of the $B^2\Sigma_u^+$ state (namely about 0.9 and 0.4). If the formalism of Fano is used, the total width is given by the formula [20]

$$\Gamma = \frac{\Gamma_1 + \Gamma_2}{1 + \pi^2 V^2} \tag{11}$$

where Γ_1 is the width due to the vibrational coupling (Eq.(7)), Γ_2 the width due to the electronic coupling (Eq.(10)) and V the coupling between the two continua. Using the same configuration interaction technique, it is possible to deduce, in the case $q = \infty$, the branching ratio for the two partial cross sections σ_1 and σ_2

$$\frac{\sigma_2}{\sigma_1} = \frac{\Gamma_2 + \Gamma_1 \; \pi^2 V^2}{\Gamma_1 + \Gamma_2 \; \pi^2 V^2} \qquad (12)$$

Now, using our MQDT method with the parameters given in Table 2, we have calculated total and partial cross sections. With the chosen values for the transition moments the peaks are very near from a Lorentzian shape. Then, from the plot of these cross sections, we can deduce approximate widths for the peaks. We verify that the peak widths are the same in the total and in each partial cross section, and are in agreement with formula (11). The intensity maxima of the peaks are given in Table 2 and their ratio follows very well the formula (12). The difference is due to the finite value of q in our case. In the last line of Table 2 we see that if the electronic coupling becomes very large, the branching ratio can be less than one. Our suggestion for explaining the anomaly which appears on Fig. 7 has been given for the case where q is large, for which the branching ratio is nearly constant in agreement with the conclusions of Starace[21]. For other values of q, the branching ratio varies very much with the energy. This example stresses out the importance of the couplings between continua which modify strongly the partial cross sections not only near the resonances but also in the region between them.

TABLE 2 : Variation of the branching ratio with the electronic coupling [(a)]

	Resonance width Γ (cm^{-1})		Branching ratio σ_2/σ_1	
V_{el}	from MQDT calc.	Eq.(11) [(b)]	from MQDT calc.	Eq.(12) [(b)]
0.1	2.5	2.5 (2.7)	11.	9.3 (40.)
0.2	8.0	8.2 (10.8)	4.	3.0 (153.)
0.5	23.0	22.0 (66.6)	0.9	0.5 (950.)

[(a)] Chosen parameters : $d\delta/\delta R = -0.1\ ao^{-1}$ $\quad \delta = 0.65$
$V = V_{el}$ $\langle v = oA \mid v = oB \rangle = V_{el} \times 0.9$ $\quad \langle v = oA \mid v = 1B \rangle = 0.4$
D_{el} $(X' - A) = 1.0$ $\quad \langle v = oX' \mid v = oA \rangle = 0.4$
D_{el} $(X' - B) = 2.2$ $\quad \langle v = oX' \mid v = oB \rangle = 0.18$ $\quad \langle v = oX' \mid v = 1B \rangle = 0.91$
(The transition moments have been chosen to approach the case $q = \infty$ valid for Eqs.(11) and (12)

[(b)] The parenthetical values are obtained from Eqs.(11) and (12) by neglecting the coupling between continua ($V = 0$)

CONCLUSION

Two features emerge from these two examples:

(i) Apparent vibrational autoionization can often be an indirect electronic autoionization process. In such a case, the "propensity rule"[8] $\Delta v = -1$ can be broken.

(ii) The results obtained by separate calculations for each channel can be strongly modified by the interchannel electronic interaction.

REFERENCES

1 L. Sanche and G.J. Schulz, Phys. Rev. A6, 69 (1972).

2 F. Gresteau, R.I. Hall, A. Huetz, D. Vichon and J. Mazeau, J. Phys. B, 12, 2925 (1979).

3 H. Lefebvre-Brion, Chem. Phys. Letters, 19, 456 (1973).

4 A.U. Hazi and M.S. Taylor, Phys. Rev., A1, 1109 (1970).

5 P.K. Pearson and H. Lefebvre-Brion, Phys. Rev., A13, 2106 (1976).

6 G. Raseev, Comp. Phys. Comm. (1980) in press.

7 F. Gresteau, R.I. Hall, A. Huetz, D. Vichon and J. Mazeau, J. Phys. B, 12, 2937 (1979).

8 R.S. Berry, J. Chem. Phys. 45, 1228 (1966).

9 C. Duzy and R.S. Berry, J. Chem. Phys. 64, 2431 (1976).

10 P.M. Dehmer and W.A. Chupka, unpublished results (1978).

11 M. Ogawa and Y. Tanaka, Canad. J. Phys., 40, 1593 (1962).

12 G. Herzberg and Ch. Jungen, J. Mol. Spect., 41, 425 (1972).

13 A. Giusti-Suzor and H. Lefebvre-Brion, Chem. Phys. Letters (1980) in press.

14 G. Raseev, H. Le Rouzo and H. Lefebvre-Brion, J. Chem. Phys., 72, 5701 (1980).

15 T.N. Rescigno, C.F. Bender, B.V. Mc Koy and P.W. Langhoff, J. Chem. Phys. 68, 970 (1978).

16 A. Lofthus and P.H. Krupenie, J. Phys. Chem., Ref. Data 6, 113 (1977).

17 M. Leoni, Thesis unpublished, Physical Chemistry Laboratory, ETH. Zurich, Switzerland (1972).

18 Ch. Jungen and D. Dill, J. Chem. Phys. 73, 3338 (1980)

19 A. Tabché-Fouhailé, K. Ito, H. Frohlich, P. Morin, P.M. Guyon and I. Nenner, to be published.

20 J.A. Beswick and R. Lefebvre, Mol. Phys. 29, 1611 (1973).

21 A.F. Starace, Phys. Rev., A16, 231 (1977).

COMPUTATIONAL MODELS FOR e^--POLYATOMICS LOW-ENERGY SCATTERING

F.A. Gianturco

Gruppo di Chimica Teorica, Nuovo Edificio Chimico
Citta Universitaria, oo189 Rome, Italy

D.G. Thompson

Department of Applied Mathematics and Theoretical Physics
The Queen's University
Belfast BT7 INN, Northern Ireland

1. INTRODUCTION

Considerable progress has been made in the last few years on the development of theoretical and computational treatments of electron molecule collisions. The main reason for this progress stems from an ever increasing need for the relevant cross sections in many applications in astrophysics, in the physics of the upper atmosphere, in laboratory plasma physics and in laser physics. In addition to their demands, recent experiments devoted to low energy measurements of electron scattering off several simple molecules are providing a new wealth of accurate data and very detailed information against which one can then fruitfully test theoretical models. Finally, the easier availability of powerful computing facilities is making it possible to obtain numerical results from new theories to an extent that was still out of reach only a few years ago. Accordingly, a number of important reviews on the subject have appeared in the last two years[1,2,3] as a clear sign of growing concern for these collisional processes.

In the present overview we will briefly discuss the major hurdles that stand in the way of a fully rigorous solution of the theoretical problem and then go on to mention some of the necessary simplifications that have been recently tested. The discussion of results is here focussed on the still very few applications that so far have been carried out for polyatomic targets where the general complexity of any computational approach is further increased by the added, structural complexity of the specific molecule involved.

The following Section contains in some detail a reminder of the theoretical framework where collisions are described as involving a molecular target in which the nuclei are constrained to fixed locations in space. This approach has gained some favour in recent years since it firstly provides a considerable simplification for the equations that are to be solved, and secondly it has been found to give reliable results for many inelastic processes (molecular excitations and/or deexcitations) where the scattering amplitude can be appropriately averaged over the nuclear coordinates.

Section III is then dealing with the various forces at play when the slow electron interacts with the molecule and with some of the methods that can be employed to represent such forces in a simple, albeit realistic way that thus defines our model approach for diatomic and polyatomic targets.

The last Section is finally presenting some of the applications to specific cases. For instance e^{-}-CH_4 scattering over a wide range of collision energies, where comparison with experiments is possible, is shown in some detail and indicates that the present approach indeed provides rather satisfactory agreement with measured data and is likely to serve as a powerful tool for general qualitative predictions in other polyatomic systems.

2. SCATTERING EQUATIONS AND CROSS SECTIONS

As we said before, one possible simplified approach is usually provided by a physical situation in which the nuclei can be held fixed during the collision. The latter reduction is invoked when the impinging electron moves through the zone of interaction faster than the typical time scale of a rotation and/or vibration of the heavier nuclei.[4] Such a picture also provides the usual first step in other approximations which remain valid when the above velocity criterion is not satisfied. The corresponding Hamiltonian then contains only electronic terms and one usually also assumes that its relativistic part can be disregarded in the processes under consideration.[5]

The familiar electrostatic Hamiltonian is then referred to a frame of reference which is rigidly attached to the molecule in question, with its origin in the nuclear centre-of-mass. This frame is usually called the molecular or body-fixed (BF) frame:

$$[H_{el}-E]\ \Psi = \{-1/2\nabla^2+H_t+\hat{V}-E\}\ \Psi = 0 \tag{1}$$

where the operators within curly brackets are, respectively, the kinetic energy operator for the scattered electron, the target electronic Hamiltonian and the interaction potential between the impinging electron and the molecule. For example in the case of a diatomic target the last term is defined as:

$$\hat{V}\,(r_1,\ldots \underline{r}_N;R) = \sum_{i=1}^{N} \frac{1}{|\underline{r}-\underline{r}_i|} - \frac{Z_A}{|\underline{r}-r_A|} - \frac{Z_B}{|\underline{r}-\underline{r}_B|} \qquad (2)$$

where Z_A, Z_B are the nuclear charges located at $\underline{r}_A$ and $\underline{r}_B$, R is the internuclear distance and the labelled coordinates $\underline{r}_1,\ldots \underline{r}_N$ identify the N bound electrons. The continuum electron is given here the coordinate $\underline{r}$, also referred to the previous BF frame. For the case of a polyatomic target the relevant interaction obviously presents, as only difference from above, a second sum over an index j that runs over all the nuclear position vectors and that replaces the last two terms on the r.h.s. of eq. (2).

The total wavefunction Ψ in eq. (1) can be expanded for each set of internuclear coordinates, in the form:

$$\Psi = A \sum_{i=1}^{n} \varphi_i\,(1,\ldots N)\,F_i\,(N+1) + \sum_{i=1}^{m} \xi_i(1,2,\ldots,N+1)\,a_i \qquad (3)$$

where the φ_i are target electronic eigenstates and any other suitable pseudostate that can represent the target response function to the perturbation caused by the incoming electron. The continuum functions F_i's describe the motion of the scattered electron (including its spin) and A is the usual antisymmetrisation operator. The ξ_i's are L^2 integrable correlation functions weighted by the multiplying coefficients a_i.

One immediately sees from the above expansion that several difficulties appear at the outstart and need to be solved before proceeding to the search for scattering observables:

i) The molecular eigenstates need to be known for one or more of the target electronic states;

ii) The distortion effects caused by the scattering process also need to be included in both the short range and long range parts of the full interaction between colliding partners;

iii) The A operator effectively introduces integrodifferential equations through the non-local nature of the bound-continuum interaction;

iv) The selection of suitable correlation functions can easily require a slow-converging sum that rapidly gets computationally out of hand unless clear selection criteria can be found for each specific case under study.

All the above points will be touched in the following section, where various model approaches are described. We now consider more specifically the derivation of the cross sections from the asymptotic form of the wavefunction of eq. (3).

We look for a solution of eq. (1) corresponding to an incident plane wave plus an outgoing spherical scattered wave and therefore need to refer back to the real laboratory frames the solutions generated via eq.s (2) and (3).

The usefulness of the BF frame appears very transparently when a standard partial wave analysis is carried out for the total wavefunction of eq. (3). To begin with, for light molecules and non relativistic electron energies we can couple the spin of the scattered electron with the spin of the target molecule to form an eigenstate of $\hat{S}^2$ and $\hat{S}_z$ corresponding to the quantum numbers S and M_S which are conserved during collision. Each of these eigenstates can obviously be in turn expanded over the set of spherical harmonics which provide a basis of a reducible representation of the molecular point group for the target molecule. Hence one can readily obtain a new set of generalized spherical harmonics which provide a symmetry-adapted basis set to be used for both continuum and bound orbitals:

$$X^{p}_{ih\ell}\,(\hat{r}_{N+1}) = \sum_{m} b^{p\mu}_{ih\ell m}\; Y^{m}_{\ell}\,(\hat{r}_{N+1}) \tag{4}$$

Here p denotes the particular irreducible representation (I.R.), μ distinguishes the component of the basis, if its dimension is greater than one, and h labels different bases of the same I.R. corresponding to the same value of ℓ; in the above expression both real and immaginary spherical harmonics can obviously be used.[6,7]

One can thus adopt a coupling scheme which is diagonal in $(p\mu SM_S)$ and then rewrite eq. (3) as follows:

$$\psi^{p\mu SM_S} = \sum_{i=1}^{n} \sum_{h\ell} \Phi^{p\mu SM_S}_{ih\ell}\,(1,\ldots N,\hat{r}_{N+1}\;\sigma_{N+1})\; r^{-1}_{N+1}\; f^{p\mu S}_{ih\ell}(r_{N+1})$$
$$+ \sum_{i=1}^{m} \xi^{p\mu SM_S}_{i}\,(1,2\ldots N+1)\; a^{p\mu S}_{i} \tag{5}$$

here σ_{N+1} is the spin variable for the continuum electron. The channel functions of this equation are defined as:

$$\Phi_{ih\ell}^{p\mu SM_S} = \sum_{M_{S_i}\, m_{S_i}} \phi_i^{p_i\mu_i\; S_iM_{S_i}}(1,\ldots N)\; X_{h\ell}^{p\mu}(\hat{r}_{N+1}) \times \eta_{1/2}^{m_{S_i}}(N+1)\;(S_iM_{S_i}\; 1/2 m_{S_i}|S_i\; 1/2\; S\; M_S) \qquad (6)$$

where the η's are electron spin functions for the free electrons and the $(S_i\; M_{S_i}\; 1/2\; m_{S_i}\; /\; S_i\; 1/2\; M_S)$ are Clebsh Gordan coefficients. The bound functions ϕ_i's describe the N-electron target wave-functions used in the expansion (3), each belonging to a specific I.R. of the molecular point group symmetry and with a specific total spin eigenstate. The main molecular axis defines the M_{S_i} direction.

Projecting the Schrödinger equation $[H_{el}-E]\;\psi^{p\mu S\; M_S} = 0$ onto the channel functions $\Phi_{ih\ell}^{p\mu S\; M_S}$ and onto the symmetry adapted correlation functions $\xi^{p\mu S\; M_S}$ gives the following infinite set of coupled integrodifferential equations for each continuum electron channel $(p\mu S)$:

$$\left[\frac{d^2}{dr^2} - \frac{\ell(\ell+1)}{r^2} + k_i^2\right] f_{ih\ell}^{p\mu S}(r) = = 2\sum_{i'=1}^{n}\sum_{h'\ell'}^{\infty}\left[V_{ih\ell,i'h'\ell'}^{p\; S}(r) + W_{ih\ell,i'h'\ell'}^{p\mu S}(r) + \Theta_{ih\ell,i'h'\ell'}^{p\mu S}(r)\right] f_{i'h'\ell'}^{p\mu S}(r) \qquad (7)$$

where we have omitted for simplicity the explicit dependence of these equations on the internuclear set of coordinates $\{\underline{R}_i\}$.

The direct potential matrix $V_{ih\ell i'h'\ell'}^{p\mu S}$ defines the coupling of two different channel functions through the operator of eq. (2). Here again the symmetry-adapted coupling scheme of eq. (5) helps to simplify the problem in the BF representation; for each electronic eigenstate of the bound electrons of the target appearing in eq. (6) one can infact write down the corresponding expansion of the potential originating from it. A typical diagonal term, for instance, is given by:

$$V^{p_i \mu_i S}(\underline{r}_{N+1}) = \sum_{h\ell} V^S_{ih\ell}(r_{N+1})\, X^{p_i \mu_i S}_{h\ell}(\hat{r}_{N+1}) \tag{8}$$

and therefore one obtains the direct matrix elements in eq. (7) as:

$$V^{p\mu S}_{ih\ell, ih'\ell'}(r)\, \delta_{pp'}\, \delta_{\mu\mu'} = \langle X^{p\mu S}_{h\ell}(\hat{r}),\, V^{p_i \mu_i S}(\underline{r}),\, X^{p'\mu' S}_{h'\ell'}(\hat{r})\rangle \tag{9}$$

which yields:

$$V^{p\mu S}_{ih\ell, ih'\ell'}(r_{N+1}) = \sum_{\substack{mm'\nu \\ gL}} b^{p\mu *}_{h\ell m}\, b^{p_i \mu_i}_{gL\nu}\, b^{p\mu}_{h'\ell' m'} \times (\ell m \ell' m' \,/\, \ell\, \ell' L \nu) \times (\ell o\, \ell' o \,/ \ell\, \ell'\, L\, o)\, V^S_{gL}(r_{N+1}) \tag{9a}$$

Outside the molecular charge distribution, say at the distance r_o, the above matrix has the asymptotic form arising from the permanent static multipoles of the target molecule:

$$V^S_{gL}(r_{N+1}) \simeq \sum_{\lambda=1}^{\lambda max} a^{\lambda}_{gL}\, v^{-\lambda-1}_{N+1} \qquad r_{N+1} > r_o \tag{9b}$$

where the first few terms have contributions arising from the permanent dipole, quadrupole, octupole etc. moments of the isolated molecule (equilibrium geometry). The non-diagonal terms are discussed below.

The second term on the r.h.s. of eq. (7) is defined by:

$$W^{p\mu S}_{ih\ell, i'h'\ell'}(r_{N+1})\, f^{p\mu S}_{i'h'\ell'}(r_{N+1})$$

$$= -N\langle \Phi^{p\mu S\, M_S}_{ih\ell}(1 \ldots N, \hat{r}_{N+1}\, \sigma_{N+1})\, |H_{el} - E| \times \Phi^{p\mu S\, M_S}_{i'h'\ell'}(1 \ldots N-1, N+1, \hat{r}_N\, \sigma_N)\, f^{p\mu S}_{i'h'\ell'}(N_N)\rangle \tag{1o}$$

where the coordinates of the scattered electron are exchanged with the coordinates of one of the target electrons. It corresponds to a non-local exchange potential operator, whose integral term vanishes exponentially at large distances.

Finally, the correlation potential $\Theta^{p\mu S}_{h\ell,h'\ell'}$ arises from the elimination of the m equations involving the correlation functions ξ's. A well known procedure to do so is to diagonalise the H_{el} operator within the space of the correlation functions, thus yielding a set of eigenfunctions $\xi^{p\mu S\ M_S}_{\alpha}$ with corresponding eigenvalues ε_{α} [8,9]. The practical problem of choosing an efficient and realistic set of m functions to describe short-range correlations is, however, still a matter of ingenuity and trial-and-error procedures.

The full solution of the integro-differential eq. (7) then provides one with a set of linearly independent vectors within each (p μ S M_S) subset and with dimensions controlled by the ($h\ell$) indeces. One can then form a general solution for the continuum function of the scattered electron by writing the following linear combination:

$$F^{p\mu S}_{ih\ell}(\underline{r}_{N+1}) = \sum_{i'h'\ell'} a^{p\mu S}_{h'\ell'} f^{p\mu S}_{ih\ell,i'h'\ell'}(r_{N+1}) X^{p\mu S}_{h'\ell'}(\hat{r}_{N+1}) \quad (11)$$

which in turn yields the total scattered function:

$$F(\underline{r}_{N+1}) = \sum_{p\mu S\ ih\ell} r^{-1}_{N+1} F^{p\mu S}_{ih\ell}(\underline{r}_{N+1}) \quad (12)$$

Using S-matrix boundary conditions the latter function has to satisfy the following asymptotic relation:

$$F(\underline{r}_{N+1}) \underset{r\to\infty}{\sim} k_i^{-1/2} \sum_{\substack{p\mu S \\ ih\ell i'h'\ell'}} a^{p\mu}_{ih\ell}[\exp\{-i(k_i r-1/2\ell'\pi)\}\delta_{h'h}\delta_{\ell'\ell}\delta_{i'i}$$

$$- \exp\{i(kr-1/2\ell'\pi)\} S^{pS}_{h\ell h'\ell'}]\ X^{p\mu}_{h'\ell'}(\hat{r}_{N+1}) \quad (13)$$

which then allows one to obtain the coefficients a's in the space fixed (SF) laboratory frame by carrying out a simple rotation and equating the ingoing and outgoing parts of eq. (13) with those of the incident plane wave. [6,10]

The corresponding scattering amplitude is therefore given by the following equation, where $\hat{k}_i$ and $\hat{r}$ are the initial and final electron directions in the molecular (BF) frame:

$$f(\hat{r};\alpha\beta\gamma) = \sum_{\substack{p\mu S \\ ih\ell, i'h'\ell'}} \frac{2\pi}{ik_i} X^{p\mu}_{ih\ell}(\hat{k}_i) X^{p\mu}_{i'h'\ell'}(\hat{r})\, i^{\ell-\ell'}$$

$$\times (S^{pS}_{i\ell h, i'\ell'h'} - \delta_{\ell\ell'}\delta_{hh'}\delta_{ii'})$$

$$= \sum_{\substack{p\mu S \\ ih\ell m \\ i'h'\ell'm'm''}} \frac{\pi^{1/2}}{ik_i} b^{p\mu *}_{ih\ell m} b^{p\mu}_{i'h'\ell'm'} (2\ell+1)^{1/2} x i^{-\ell-\ell'}$$

$$\times \mathcal{D}^{\ell *}_{om}(\alpha\beta\gamma)\mathcal{D}^{\ell'}_{m''m'}(\alpha\beta\gamma)\, Y^{m''}_{\ell'}(\hat{r}')$$

$$\times (S^{pS}_{ih\ell, i'h'\ell'} - \delta_{\ell\ell'}\delta_{hh'}\delta_{ii'}) \qquad (14)$$

The last expression now depends upon the orientation $(\alpha\beta\gamma)$ of the target molecule with respect to the incident beam direction and the final electron direction $\hat{r}'$ is referred to the SF frame of reference[11].

The differential cross section is then obtained by averaging over all molecular orientation of the rigid target:

$$\frac{d\sigma}{d\omega} = \frac{1}{8\pi^2}\int |f(\hat{r}';\alpha\beta\gamma)|^2\, d\alpha\, d\beta\, d\gamma \qquad (15)$$

or, alternatively[10]:

$$I(\theta) = \frac{1}{4k_i^2}\sum_L A_L(k_i)\, P_L(\cos\theta) \qquad (16)$$

where θ is the angle formed with the incident beam of electrons, in the laboratory experiment, by the outgoing electron beam. The A_L coefficients relate in turn with all the angular factors and S-matrix elements that one averages over in the integration (15).[6]

Finally, the total cross section is:

$$\sigma_{TOT} = \frac{\pi}{k_i^2} A_o(k_i) = \int I(\theta)\sin\theta\, d\theta\, d\varphi \qquad (17)$$

and the momentum transfer cross section is:

$$\sigma_m = \frac{\pi}{k_i^2}(A_o - 1/3\, A_1)$$

$$= \int I\,(1-\cos\theta)\sin\theta\, d\theta\, d\phi \qquad (18)$$

3. COMPUTATIONAL MODELS

3.1. The Direct Static Potential

When analysing the structure of eq. (7) one quickly finds that the three contributions to the full interaction which appears on its r.h.s. play quite distinct roles in elucidating the forces at play in the scattering process.

To begin with, the direct interaction defines the following matrix elements:

$$V^{p\mu S}_{ih\ell,i'h'\ell'}(r_{N+1}) = \langle \Phi^{p\mu S}_{ih\ell}{}^{M_S}(1,\ldots N,\hat{r}_{N+1}\,\sigma_{N+1})| \times \hat{V}(\underline{r}_1,\ldots\underline{r}_{N+1};\,\underline{R}) \times |\Phi^{p\mu\ S\ M_S}_{i'h'\ell'}(1,\ldots N,\hat{r}_{N+1}\sigma_{N+1})\rangle \tag{19}$$

where the integration is taken over all (N+1) electron coordinates except the radial coordinate of the (N+1)th electron. For each of the electronic states of the molecule that are included in the expansion (3) one therefore obtains matrix elements like (19). The diagonal contributions correspond to the undistorted interaction for each of the above states; when the latter describe real physical situations of the molecular target one is simply obtaining the direct static potentials caused by the 'frozen' target electronic configuration and felt by the impinging electron. For collisions that take place at relatively low (near-thermal) energies the most important of such potentials is obviously the ground state static potential from the Φ_{GS} function of eq. (3):

$$V_{GS}(\underline{r}_{N+1}) = \langle \Phi_{GS}(1,\ldots N)|\,\hat{V}(\underline{r}_1,\ldots\underline{r}_{N+1};\,\underline{R})\,|\Phi_{GS}(1,\ldots N)\rangle \tag{2o}$$

where the integration is carried out over the space and spin coordinates of the N target electrons. The corresponding expansion (8) can then be written for the ground state, A_1 configuration of a closed shell polyatomic target in the following way:

$$V_{GS}(\underline{r}_{N+1}) = \sum_{h\ell} V^{A_1}_{h\ell}(r_{N+1})\,X^{A_1}_{h\ell}(\hat{r}_{N+1}) \tag{21}$$

where the expansion is performed around the centre of mass of the molecule in question and the radial coefficients can be obtained by numerical quadrature over the charge distributions yielded by SCF calculations of the MO-LCAO type for the target bound states.[7,12]

The question of convergence obviously arises immediately and has been extensively examined in the recent literature by our group[7,10] and others[13-15]. Suffice it to say that the existence of nuclei with large Z which are located away from the centre of mass provides one of the hardest situations where to reach acceptable convergence. On the other hand, molecular targets where one heavy atom is surrounded by one or more light atoms (e.g. H atoms) can be treated much more efficiently and provide a reliable level of convergence even with relatively few terms in the expansion(21).[16] A further test of the accuracy reached in describing the target molecule charge distribution is provided by the asymptotic coefficients of eq. (11) that can be compared with the static, permanent molecular multiples yielded by standard quantum chemistry codes.

3.2. The Polarization Potential

In addition to the diagonal matrix elements, we also have off-diagonal potentials. The most important of these couple electronic states between which optical dipole transitions are allowed. Higher order multipoles are also possible, albeit with smaller contributions.

The asymptotic form of the leading term usually goes as r^{-2}, with coefficients which are proportional to the oscillator strength of the transition from state i to state i'. The result of coupling an excited, dipole allowed electronic state with the target ground state is to provide, in 2nd order, an effective potential that goes as r^{-4} and describes the polarization potential seen by the incident electron interacting with that particular state.[17]

One therefore sees that each i state included in the expansion provides an improved description of the target response to the perturbation caused by the impinging electron. Of course the full polarization potential of the ground state can only be obtained by coupling all allowed target eigenstates including the continuum states. On the other hand, the r^{-4} component of the polarization function can also be included via a finite sum over suitably chosen pseudostates.[18] This has been widely used in electron atom scattering,[19] while for molecular targets only the H_2 case has been approached in this way.[20]

It therefore follows that model calculations need to be performed in order to obtain a simpler, but still realistic, account of the molecular response function effects on the computed phase shifts or on the observed cross sections. As far as the dipole contribution with asymptotic r^{-4} behaviour is concerned, this aspect can be implemented by a polarization potential which, for a diatomic target, is given as follows for large r values:

$$V_P(\underline{r};\ R) \underset{r\to\infty}{\sim} - \frac{\alpha_o(R)}{2\ r^4} - \frac{\alpha_2(R)}{2\ r^4} P_2\ (\hat{r}\ \hat{R}) \tag{22}$$

where α_o and α_2 are defined with respect to the polarizability components parallel and perpendicular to the molecular axis.[21] The latter data can either be obtained from experiments or computed from an SCF wavefunction of quality comparable to the one used to yield the direct static potential of above.[22]

For molecules with C_{2y} symmetry eq. (22) acquires the following form for an A_1 electronic state:

$$V_p\ (\underline{r};\ \underline{R}) \underset{r\to\infty}{\sim} - \frac{\alpha_o(\underline{R})}{2\ r^4} - (\frac{4\pi}{5})^{1/2} \frac{\alpha_2}{2r^4} X_{20}^{A_1}\ (\hat{r})$$

$$- (\frac{4\pi}{15})^{1/2} \frac{\alpha'_2}{2r^4}\ X_{22}^{A_1}\ (\hat{r}) \tag{23}$$

where the polarizabilities along the three principal axes of inertia of the molecule, which are the quantities usually quoted in the literature, can be easily expressed in terms of α_o, α_2 and α'_2.[10]

The problem of generating the polarization potential when charges begin to overlap is however still open. For electron molecule scattering at energies even as low as ∿1.o eV the sampling of the molecular charge distribution by the electron beam is such that the simple asymptotic relations of before are far from valid and one therefore needs to include short-range polarization contributions. The latter in turn cannot be given by the correlation functions included in the Θ's of eq. (7) that are instead required to deal with the orthogonalisation of the continuum electron to the bound electron wavefunctions.

For systems where a strong resonance structure is likely to occur (e.g. the $^2\pi_g$ configuration of N_2), part of the target polarizability can be included without resorting to pseudostate expansion, but rather by computing the N-electron MO's of the M resonant state via some sort of stabilization method,[23] thereby treating hopefully the most important part of the distortion.[24]

Alternatively, a widely used procedure has been to define an ad hoc correction to the asymptotic expressions (22) and (23) whereby the correct behaviour at the origin is forced on them via a cutoff function $F(r,r_c)$:

$$V_{Pol}(\underline{r};\underline{R}) = V_p\ (\underline{r};\underline{R}) \times F_j(r,r_c) \tag{24}$$

where r_c is a free, adjustable parameter and F_j is either:

$$F_1\ (r,r_c) = 1 - \exp\ [-(r/r_c)^6] \tag{25a}$$

or:

$$F_2\ (r,r_c) = [1 - \exp\ (-r/r_c)]^6 \tag{25b}$$

The selection for the best r_c value has been usually done by making some best fit of a computed cross section to experiments[25,26] although this is not a conceptually satisfactory procedure and a more detailed analysis of the possible physical meaning of r_c in simple targets is therefore still needed. On the other hand, we found that for several polyatomic targets many different observables could be satisfactorily understood by essentially choosing one r_c value for all of them in each molecular case.[10] This point will be further examined when discussing results.

3.3 The Exchange Potential

The matrix elements of eq. (7) that are explicitely shown in eq. (10) arise from the antisymmetrising operator A which appears in the eq. (3) for the total wavefunction. Its computational evaluation essentially involves an additional integration over the radial function $f^{p\mu S}_{i'h'\ell'}(r)$:

$$W^{p\mu S}_{ih\ell,i'h'\ell'}(r)\ f^{p\mu S}_{i'h'\ell'}(r)$$

$$= \sum_{\alpha} \int_0^{\infty} K^{p\mu S}_{ih\ell,i'h'\ell'}(r,r')\ f^{p\ S}_{i'h'\ell'}(r')\ r' \tag{26}$$

where the Kernel integrals can be expressed in terms of the radial orbitals arising from the single centre expansion of the occupied target molecular orbitals, their number being controlled by the index α. One see immediately that the exchange potential is son-local, a cause of difficulties when solving eq. (7), and it has the range of the occupied target orbitals which vanish exponentially. The last property means that for any choice of the target eigenstates and pseudostates $\{i\}$, one can select a radial value $r_{N+1} = r_o$ beyond which the exchange potential is negligible and the direct static potentials are given by eq. (11). This value could also correspond to some classical turning point for very low energy impinging electrons.

Considerable effort has been given to the development of alternative models which avoid computing the integrals of eq. (26), as we will discuss below, since the numerical solutions of the correct integro-differential equations have only been attempted in simple molecules[27,28] and have proven to be even there a formidable problem. Its treatment in a single-centre expension formulation requires in fact a large number of terms, thus making the accuracy of the final results strongly dependent on the level of convergence achieved. On the other hand, polyatomic targets with one heavy atom and several light atoms around it constitute the ideal candidates for testing possible rapid convergence of expansions around the heavy atom itself.[7] The results of the following section will in fact emphasize just this very point.

To begin with, the representation of exchange by a local potential has provided in recent years one of the most successful and widely used approaches: the use of the free-electron gas model was in fact introduced by Hara[29] and discussed recently by many others [30-32], while a semiclassical derivation has also been put forward by Riley and Truhlar[30]. The accuracy of these simplifications has been hard to assess since these prescriptions rely on the knowledge of the target wavefunction (undistorted) that is often lacking, even for simple targets, in those regions of space where the model is most sensitive to details, while often having to further resort to additional approximations to treat the polarization potential.

One important property of the continuum functions (in the static + exchange approximation) that needs to be ensured in an exact solution of the problem is their being orthogonal to those bound orbitals of the same symmetry which are fully occupied in the molecular ith state under consideration. This is not guaranteed by approximate exchange models like those mentioned above, while it is likely to have a dramatic effect on computed phaseshifts.

The imposition of this constraint, which gives an inhomogeneous term in the coupled differential equations (7), which thus are not anymore integro-differential, was introduced by Burke and Chandra[11] for diatomic molecules and extended to polyatomic targets by Gianturco and Thompson.[25,10] It appeared from their results that this term indeed represents a substantial part of the exchange potential, while however failing to properly describe the correct physical situation for those continuum states of highly localized nature (resonant states) which belong to a molecular symmetry (I.R.) that does not contain occupied orbitals in the chosen ith functions which describe the target. Recent, further calculations on polar diatomics[26,33] have thus shown that the use of a local form of the exchange potential plus the constraint of orthogonalization of the continuum orbitals to the target states improves the quality of the results and helps to eliminate spurious resonances often introduced by local exchange only. This approach is presently being introduced and extended to polyatomic molecules.[34]

4. RESULTS OF CALCULATIONS

Several aspects of the physical models briefly described in the previous section need obviously to be tested against experimental findings or against each other to assess their relative performances and feasibility. This task has been taken up in the recent literature by many authors, including some of the ones quoted before, with regard to each of the models and we will not repeat their numerous conclusions in the present review.

It is still of some interest, however, to point out the obvious weaknesses of the above models and to stress the correct quality of the physics that they still manage to describe, albeit qualitatively, in many different situations.

The treatment of scattering problems at the static-exchange level of approximation, for instance, disregards the dramatic effects that originate from the distortion of the molecular charge distribution by the electric field of the scattering electron. The energy of the perturbed molecule is in fact lower than that of the undistorted target and therefore gives rise to an additional attractive term in the potential energy as we discussed in the previous section.

As an example of comparison, Figure 1 reports the static potential (first two terms in the expansion of eq. (21)) generated by a near Hartree-Fock, SCF calculation for the ground state wavefunction of the N_2 molecule ($^1\Sigma_g$) and the corresponding behaviour of the asymptotic form of the polarization potential for the spherical coefficient given by the same wavefunction.[35] One clearly sees the strong changes that such a potential is likely to bring about, although the problem of defining its correct form at intermediate r_{N+1} value is by no means a simple one. The use of analytic cut-off functions as in eq. (24) for only the lowest-order contributions of a multipolar expansion of V_{Pol} has been in fact, questioned with regard to adiabatic calculations,[36] although no alternative viable strategies have been put forward for general cases. The inclusion of the second order effects yielded by polarization is thus generally accepted as essential to meaningful comparison with experiments, although nothing better than the stratagem of eq. (24) has been found viable for polyatomics.

Another essential ingredient of the models, as discussed before, is the inclusion of exchange potentials. At very low collision energies, however, the effect of such a potential is often surmised to be less important with respect to other long-range potentials, if they exist.[37] The HCl molecule with its strong dipole moment (∿0.47 a.u.) is a case in point and various calculations

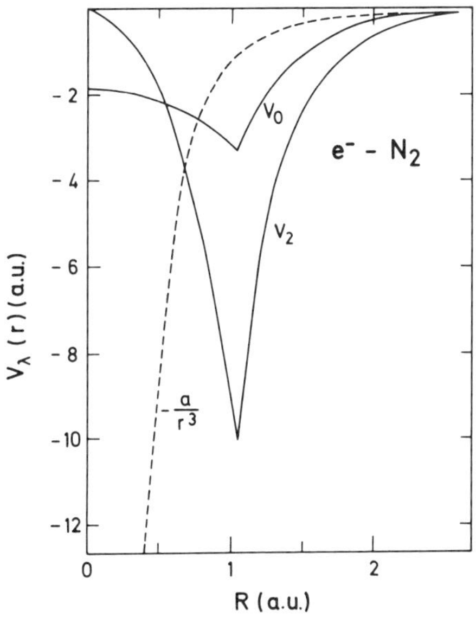

Fig. 1 - Static potential multipolar expansion coefficients (V_0 and V_2) from the ground state configuration of the N_2 target. The spherical part of the asymptotic form of its computed polarization potential is also shown (dashed line).

of the rotationally summed, vibrationally inelastic integral cross section are compared with experiments, as functions of collision energy, in Fig. 2. The continuous curve labelled FBA points out the failing of the simple Born approximation in giving both the shape and the order of magnitude of the measured cross sections.[38] The calculations reported by a dashed line (I.T.) correspond to body-fixed, close coupling calculations that employ a model potential and no exchange.[37] The method obviously fails at low collision energies and has been therefore extended by spacefixed calculations (GR) that include static and polarization potential terms in the close coupling approach, but no exchange contributions.[39] The latter reproduce very well the measured points at collision energies around 2oo meV and below, thus indicating the dominating role of long range forces for special targets or at specific collision energies.

Differential cross sections (DCS), partial or total, are usually considered a rather sensitive test of computational models, since the interference pattern that they indicate stems from the interplay of short-range and long-range forces as functions of the impact parameters and collision energies which can be associated to the impinging electron. At high values of k_1^2, for instance, the use of exchange potentials and higher order contributions to the static-plus-polarization terms is an all-important exercise as the results of Figure 3 briefly indicate for the case of the CO_2 molecule.

The computed curves labelled OTI and OT XVIII refer to a full quantum calculation with model exchange and model polarization, plus a simplified description of the static interaction.[4o] The third curve (-.-.), labelled Glauber, employs a semiclassical model with only lower-term contributions to static and polarization interactions.[41] The large differences between these models are clearly shown by the DCS results and allow a very straightforward assessment of their relative merits.

A polyatomic system which has been studied recently by several experimental groups[42-44] and which constitutes an interesting test of theoretical models on polyatomics is the methane molecule.

One of the most marked features of the above system is the existence of a pronounced Ramsauer-Townsend minimum below 1.0 eV that has been theoretically associated to the dominant interference pattern between short range repulsive and long range attractive branches of the potential for s-wave scattering in the A_1 symmetry.[25] Recent calculations on the same system[45] employed a multiple scattering model to get the static potential and used a free electron gas model to yield the exchange contributions, while the polarization effects were treated via the usual cut-off functional form as in

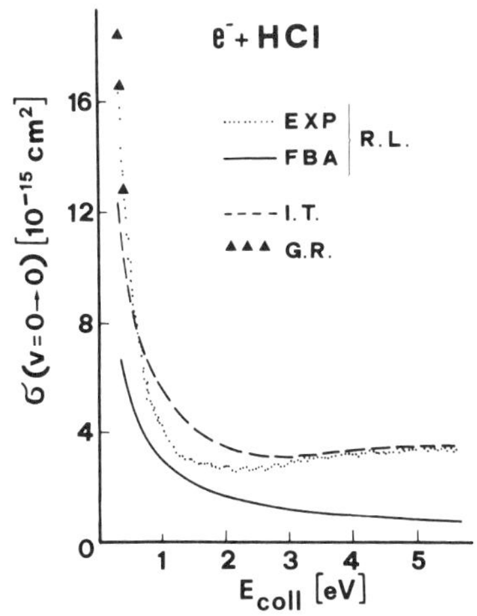

Fig. 2 - Rotationally summed, vibrationally elastic cross sections for HCl targets as functions of collision energy. The small dots are the experimental points (EXP) while the other three curves report calculated cross sections. See text for meaning and references.

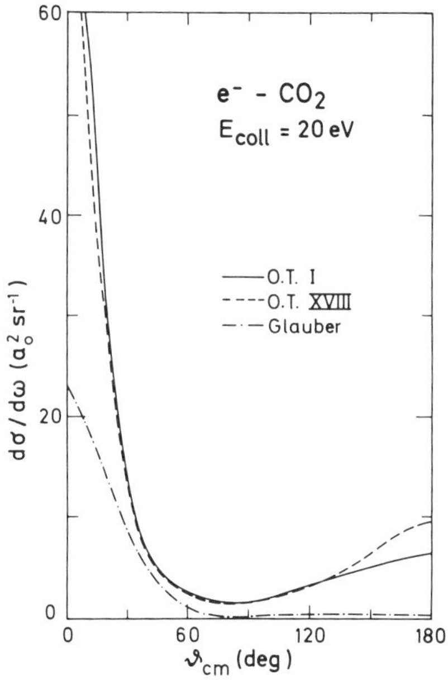

Fig. 3 - Differential cross sections at a collision energy of 20 eV for e^- - CO_2 scattering. The three curves refer to different computational models, as reported in the main text.

eq.s (25). Their results were essentially in agreement with our previous calculations and also confirmed the presence of a broad shape resonance in the 5 eV region, the latter being dominated by the ℓ = 2 partial wave (T_2 symmetry), as suggested by our previous calculations[25] and confirmed by our further extension of them.[10]

The corresponding DCS for the CH_4 molecule is also a sensitive tool for sampling resonant regions above the centrifugal barriers of the higher partial waves,[46] and the recent experimental measurements of their absolute values over a wide range of energy[47] are an interesting set of data against which theoretical models can be tested.

We have therefore performed our calculations by using the ground state configuration for the target molecule and by employing the same partial wave expansion already reported by us.[10] The need to use a cutoff parameter for the polarisation potential, a term which is usually strongly affecting final cross sections, was solved here by adopting a value kept constant across the examined energy range and which had provided encouraging fits of various experiments[10] (r_0 = 0.88 a.u.). It was also used only for the spherical part of the polarisability, since the ℓ=2 term does not contribute to the A_1 symmetry expansion. This is another instance where the use of symmetry adapted partial wave expansions provides diagonal representations for each $\{p\mu\}$ subset hence favours in some cases a faster convergence with fewer contributing terms.

The very low energy results (E = 2 eV) are reported in Figure 4, where the experimental data[47] at 3 eV are also shown. Since both theory and experiments use absolute values, the agreement of the overall shape is really excellent. It is also worth noting that the corresponding DCS for Ar at the above collision energies are very similar to those shown in our Figure, indicating that the scattering experiment is only sampling the outer range of the target structure, with little contribution from its molecular shape.

Further results are also shown in Figure 5, for collision energies of 5 and 10 eV. The comparison with calculations is only shown here with the data at 5 eV, while the data at 10 eV appear analysed in more detail in the following Figure 6. The experiments of Fig. 5 show that the shape of the DCS below 10 eV is still very similar to the Ar case,[47] while at higher energies the angular distributions are markedly different for the molecular target, since its polycentre nature begins to be sampled by the impinging electrons.

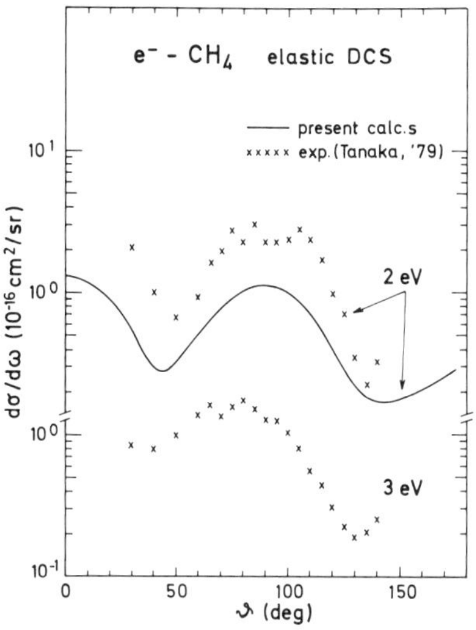

Fig. 4 - Differential (vibrationally elastic) total cross sections measured at 2 and 3 eV of collision energy. The computed results (not normalized) are given by the continuous line.

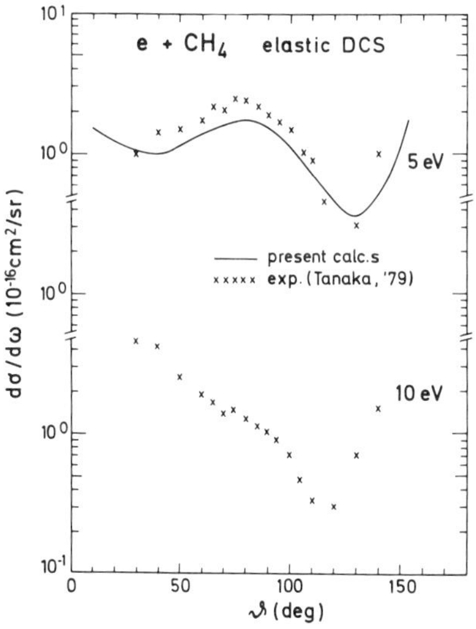

Fig. 5 - Same as in Figure 4 but for the higher collision energies of 5.0 and 10.0 eV.

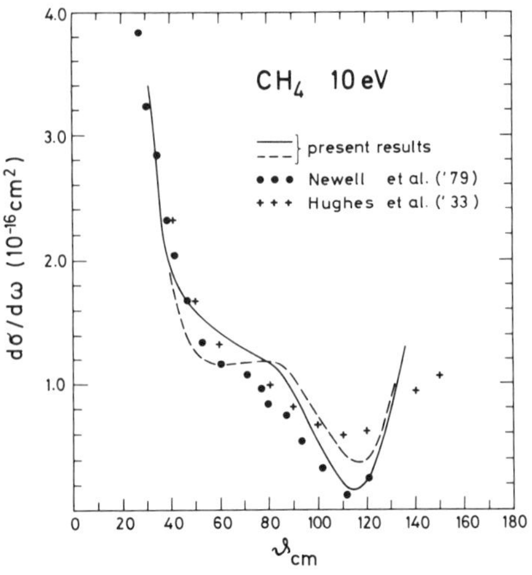

Fig. 6 - Differential (elastic) cross sections for e´ - CH_4 scattering, on a linear scale, at 10.0 eV. See text for meaning of symbols.

The results reported in Fig. 6 show two different choices of cut-off parameter for the polarization potential, one being the r_0=0.88 u.a. that was used for all the present calculations (dashed line) and the other corresponding to a weaker potential with r_0=0.92 (continuous line) that appears to get surprisingly well the recent experimental data of Newell.[48] The previous choice of r_0, on the other hand, follows more closely the older data of Hughes.[49] Both are, however, in very good agreement with measured data.

Finally, higher energy results are compared with our calculations in Figure 7, whereby showing that the present model is still in good agreement with measurements in spite of the various limitations that its ingredients contain and which we have discussed above in the previous sections. Obviously, more work needs to be done on polyatomic targets, hopefully parallel to the increase in sophistication for diatomics that is presently taking place, but one should also keep in mind that the increase in molecular complexity brings about the additional help of symmetry-adapted representations that here become much more efficient selectors of relevant contributions to physical phaseshifts. The case of the methane molecule discussed here should, hopefully provide a clear example of such behaviour.

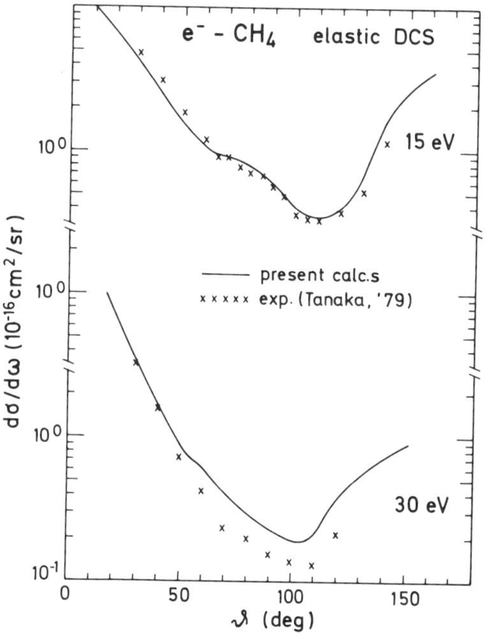

Fig. 7 - Same as in Figs. 4 and 5 but for the higher collision energies of 15.0 and 30.0 eV.

ACKNOWLEDGEMENTS

We are grateful to Dr. Tanaka for sending us his experimental results prior to publication, to the ZIF Bielefeld Universität for financial support during the Meeting where our results were presented and to the MPI für Strömungsforschung, Göttingen, for the preparation of the Figures. A NATO travelling grant awarded for this collaborative research is also here acknowledged.

REFERENCES

1. N.F. Lane, Rev. Mod. Phys. 52:29 (198o).
2. T.N. Rescigno, V. McKoy and B.I. Schneider "Electron Molecule and photon-molecule Collisions", Plenum Press, New York and London (1979).
3. S.C. Brown, "Electron Molecule Scattering", John Wiley, New York (1979).
4. e.g. see: P.G. Burke and A.L. Sinfailam, J. Phys. B. 3:641 (197o).
5. S. Chung, C.C. Liu, Phys. Rev. A17:1874 (1978).
6. P.G. Burke, N. Chandra and F.A. Gianturco, J. Phys. B5:2212 (1972).
7. F.A. Gianturco and D.G. Thompson, Chem. Phys. 14:111 (1976).
8. H. Feshbach, Ann. of Phys. 5:357 (1958).
9. H. Feshbach, Ann. of Phys. 19:287 (1962).
1o. F.A. Gianturco and D.G. Thompson, J. Phys. B 13:613 (198o).
11. P.G. Burke and N. Chandra, J. Phys. B 5: 1696 (1972).
12. G.B. Schmid, D.W. Norcross, L.A. Collins, Comp. Phys. Comm. 21:79 (198o).
13. L.A. Collins and D.W. Norcross, Phys. Rev. A 18:467 (1978).
14. M.E. Riley and D.G. Truhlar, J. Chem. Phys. 65:792 (1976).
15. M.A. Morrison and L.A. Collins, Phys. Rev. A17:918 (1978).
16. F.A. Gianturco and D.G. Thompson, J. Phys. B1o:L21 (1977).
17. L. Castillego, I.C. Percival and M.J. Seaton, Proc. Roy. Soc. A254:259 (196o).
18. R. Damburg, E. Karule, Proc. Phys. Soc. 9o:677 (1967).
19. M. Le Dourneuf, Vo Ky Lan, P.G. Burke, Comm. Atom. Molec. Phys. 7:1 (1977).
2o. B.I. Schneider, in "Electronic and Atomic Collisions", G. Watel Ed., North Holland Publ. Co., Amsterdam (1978).
21. F.A. Gianturco, U.T. Lamanna, J. Phys. B 12:742 (1979).
22. F.A. Gianturco and C. Guidotti, J. Phys. B11:L385 (1978).
23. H.S. Taylor, Adv. Chem. Phys. 18:91 (197o).
24. B.I. Schneider, M. Le Dourneuf, Vo Ky Lan, Phys. Rev. Lett. 43:1923 (1979).
25. F.A. Gianturco and D.G. Thompson, J. Phys. B 9:L383 (1976).
26. L.A. Collins, R.J.W. Henry, D.W. Norcross, J. Phys. B13:2299 (198o).
27. G.D. Buckley and P.G. Burke, in "Symposium on Electron Molecule Collisions", I. Shimamura and M. Matsuzawa Eds., U. of Tokyo, Tokyo (1979).
28. L.A. Collins, W.D. Robb, M.A. Morrison, J. Phys. B11:L777 (1978).
29. S. Hara, J. Phys. Soc. Japan, 22:71o (1967).
3o. M.E. Riley and D. Truhlar, J. Chem. Phys. 63:2182 (1975).
31. P. Baille and J.W. Darewych, J. Chem. Phys. 67:3399 (1977).
32. M.A. Morrison and L.A. Collins, Phys. Rev. A17:918 (1978).
33. L.A. Collins, W.D. Robb and D.W. Norcross, Phys. Rev. A2o:1839 (1979).
34. S. Salvini and D.G. Thompson, private communication.
35. F.A. Gianturco, umpublished results.

36. M.A. Morrison and P.J. Henry, Phys. Rev. A2o:74o (1979).
37. Y. Itikawa, K. Takayanagy, J. Phys. Soc. Japan 26:1254 (1969).
38. F. Linder in "Electronic and Atomic Collisions" G. Warel Ed., N. Holland, Amsterdam (1978).
39. F.A. Gianturco and N.K. Rahman, J. Phys. B11:727 (1978).
4o. K. Onda and D.G. Truhlar, J. Phys. B12:283 (1979).
41. F.A. Gianturco, U. Lamanna and S. Salvini: Int. J. Quantum Chem., S.13:529 (198o).
42. D. L. McCockle, L.G. Christopharou, D.V. Maxey and J.G. Carter, J. Phys. B 11:3o67 (1978).
43. E. Barbarito, M. Basta, M. Calicchio and G. Tessari; J.Chem.Phys. 71:54 (1979).
44. K. Rohr, J. Phys. B13:4897 (198o).
45. Zs. Varga, I. Gyémánt and M.G. Benedict, Acta Phys. Chem., Szeged, XXV:85 (1979).
46. J.L. Dehmer, J. Siegel and D. Dill, J. Chem. Phys. 69:52o5 (1978).
47. T. Okada, M. Kubo, T. Yamamoto, T. Suzuki and H. Tanaka, At. Coll. Res. Japan, Progr. Rep. 6:17 (1979).
48. W.R. Newell, D.F.C. Brewer, A.C.H. Smith, Abstracts of contributed papers, XI ICPEAC, Kyoto (1979).
49. A.L. Hughes, J.M. McMillen, Phys. Rev. 44:876 (1933).

DYNAMICAL THEORY OF RESONANT ELECTRON-MOLECULE SCATTERING NEAR THRESHOLD

W. Domcke and L.S. Cederbaum

Theoretische Chemie
Universität Heidelberg
D-69oo Heidelberg, Germany

I. INTRODUCTION

Resonances play an important role in the electronic and vibrational excitation of molecules by electron impact[1,2] and in dissociative attachment and recombination processes.[3] The nuclear motion in the resonance state is usually described within the local complex potential approach.[3,4] The local approximation is expected to break down, however, for resonances close to threshold.[3] In recent years the observation of pronounced threshold peaks in the vibrational excitation functions of many polar and certain non-polar molecules[5] has stimulated renewed interest in resonance and threshold effects in electron-molecule collisions.[6-1o]

The aim of this contribution is two-fold. We develop a theory of resonance scattering which includes exactly the effects of vibrational motion. This is of particular interest for resonances close to threshold where the local complex potential approximation is invalid, especially in the presence of long-range potentials. In addition, we will be concerned with the analytic properties of the fixed-nuclei S-matrix, in particular with the resonances, virtual state and bound state poles of S and their movement with the internuclear distance. We shall see that the energy dependence of the resonance width at threshold determines the analytic properties of the S-matrix and governs also the importance of dynamic (in particular non-adiabatic) effects.

II. THE HAMILTONIAN AND THE CALCULATION OF THE CROSS SECTION

It is well-known that resonances can be described as discrete states embedded into and interacting with a continuum.[11-14] Assuming a single discrete electronic state $|d\rangle$ and a single electronic continuum $|k\rangle$ and including the vibrational degrees of freedom we write for the Hamiltonian[15]

$$H = H_o + V + W \tag{1a}$$

$$H_o = |d\rangle\varepsilon_d\langle d| + \sum_k |k\rangle\varepsilon_k\langle k| + \tilde{H}_o \tag{1b}$$

$$\tilde{H}_o = \frac{-1}{2\mu}\frac{d^2}{dR^2} + V_o(R) = T_N + V_o(R) \tag{1c}$$

$$V = \sum_k V_{dk}|d\rangle\langle k| + \text{h.c.} \tag{1d}$$

$$W = \sum_{kk'} W_{kk'}|k\rangle\langle k'|. \tag{1e}$$

We have suppressed the translational and rotational degrees of freedom and assumed a single vibrational coordinate R for simplicity. These restrictions can easily be relaxed if necessary. The continuum states are taken to be box-normalized and orthogonal to the discrete state

$$\langle k|k'\rangle = \delta_{kk'} \tag{2a}$$

$$\langle d|k\rangle = 0. \tag{2b}$$

The basis states in eq. (1) are of course multielectron states, but since we will not be concerned with electronic excitation of the target, it suffices to characterize them by the single index d or k.

$V_o(R)$ denotes the electronic potential energy of the target molecule and T_N the nuclear kinetic energy. H_o is thus the vibrational Hamiltonian of the target. The interaction V mixes $|d\rangle$ with the continuum and converts the discrete state into a resonance. The interaction W acts only in the continuum and represents the non-resonant background scattering. In principle, the influence of background scattering on the resonance could be included by a re-definition of the basis states $|k\rangle$, i.e., by pre-diagonalizing the Hamiltonian in the continuum. We know, however, that long-range potentials will introcude rapid variations of the continuum density of states with energy near threshold. We therefore take W of eq. (1e) to represent a typical long-range potential such as the Coulomb, dipole, or polarization potential. This allows us to treat the influence of long-range potentials on the resonant scattering process explicitly, while all short-range direct scattering potentials are assumed to be included in the definition of the basis states $|k\rangle$. This parametrization of the Hamiltonian bears some resemblance to the quantum-defect theory of atomic physics.[16,17]

It is a well-established experimental fact that vibrational excitation via resonances is much more effective than vibrational excitation by electron impact in non-resonant energy regions.[1] We have incorporated this into eq. (1) by assuming ε_k and $W_{kk'}$ to be independent of the vibrational coordinate R. There is then no direct coupling between electrons in the continuum and the nuclear motion. In the resonant energy region this coupling is accomplished by the mixing of the continuum states with the discrete state $|d\rangle$ whose energy ε_d is assumed to depend on the internuclear distance R. The discrete state-continuum matrix element V_{dk} may also be a function of R.

To complete the definition of the Hamiltonian (1) we introduce the adiabatic or Born-Oppenheimer (BO) approximation for the basis states. We assume that the electronic wave function of the discrete state

$$\phi_d(\underline{r},R) = \langle \underline{r} \,|\, d\rangle,$$

where $\underline{r}$ denotes collectively the electronic coordinates, depends sufficiently weakly on the nuclear coordinate R such that

$$[T_N, \phi_d(\underline{r},R)]_- = 0 \tag{3}$$

to a good approximation. Eq. (3) is identical with the well-established BO approximation for bound electronic states, which is based on the fact that the energy separation of electronic states is large compared to the spacing of vibrational energy levels. Since the energy separation of discrete states $|d\rangle$ giving rise to different electronic resonances is of the same order of magnitude as the energy separation of bound electronic states (typically a few electron volts), the BO approximation should work well for these discrete states. It should be stressed that the BO approximation is not necessarily a good approximation for the resonance wave function obtained by diagonalizing $H_0 + V$ for each R (neglecting T_N), since the degree of mixing of the basis states $|d\rangle$ and $|k\rangle$ may vary rapidly with internuclear distance. The nuclear kinetic energy operator will then not commute with the adiabatic resonance wave function although it commutes with $\phi_d(\underline{r},R)$. The basis states in eq. (1) are analogous to the "diabatic" basis states frequently used to describe atomic collisions[18] and vibronic spectra.[19]

Starting from the Hamiltonian (1) the scattering cross section can be calculated exactly using the standard formalism of scattering theory.[20] Let us first discuss the pure resonance scattering problem, $H = H_o + V$, without any long-range potential W. We define the initial and final asymptotic states

$$|i\rangle = |k_i\rangle|0\rangle, \quad |f\rangle = |k_f\rangle|v\rangle \tag{4}$$

where $|0\rangle$ and $|v\rangle$ denote the vibrational ground state and the v-th excited vibrational state of the target molecule, respectively. k_i and k_f indicate the initial and final momenta of the scattered electron and it is assumed that the channels corresponding to excited electronic states of the target molecule are closed. The BO factorization of the electronic and vibrational parts of the wave function implied by eq. (4) is an excellent approximation for the electronic ground states of the overwhelming majority of molecules. The differential cross section for the excitation of the v-th vibrational level is given by (in atomic units, $\hbar = c = e = m_e = 1$)

$$d\sigma_v/d\Omega = (\Omega_N/2\pi)^2(k_f/k_i)|\langle f|T|i\rangle|^2 \tag{5}$$

with

$$T = V + V(E_t - H_o + i\eta)^{-1}T$$

$$= V + V(E_t - H_o + i\eta)^{-1}V + \ldots \tag{6}$$

where $E_t = E_i + \langle 0|\tilde{H}_o|0\rangle$ denotes the total energy and $E_i = k_i^2/2$ is the kinetic energy of the incident electron. η is a positive infinitesimal. Ω_N is the normalization volume which drops out of the final expressions. Taking account of eq. (3) we can integrate over the electronic coordinates in each term arising from the expansion (6). The resulting infinite series which still contains vibrational operators can be summed exactly, giving[15]

$$d\sigma_v/d\Omega = (\Omega_N/2\pi)^2(k_f/k_i)|\langle v|V_{k_f d}(E_t - \mathcal{H})^{-1}V_{dk_i}|0\rangle|^2 \tag{7a}$$

$$\mathcal{H} = \tilde{H} + \Delta(E_t - \tilde{H}_o) - 1/2\, i\Gamma(E_t - \tilde{H}_o) \tag{7b}$$

$$\tilde{H} = \tilde{H}_o + \varepsilon_d(R) \tag{7c}$$

$$\Gamma(E) = 2\pi \sum_k V_{dk}\delta(E - \varepsilon_k)V_{kd} \tag{7d}$$

$$\Delta(E) = (2\pi)^{-1}P\int dE'\Gamma(E')/(E - E') \tag{7e}$$

where P denotes the Cauchy principal value. $\tilde{H}$ is the Hamiltonian for the vibrational motion when an electron occupies the discrete state $|d\rangle$. The non-unitary energy-dependent operator $\mathcal{H}$ describes the time evolution of the vibrational motion in the resonance state. Through the coupling to the continuum the discrete state has acquired a width $\Gamma(E_t-\tilde{H}_o)$ and has been shifted by $\Delta(E_t-\tilde{H}_o)$. The width and the level shift are not simply functions of the energy E_t and the internuclear distance R, but depend also - via H_o - explicitly on the kinetic energy of the vibrational motion. The dependence of the resonance position and width on the nuclear kinetic energy is, by definition, a non-adiabatic effect. It should be emphasized that eq. (7) is a completely general result depending only on the idea of a quasi-diabatic discrete state interacting with a continuum. The specific form of the wave functions of the discrete and continuum states is not of relevance.

Equation (7a) can be written in a more compact form when V_{dk} is weakly dependent on R and can be taken out of the vibrational matrix element. Integrating over the angles we obtain for the integral excitation functions

$$\sigma_v = \nu\pi/k_i^2 \Gamma(E_i)\Gamma(E_f) |\langle v|(E_t - \mathcal{H})^{-1}|0\rangle|^2 \tag{8}$$

where E_f is the final kinetic energy of the scattered electron and ν counts the spatial degeneracy of the discrete state $|d\rangle$. Eq. (8) together with eqs. (7b-e) represent the exact formal solution of the multi-channel scattering problem of an electronic resonance coupled to the vibrational degrees of freedom. In the fixed-nuclei limit, $T_N \to 0$, eq. (8) reduces to the well-known isolated resonance Breit-Wigner formula[20] with energy-dependent width.

III. INCLUSION OF LONG-RANGE POTENTIALS

The importance of non-adiabatic effects, formally described by the operator nature of the decay width and the level shift in the effective Hamiltonian $\mathcal{H}$ of eq. (7b), is seen to be governed by the energy dependence of the functions $\Gamma(E)$ and $\Delta(E)$ defined in eqs. (7d,e). The strongest energy dependence of $\Gamma(E)$ is expected to occur near threshold where $\Gamma(E)$ is forced to approach zero. From the general theory of threshold laws[21] we know that the energy dependence of Γ near threshold is determined by the centrifugal potential. Introducing a partial wave representation of the continuum, the width function (7d) can be written as

$$\Gamma(E) = \sum_{l=o}^{\infty} \Gamma_l(E). \tag{1o}$$

In the absence of long-range forces the threshold behavior of $\Gamma_l(E)$ is given by[21]

$$\Gamma_l(E) \sim E^{(2l+1)/2} \tag{11}$$

It is thus clear that the lowest l allowed by the symmetry selection rules dominates the energy dependence of the width at threshold.

To discuss the influence of long range potentials we go back to eq. (5) and calculate the scattering cross section using the full Hamiltonian (1), i.e. $H = H_o + V + W$, where W represents some long-range direct scattering potential. It is assumed that W does not depend on the internuclear distance. In analogy to equation (6) the T-operator is now given by

$$\begin{aligned} T &= (V+W)+(V+W)[E_t-H_o+i\eta]^{-1}T \\ &= (V+W)+(V+W)[E_t-H_o+i\eta]^{-1}(V+W)+\ldots \end{aligned} \tag{12}$$

By a re-arrangement of the terms in the infinite series for $\langle f|T|i\rangle$ we can, in each order in V, sum all terms up to infinite order in W. Introducing the exact scattering states in the potential W (it is convenient to change now to energy-normalized continuum states)

$$\begin{aligned} |\bar{k}^{(\pm)}\rangle &= |k\rangle + W\,G_o^{(\pm)}|\bar{k}^{(\pm)}\rangle \\ &= |k\langle + W\,G_o^{(\pm)}|k\rangle + W\,G_o^{(\pm)}\,W\,G_o^{(\pm)}|k\rangle + \ldots \end{aligned} \tag{13a}$$

where

$$G_o^{(\pm)} = (E_t - H_o \pm i\eta)^{-1} \tag{13b}$$

we obtain in zeroth order in V the non-resonant scattering amplitude

$$T_{fi}^{(D)} = \langle k_f|W|\bar{k}_i^{(+)}\rangle\delta_{v,0}. \tag{14}$$

The resonant scattering amplitude is obtained by summing the remaining terms to infinite order in V

$$T_{fi}^{(R)} = \langle v|\bar{V}_{k_f d}(E_t - \bar{\mathcal{H}})^{-1}\,\bar{V}_{dk_i}|0\rangle \tag{15a}$$

$$\bar{\mathcal{H}} = \tilde{H} + \bar{\Delta}(E_t - \tilde{H}_o) - 1/2\; i\;\bar{\Gamma}(E_t - \tilde{H}_o) \tag{15b}$$

where

$$\bar{V}_{k_f d} = \langle \bar{k}_f^{(-)} | V | d \rangle, \tag{16a}$$

$$\bar{V}_{dk_i} = \langle d | V | k_i^{(+)} \rangle \tag{16b}$$

and $\bar{\Gamma}(E)$ and $\bar{\Delta}(E)$ are given by eqs. (7d,e) with V_{dk} replaced by $\bar{V}_{dk}$. Eqs. (14-16) are exact and include the long-range potential W to all orders.* Apart from the appearance of a background scattering amplitude (which contributes only in the elastic channel owing to the assumed R-independence of W) the equations derived in Section II remain completely unchanged. The direct scattering potential W just leads to a "renormalization" of the width function $\Gamma(E)$. To describe this renormalization near threshold, it is convenient to introduce the Jost function[20] for the potential W. Owing to the localization of $\phi_d(\underline{r})$ in space, $\langle \underline{r} | k^{(\pm)} \rangle$ can be replaced by the free solution $\langle \underline{r} | k \rangle$ divided by the Jost function f(k) for energies close to threshold.[20] The renormalized width function is thus given by

$$\bar{\Gamma}(E) = (E)/ f(k)^{2}. \tag{17}$$

As is well-known,[20] the Jost function is non-analytic at threshold for long-range potentials. In the case of the r^{-4} potential the threshold expansion of f(k) contains a term proportional to $k^2 \ln k$, which makes f(k) non-analytic at threshold, but does not alter the leading term of the expansion for $l = 0$.[22,23] The potential of a point dipole

$$W = - Dr^{-2} \cos\theta \tag{18}$$

is of particular interest for electron-polar molecule scattering. Being anisotropic, the dipole potential does not commute with the rotational kinetic energy operator. Therefore, any rigorous treatment of dipole effects has to include explicitly the rotational degrees of freedom.[24] The consideration of a fixed (non-rotating) dipole is nevertheless of interest for comparison with fixed-nuclei *ab initio* calculations. The Jost function corresponding to the s-wave component of the scattering wave function in the dipole field reads for $k \to 0$ and $D < D_{crit}$.[25]

$$f(k) \sim (-ikr_o)^{1/2-\alpha} \tag{19}$$

---

* Eqs. (15,16) hold also for R-dependent long-range potentials W if $|k^{(\pm)}\rangle$ and thus $\bar{V}_{dk}$ are considered as operators in R-space. Eq. (13) shows that these quantities are functions of $E_t - \tilde{H}_o$.

where

$$\alpha = (\gamma + 1/4)^{1/2}. \tag{20}$$

here r_o denotes a short-range cut-off radius and γ is dimensionless reduced dipole moment given by the lowest eigenvalue of an infinite-dimensional tridiagonal matrix.[26] $\gamma = -1/4$ corresponds to $D = D_{crit}$, which is the minimum dipole moment required to bind an electron in the dipole field.[26] Combining eqs. (11,17,19) we have for the energy dependence of the width function at threshold

$$\bar{\Gamma}(E) \sim E^{\alpha} \tag{21}$$

with $0 < \alpha < 1/2$. Thus the square root behavior of $\Gamma(E)$ typical for s-wave scattering is converted to a step tunction behavior when the dipole moment approaches the critical value.

A super-critical non-rotating dipole possesses an infinite number of bound states.[26] As a consequence, $\bar{\Gamma}(E)$ exhibits an infinite series of δ-function peaks at negative energies and $\bar{\Delta}(E)$ exhibits simple poles at these energies, as follows from eq. (7e). A similar result is obtained for the attractive Coulomb potential where $\bar{\Gamma}(E)$ exhibits δ-function peaks at the Rydberg energy levels. The bound states of the dipole or Coulomb potentials thus appear as a continuation of the continuum to negative energies as is familiar from quantum-defect theory.[16,17]

IV ANALYTIC PROPERTIES OF THE FIXED-NUCLEI S-MATRIX

In the fixed-nuclei limit, $T_N \to 0$, the operator $E_t - \tilde{H}_o$ appearing in the argument of Γ and Δ in eq. (7b) reduces to

$$E' = E_t - V_o(R), \tag{22}$$

where E_t is the total energy and E' is the kinetic energy of the scattered electron. Considering E_t as fixed, E' becomes a function of the internuclear distance R, since the threshold itself is a function of R. Introducing E' as the energy variable, the operator expression (8) reduces to the standard Breit-Wigner formula[20]

$$\sigma(E') = \frac{2\pi\nu}{E'} \frac{[\Gamma(E')/2]^2}{[E'-\varepsilon_d(R)-\Delta(E')]^2 + [\Gamma(E')/2]^2} . \tag{23}$$

The corresponding S-"matrix" reads

$$S(E') = [1+iK(E')][1-iK(E')]^{-1} \tag{24}$$

with the real K-"matrix"

$$K(E') = \tan\delta(E') = -\frac{\Gamma(E')/2}{E'-\varepsilon_d(R)-\Delta(E')} \quad . \tag{25}$$

$\delta(E')$ denotes the fixed-nuclei phase shift.

Eq. (25) for the phase shift allows us to make contact with fixed-nuclei ab initio calculations. Given the ab initio calculated phase shift as a function of energy an internuclear distance, one may fit the resonant part using expression (25) to determine the functions $\Gamma(E')$ and $\varepsilon_d(R)$ which are required as the input data for the dynamical calculation of the cross sections via eqs. (7,8). It should be mentioned that an additional dependence of $\Gamma(E')$ on R represents no problem in the fixed-nuclei limit. It is only the dynamical calculation which becomes more complicated when Γ is a function of both energy and internuclear distance.

The usual criterion for the occurence of a resonance in the scattering cross section is an increase of the phase shift by approximately π. The position of the resonance may be defined as the energy where $\delta(E)$ (mod π) is equal to $\pi/2$, which means that K(E) has a singularity. It follows from eq. (25) that the singularities of the K-matrix are given by the solutions E'_r of the equation

$$E'_r - \varepsilon_d(R) - \Delta(E'_r) = 0. \tag{26}$$

A second possibility to look for resonances is to search the poles of the S-matrix in the complex momentum or energy plane. To find the analytic continuation of S we consider k and $z = k^2/2$ as komplex variables and define

$$F(z) = \frac{1}{2\pi} \int dE \, \frac{\Gamma(E)}{z-E} \quad . \tag{27a}$$

We then have for z on the real energy axis

$$\mathrm{Re}F(E \pm i\eta) = \Delta(E) \tag{27b}$$

$$\mathrm{Im}F(E \pm i\eta) = \mp \Gamma(E)/2 \tag{27c}$$

F(z) is a single-valued function on the z-plane cut along the positive real axis. The cut on the real axis is a consequence of the mapping $z = k^2/2$. When considered as a function of k, F(k) is an analytic single-valued function on the k-plane in the absence of long-range forces. We may now define the S-matrix in the complex k-plane as

$$S(k) = \mathfrak{f}(-k)/\,\mathfrak{f}(k) \tag{28}$$

with

$$\mathcal{f}(k) = -\frac{1}{2}k^2 + \varepsilon_d(R) + F(k). \qquad (29)$$

$\mathcal{f}(k)$ is essentially the Jost function for the resonance scattering problem. The zeros of $\mathcal{f}(k)$ determine the poles of S(k). $\mathcal{f}(k)$ should not be confused with the Jost function f(k) for scattering from the long-range potential W introduced in Section III. It is easy to show that S(k) of eq. (28) reduces to eq. (24) for k on the positive real axis and that $\mathcal{f}(k)$ and S(k) fulfill the usual symmetry requirements.[20]

Eqs. (27,28,29) provide us with the analytic continuation of the Breit-Wigner resonance formula for arbitrary energy dependence of the width Γ. This allows us to discuss resonance poles, bound state poles and virtual state poles and their movement with the internuclear distance R in a very general way. To illustrate how this works we consider the following simple parametrization of the width function

$$\Gamma(E) = A(E/B)^{\alpha}(2 - E/B)^{\alpha} \qquad (30)$$

where A,B and α are free parameters. Eq. (30) yields $\Gamma(E) \sim E^{\alpha}$ at threshold and α may be called the threshold exponent. A is the maximum value of the width and 2B determines the energy range for which Γ(E) is different from zero. Γ(E) has to approach zero at high energies to guarantee the existence of the Hilbert transform in eq. (7e). The detailed form of Γ(E) at high energies is of course irrelevant for the scattering near threshold and we choose the simple symmetric form of eq. (30) for convenience. Direct evaluation yields

$$F(z) = \frac{A}{2\sqrt{\pi}}\left\{ -\frac{\Gamma(\alpha)}{\Gamma(\frac{1}{2}+\alpha)}\, {}_2F_1\left(\tfrac{1}{2},-\alpha;1-\alpha;1-(z/B-1)^2\right)\right.$$
$$+\frac{\Gamma(1+\alpha)\Gamma(-\alpha)}{\Gamma(1/2)}(z/B-1)[(z/B-1)^2-1]^{\alpha}$$
$$\left. {}_2F_1\left(1+\alpha,\tfrac{1}{2};1+\alpha;1-(z/B-1)^2\right)\right\} \qquad (31)$$

where Γ(a) and ${}_2F_1$ (a,b;c;z) denote the Gamma function and the hypergeometric function, respectively.[27] The Hilbert transform Δ(E) follows from eq. (27b).

For half-integral values of α , α = (2l + 1)/2, eq. (30) represents the threshold behavior characteristic for the scattering of electrons with angular momentum l in the absence of long-range forces (cf. eq. (11)). A (non-rotating) permanent dipole leads to non-half-integral values of α as discussed in Section III. In the following we consider the two special cases α = 1/2 (s-wave scattering) and α = 0 (s-wave scattering with critical dipole moment). In both cases F(k) is given by simple expressions, namely (C^2 = 4B)

$$F(k) = -A/C^2[C^2/2 - k^2 + ik(C^2 - k^2)^{1/2}] \tag{32}$$

for $\alpha = 1/2$ and

$$F(k) = A/(2\pi) \ln[k^2/(k^2-C^2)] \tag{33}$$

for $\alpha = 0$. The latter case is discussed also in Ref. 25 and the reader is referred to this paper for mathematical details. The transition from $\alpha = 1/2$ to $\alpha = 0$ with gradually increasing dipole moment has also been discussed in detail,[28] but cannot be outlined here due to limitations in space.

The poles of the fixed nuclei S-matrix in the complex k-plane and their movement with the internuclear distance R obtained from the solution of

$$-\frac{1}{2}k^2 + \varepsilon_d(R) + F(k) = 0 \tag{34}$$

are shown in Fig. 1 and Fig. 2 for $\alpha = 1/2$ and $\alpha = 0$, respectively. In both cases A = 3eV, B = 7eV and ε_d has been taken as a linear function of R

$$\varepsilon_d(R) = \varepsilon_d(R_o) + \varepsilon_d'(R_o)(R-R_o) \tag{35}$$

with $\varepsilon_d(R_o) = 3.2\text{eV}$, $\varepsilon_d'(R_o) = 8.33\text{eV/Å}$. Fig. 1 shows two resonance poles of S in the lower half of the complex k-plane situated symmetrically to the imaginary axis. With increasing R the poles move inward essentially parallel to the real axis and meet on the negative imaginary axis. With further increasing R one pole moves down the imaginary axis, while the other pole moves upwards and crosses the origin to become a bound state pole. The poles on the negative imaginary axis are usually called virtual state poles.[20,29] The pattern of pole motion with the parameter R shown in Fig. 1 is in full accord with the general analytic theory of s-wave scattering.[29] This result illustrates that the standard Breit-Wigner formula can be used for an exact analytic description of a resonance passing through threshold provided one includes the proper threshold energy dependence of $\Gamma(E)$ as well as the energy-dependent level shift $\Delta(E)$.

Fig. 2 illustrates the modification of the analytic properties of the S-matrix by a strong dipole potential. There is now a branch point at the origin of the k-plane and we have to cut the k-plane along the negative imaginary k-axis. The branch cut is a consequence of the non-analyticity of the dipole Jost-function (19) at k=0 and thus ultimately a consequence of the infinite range of the dipole potential. The branch cut prevents the resonance poles from reaching the imaginary axis. The poles are deflected downwards and run away into the lower half of the k-plane as shown in Fig. 2. There exists

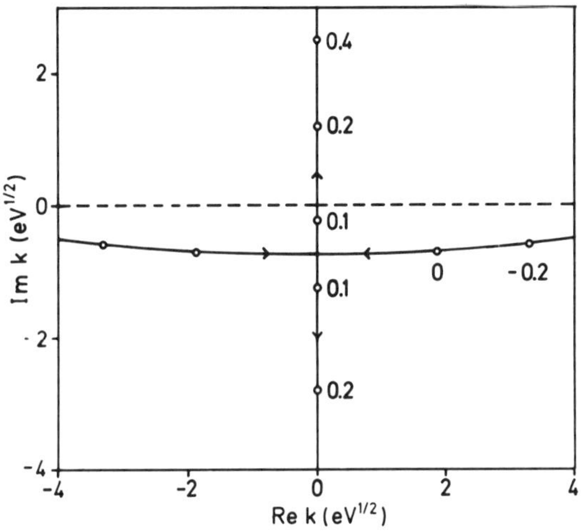

Fig. 1 Trajectories of the poles of the fixed nuclei S-matrix in the complex momentum plane for $\alpha = 1/2$ (s-wave scattering). The numbers give the values of the corresponding internuclear distance R to indicate the movement of the poles with R.

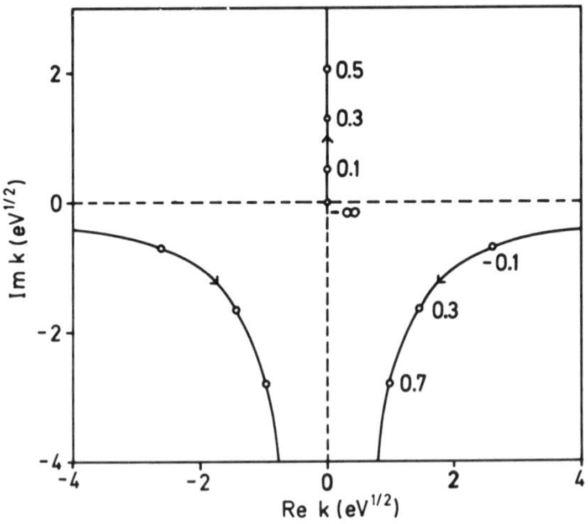

Fig. 2 Trajectories of the poles of the fixed-nuclei S-matrix in the complex momentum plane for $\alpha = 0$ (w-wave scattering and critical dipole potential).

also a bound state pole which starts at the origin for $R = -\infty$ and moves up the imaginary axis with increasing R. Note that the virtual state pole, which connects the resonance poles with the bound state pole in Fig. 1, has disappeared. The threshold peaks in the vibrational excitation functions of polar molecules can thus not be explained by a virtual state close to threshold as proposed previously.[7,8]

V. AN EXACTLY SOLVABLE MODEL

Having seen that the implementation of the threshold laws for $\Gamma(E)$ into the Breit-Wigner formula provides us with an analytically exact theory in the fixed-nuclei limit, we return now to the dynamical calculation of the cross sections. We know already that the energy dependence of Γ and Δ governs the importance of non-adiabatic effects. We expect, therefore, non-adiabatic effects to be of particular importance in the presence of long-range potentials which cause Γ and Δ to be rapidly varying functions of energy near threshold.

To evaluate the operator expressions (8) we represent the effective Hamiltonian H of eq. (7b) in the basis of the eigenstates of the target vibrational Hamiltonian $\tilde{H}_o$. The advantage of this representation is that the complicated operators $\Gamma(E - \tilde{H}_o)$ and $\Delta(E - \tilde{H}_o)$ are diagonal, independently of the form of $\Gamma(E)$ and $\Delta(E)$. The evaluation of the resolvent in eq. (8) then reduces to the inversion of a complex symmetric matrix. One can even obtain explicit formulas for the cross sections when adopting a simple model for the potential energy curves of the target and the discrete state. Choosing a harmonic potential energy curve for the target

$$V_o(R) = \frac{1}{2} V''_o (R-R_o)^2 \tag{36}$$

and taking $\varepsilon_d(R)$ as a linear function of T as already introduced in the preceding section (cf. eq. (35)), we have

$$\tilde{H}_o = -\frac{1}{2}\omega d^2/dQ^2 + \frac{1}{2}\omega Q^2 \tag{37a}$$

$$\tilde{H} = \tilde{H}_o + \varepsilon_d(R_o) + \sqrt{2}\kappa Q \tag{37b}$$

where Q is the dimensionless vibrational coordinate given by $Q = (\mu\omega)^{1/2}(R-R_o)$, R_o is the equilibrium internuclear distance of the target molecule, ω the harmonic vibrational frequency and κ is given by $(2\mu\omega)^{-1/2}\,\varepsilon'_d(R_o)$. We choose $\mu = 1$ and $\omega = 0.4$eV for the ensuing calculation.

With the simple Hamiltonians (37), H of eq. (7b) is tridiagonal in the representation of the target vibrational states. Introducing the matrix elements of the resolvent

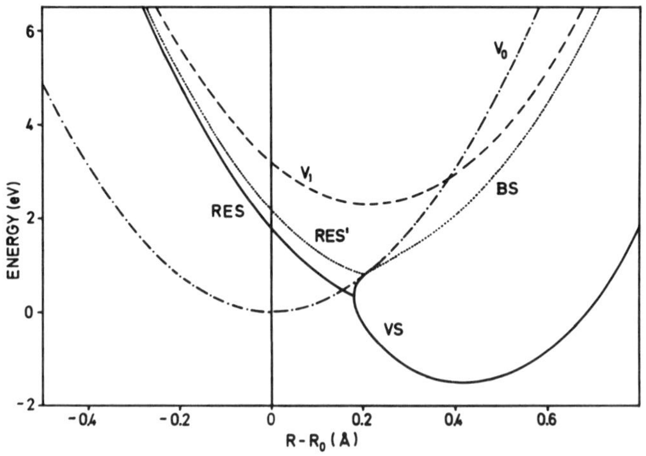

Fig. 3. Fixed-nuclei potential energy curves for $\alpha = 1/2$ (s-wave scattering). V_0 and V_1 are the potential curves of the target and the discrete state, respectively. RES and RES' represent the real part of the resonance pole of the S-matrix and the singularity of the K-matrix, respectively. BS and VS are the bound state and virtual state poles of the S-matrix.

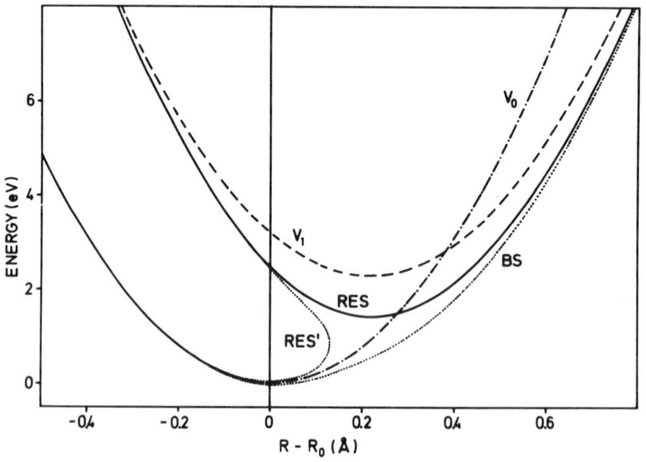

Fig. 4. Fixed-nuclei potential energy curves for $\alpha = 0$ (s-wave scattering and critical dipole potential). See caption of Fig. 3 for the explanation of the symbols.

$$R_{m,o}(E_i) = \langle m|(E_t - H)^{-1}|o\rangle, \tag{38}$$

the tridiagonality of $\langle n|H|n'\rangle$ leads to a continued fraction expression for the $R_{v,o}(E_i)$[10,30]

$$R_{v,o}(E_i) = \sqrt{v}\,\kappa R_{v-1,0}(E_i)\_/\overline{}E_i - \tilde{E}_v - (v+1)\kappa^2\_/\overline{}E_i - $$

$$- \tilde{E}_{v+1} - (v+2)\kappa^2\_/\overline{}E_i - \tilde{E}_{v+2} - \ldots \tag{39}$$

where

$$\tilde{E}_v = \varepsilon_d(R_o) + v\omega + \Delta(E_i - v\omega) - \frac{1}{2}\,i\Gamma(E_i - v\omega) \tag{40}$$

and for $v = 0$, $\sqrt{v}\kappa R_{v-1,0}(E_i)$ is defined as 1. The cross sections are finally given by

$$\sigma_v = \nu\pi/k_i^2\Gamma(E_i)\Gamma(E_i - v\omega)|R_{vo}(E_i)|^2. \tag{41}$$

Eqs. (39-41) give the exact multi-channel scattering cross sections for the model considered. For the interpretation of these cross sections it is useful to consider the fixed-nuclei potential energy curves obtained by drawing the energy of the bound state and virtual state poles and the real part of the energy of the resonance poles of the S-matrix as a function of internuclear distance together with the target potential energy curve $V_o(R)$. Fig. 3 and Fig. 4 show these potential energy curves for the two examples introduced in Sec. IV, namely pure s-wave scattering ($\alpha = 1/2$) and s-wave scattering with a critical dipole potential ($\alpha = 0$). The dashed-dotted line represents the harmonic target potential energy curve $V_o(R)$, the dashed line the discrete state potential curve $V_1(R) = V_o(R) + \varepsilon_d(R)$. The full lines give the virtual state pole (VS) and the real part of the resonance pole (RES) of the S-matrix. The dotted line gives the resonance energy determined from the singularity of the K-matrix (RES') and the bound state pole of the S-matrix (BS). Both are given by the solution of eq. (26) with $E_r' > 0$ and $E_r' < 0$, respectively. Note that the real part of the resonance pole of the S-matrix and the singularity of the K-matrix do not coincide, especially near threshold. For pure s-wave scattering the resonance pole of S moves through threshold and decays into two virtual states, one of which becomes a bound state subsequently (see Fig. 3). A strong dipole potential changes the situation profoundly, as shown by Fig. 4. Now a bound state pole and a resonance pole exist for all internuclear distances. A more detailed discussion of the potential energy curves of Fig. 4 can be found in Ref. 25.

Exact vibrational excitation cross sections corresponding to the fixed-nuclei potential curves in Figs. 3 and 4 are shown in Fig. 5 and Fig. 6, respectively, for v → 1 and v = 0 → 2 excitation. Fig. 5 represents pure s-wave scattering. The cross sections show oscillations reminiscent of the so-called "boomerang-effect",[4] but with clearly non-Lorentzian line shape due to the strong energy dependence of Γ in the s-wave case. At the opening of new channels the cross sections are additionally distorted by Wigner cusp structure. In Fig. 6, which represents resonant s-wave scattering in the presence of a strong dipole potential, the oscillations

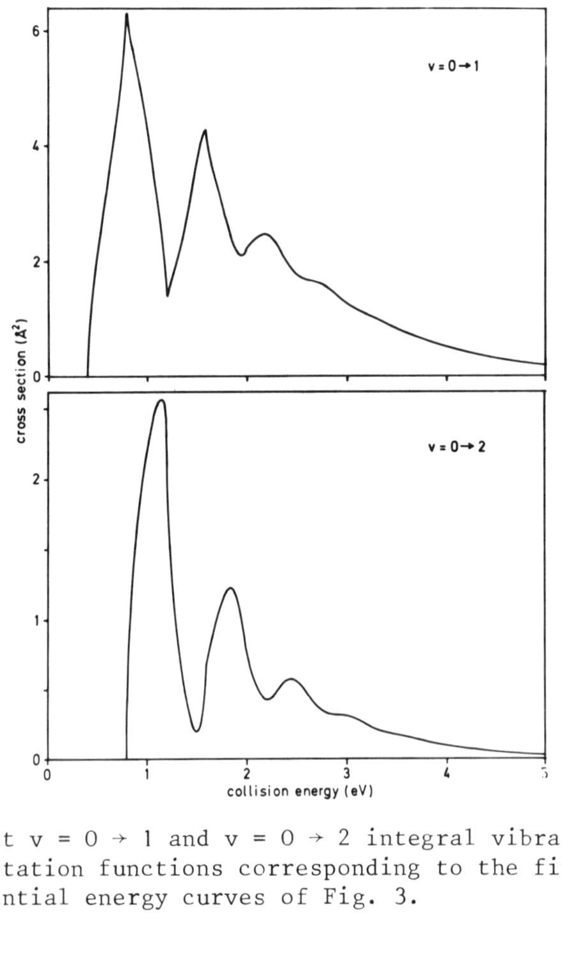

Fig. 5 Exact v = 0 → 1 and v = 0 → 2 integral vibrational excitation functions corresponding to the fixed-nuclei potential energy curves of Fig. 3.

have largely disappeared. The characteristic feature now is the steep onset and the strong peak at threshold. The cross sections in Fig. 6 resemble qualitatively the excitation functions in HCl measured by Rohr and Linder.[5] The absolute magnitude of the calculated cross sections is in good agreement with this experiment. It should be mentioned that strong threshold peaks can occur also in pure s-wave scattering[10] if the virtual state to bound state

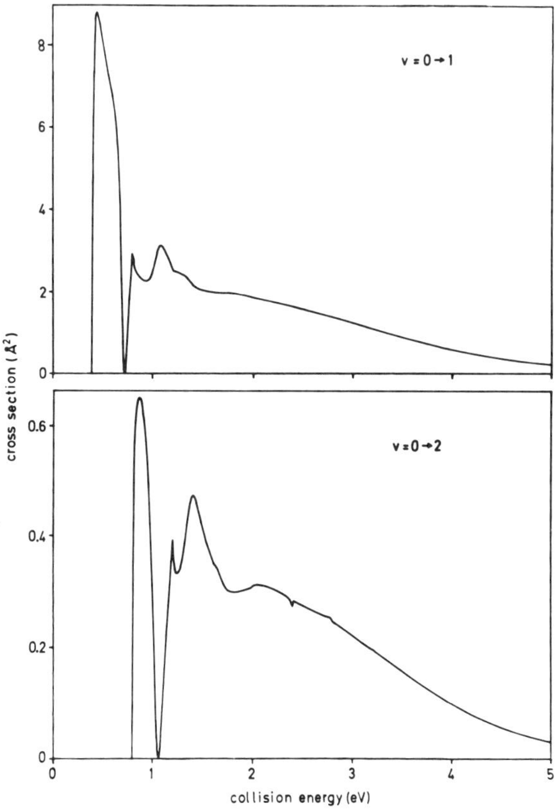

Fig. 6 Exact v = 0 → 1 and v = 0 → 2 integral vibrational excitation functions corresponding to the fixed-nuclei potential energy curves of Fig. 4.

transition (often called zero-energy resonance[20]) occurs within the Franck-Condon zone of the target molecule. With a strong dipole potential, on the other hand, threshold peaks are obtained for a wide range of the vertical position of the resonance. This may explain why threshold peaks in the vibrational excitation functions seem to be a fairly common phenomenon for polar molecules.

VI. CONCLUSIONS

We have derived an operator version of the wellknown Breit-Wigner resonance formula which includes exactly - within the propositions of the theory - the effects of nuclear motion. We have seen, furthermore, that the implementation of the threshold laws for the resonance width $\Gamma(E)$ yields an analytically exact description of resonances, virtual states and bound states in the fixed-nuclei limit. It is clear that both generalizations of the Breit-Wigner formula together provide us with an analytic multi-channel theory of low-energy resonant electron-molecule scattering including the effect of long-range forces.

We have sketched the theory in its simplest form for the sake of brevity and clarity. The approach is amenable to extension in many respects. One may include several electronic channels and consider several (possibly overlapping) electronic resonances. The resonance width will in general be a function of both $E_t - \hat{H}_o$ and R, which complicates the evaluation of the operator formula (7a). A dependence of the long range potential W on the internuclear distance renders the determination of the threshold onset of $\Gamma(E)$ a nontrivial problem. Finally one should include the rotational degrees of freedom for the proper description of anisotropic electron-molecule interactions, in particular the long-range dipole potential.

Sufficiently generalized, the operator Breit-Wigner formula should be a useful tool to convert <u>ab initio</u> calculated fixed-nuclei electron-molecule scattering information into the desired multichannel vibrational excitation cross sections. From the fixed-nuclei phase shift as a function of energy and internuclear distance one can extract, in principle, the functions $\varepsilon_d(R)$ and $\Gamma(E,R)$ entering the dynamical calculation. The feasibility of this approach for realistic systems remains to be tested.

REFERENCES

1. G.J. Schulz, Rev. Mod. Phys. 45:423 (1973).
2. N.F. Lane, Rev. Mod. Phys. 52:29 (1980).
3. J.N. Bardsley, J. Phys. B 1:349 (1968).
4. D.T. Birtwistle and A. Herzenberg, J. Phys. B4:53 (1971).
5. K. Rohr and F. Linder, J. Phys. B9:2521 (1976); K. Rohr, XI. ICPEAC Sat. Meeting, Symp. on Electron-Molecule Collisions, invited papers (Tokyo 1979).
6. H.S. Taylor, E. Goldstein and G.A. Segal, J. Phys. B11:2253 (1977).
7. R.K. Nesbet, J. Phys. B10:L739 (1977).
8. L. Dubé and A. Herzenberg, Phys. Rev. Lett. 38:820 (1977).
9. F.A. Gianturco and N.K. Rahman, Chem. Phys. Lett. 48:380 (1977).
10. W. Domcke, L.S. Cederbaum and F. Kaspar, J. Phys. B12:L359 (1979).
11. P.A.M. Dirac, Z. Physik 44:585 (1927).
12. V. Weisskopf, Ann. Physik 9:23 (1931).
13. H. Feshbach, Ann. Physics 5:357 (1958).
14. U. Fano, Phys. Rev. 124:1866 (1961).
15. W. Domcke and L.S. Cederbaum, Phys. Rev. A 16:1465 (1977).
16. M.J. Seaton, Proc. Phys. Soc. London 88:801 (1966).
17. C. Greene, U. Fano and G. Strinati, Phys. Rev A 19:1485 (1979).
18. M.S. Child in "Atom-Molecule Collision Theory", ed. R.B. Bernstein, Plenum Press, New York, 1979.
19. H.C. Longuet-Higgins, Adv. Spectrosc. 2:429 (1961).
20. J.R. Taylor, "Scattering Theory", Wiley, New York, 1972.
21. E.P. Wigner, Phys. Rev. 73:1002 (1948).
22. T.F. O'Malley, L. Spruch and L. Rosenberg, J. Math. Phys. 2:491 (1961).
23. T.F. O'Malley, Phys. Rev. 137: A 1668 (1965).
24. W.R. Garrett, Phys. Rev. A3:961 (1971).
25. W. Domcke and L.S. Cederbaum, J. Phys. B, 14:149 (1981).
26. O.H. Crawford, Proc. Phys. Soc. 91:279 (1967).
27. M. Abramowitz and I.A. Stegun, "Handbook of Mathematical Functions", Dover, New York, 1965.
28. W. Domcke, to be published.
29. R.G. Newton, "Scattering Theory of Waves and Particles", Mc-Graw-Hill, New York, 1966, Sect. 12.
30. W. Domcke and L.S. Cederbaum, J. Phys. B 13:2829 (1980).

ELECTRON IMPACT IONIZATION OF POSITIVE IONS

H. Jakubowicz and D. L. Moores

Dept. of Physics and Astronomy
University College London
Gower Street
London, WC1E 6BT

INTRODUCTION

Electron impact ionization cross-sections are urgently required for the study of both laboratory and astrophysical plasmas. Furthermore, it is important that the values used are reliable since the use of ionization balance curves based on different approximate cross-sections may lead to conflicting conclusions about the physical properties of the plasma being studied (Cheng et al. (1979)). Unfortunately, ionization remains one of the outstanding problems in atomic physics and as yet there is no satisfactory procedure for describing the process (reviews of the theory of ionization can be found in Rudge (1968) and Peterkop (1977)). The source of the difficulties is the long-range nature of the Coulomb potential which ensures that the two continuum electrons continue to interact with the residual ion and each other even out at infinity. A complete treatment of the ionization process therefore requires a full solution of the three-body problem in the asymptotic region. In practical calculations it is usual to neglect the correlation between the continuum electrons and hence errors are introduced into the results through the use of an approximation to the true ionization amplitude; this is further compounded by the resulting need to approximate exchange between the continuum electrons. Finally, it is necessary to approximate the target states for complex ions and further errors may be introduced through the use of inaccurate wavefunctions for the target. The most commonly used quantum mechanical approximation which has been applied in the calculation of ionization cross-sections for positive ions is the Coulomb-Born approximation in which the ionization process is depicted as a bound-free transition provoked by electron impact; its validity and that of its various exchange derivatives has been discussed by Jakubowicz and Moores (1980). The essence of the argument used in support of the Coulomb-Born approximation for positive ions is that as the charge of the target ion increases the correlation between the continuum electrons becomes less important and the Hamiltonian is dominated by the potential due to the screened charge of the nucleus and the kinetic energy of the ionizing electron. It can thus be expected that the true ionization amplitude is well represented by the Coulomb-Born approximation and hence the main source of uncertainty lies in the methods used to include exchange. The results of the Coulomb-Born exchange approximation for ionization of He^+ (Rudge and Schwartz (1966)) are, however, in excellent agreement with experiment (Dolder, Harrison and Thonemann (1961) and Peart, Walton and Dolder (1969)) and it is expected that the accuracy of the approximation will improve with increasing charge of the ion. The Coulomb-Born exchange approximation may therefore be an adequate method for caculating ionization cross-sections for positive ions provided that accurate wavefunctions are used to describe the target ion.

In this paper we will describe a method of calculating ionization cross-sections for complex positive ions using close-coupling wavefunctions to describe the states of the (N+1)-electron target ion and the N-electron residual ion plus ejected electron within the Coulomb-Born approximation (Jakubowicz and Moores (1980)). The close-coupling method is capable of producing accurate wavefunctions both in the case of bound states and for continuum states when the energy of the continuum electron is small. Close-coupling wavefunctions are therefore particularly appropriate for the ionization process since it is known that the ejected electron moves much more slowly than the ionizing electron in the majority of ionization events for incident electron energies away from threshold. This form of Coulomb-Born approximation enables the structural effects of inner-shell ionization, simultaneous excitation and ionization together with autoionization to be accounted for directly in the calculation of the ionization cross-section by including appropriate N-electron target states in the close-coupling expansion. These additional processes can give rise to significant changes in the ionization cross-section (Moores and Nussbaumer (1970)) but until now have only been accounted for in theoretical calculations by the addition of the appropriate cross-sections, obtained from independent calculations, to the naked cross-section describing the knockout process alone. Finally, by using accurate target wavefunctions it is possible to more clearly assess the limitations inherent in the use of the Coulomb-Born approximation and the exchange approximations by comparison of the theoretical results with experimental data. Results will be presented for a variety of beryllium-like and lithium-like ions including, for the first time, the effects of autoionization directly in the calculation of the cross-sections for the lithium-like ions.

THEORY

The Coulomb-Born approximation has been reviewed by Rudge (1968) and Jakubowicz and Moores (1980). We will consider three distinct exchange approximations: the Coulomb-Born approximation (CB), the Coulomb-Born no exchange approximation (CBOX) and the Coulomb-Born exchange approximation (CBX). In the Coulomb-Born approximation (CB), which is also sometimes referred to as the 'full range' Coulomb-Born approximation, exchange is ignored from the outset and the two continuum electrons are treated as being distinguishable. In the resulting expression for the total ionization cross-section the integration over ejected electron energies is over the complete energy range; however, because the magnitude of the exact exchange amplitude describing exchange in the continuum is related to the direct amplitude by $g(\underline{k},\underline{\chi})=f(\underline{\chi},\underline{k})$, where $\underline{k}$ is the momentum of the ionizing electron and $\underline{\chi}$ the momentum of the ejected electron, CB has a contribution from events which can only occur by exchange (for which $\chi>k$). Thus, although CB pretends to ignore exchange, it has a contribution from events which can occur by exchange alone and in practice it is found to severly overestimate the cross-section. A better and more consistent approximation is the Coulomb-Born no exchange approximation (CBOX) in which the exchange terms are set equal to zero and the resulting integration over ejected electron energies in the total cross-section only carried out over half the complete energy range. This is equivalent to summing only over events in which the ejected electron is the slower. CBOX is also sometimes referred to as the 'half-range' Coulomb-Born approximation or the 'modified' Coulomb-Born approximation. Formally CBOX corresponds to the Coulomb-Born no exchange approximation to electron impact excitation of ions and is equivalent to putting the exchange amplitude equal to zero. Whereas this is physically unsatisfactory, the exchange and interference terms in the total cross-section tend to cancel each other and hence CBOX can be expected to give reliable estimates of cross-sections. This is perhaps most easily understood by noting that the exchange amplitude attains its maximum value when it is identical to the direct amplitude, which is when the continuum electrons have equal and opposite momenta. In this case the exchange and interference terms cancel and an expression corresponding to CBOX is obtained. CBOX is also particularly attractive because it is relatively easy to calculate. The major problem with CBOX is that, because it does not completely account for the full range of possible ejected electron energies, it cannot correctly describe the thresholds for processes such as inner-shell excitation and autoionization which lie above the energy required for ionization of the ground state. These contribute only through the exchange amplitude at first and do not therefore appear in CBOX until twice their threshold energies. This point is of particular relevance in the case of ionization of lithium-like ions (see below). Approximations which attempt to account for the interference between the direct and exchange ionization processes are known as 'Coulomb-Born exchange' approximations (CBX).

The exchange amplitude can be written as

$$g_{CB}(\underline{k},\chi) = | f_{CB}(\chi,\underline{k}) | e^{i\delta(\underline{k},\chi)} \quad (1)$$

where $\delta(\underline{k},\chi)$ is the relative phase of the direct and exchange amplitudes. In order to evaluate the interference term in the Coulomb-Born approximation it is necessary to make some choice for the phase $\delta(\underline{k},\chi)$. In this work we adopt an approximation which is such that, in the limit of an infinite nuclear charge, the interference term is a maximum. There is no rigorous justification for this or any other choice of phase but, as in the argument used above in the discussion of CBOX, it can be seen that exchange is most significant when the exchange and interference terms cancel, which is precisely when the interference is a maximum. In addition, this form of CBX gives the correct result for the case of an infinitely charged nucleus and, because it sets a lower limit to the ionization cross-section, enables the importance of exchange to be assessed quantitatively by comparison with CB and CBOX results. Previous calculations which have used this type of approximation to the phase include those of Rudge and Schwartz (1966a and b), Burgess and Rudge (1963) and Younger (1980a and b). Clearly CB and CBOX ionization cross-sections are automatically generated in the calculation of a CBX cross-section. All of the Coulomb-Born approximations, CB, CBOX and CBX, can be modified to allow for screening effects by using distorted Coulomb waves in place of Coulomb waves; these approximations will be distinguished by adding the prefix 'D' to the appropriate abbreviation.

In this work the (N+1)-electron system both before and after the ionizing collision will be described by close-coupling wavefunctions of the form discussed by Seaton (1974) and Jones (1974) and are calculated using the program IMPACT (Crees, Seaton and Wilson (1977)). The wavefunction of the initial state, Ψ_o, is given by

$$\Psi_o(\alpha_o S_o L_o M_{S_o} M_{L_o} \pi | \underline{x}_1 \ldots \underline{x}_{N+1}) = \sum_{i=1}^{N_c} \Theta_i + \sum_{j=1}^{N_b} c_j \Phi_j \quad (2)$$

where S, L, M_S, M_L are the spin and orbital angular momentum quantum numbers, π is the parity, α describes all other quantum numbers of the target and $\underline{x}_i$ describes the space and spin coordinates of the i-th electron. Eq (2) is a linear combination of N_c free channel functions Θ_i, which are antisymmetrised vector-coupled products of N-electron wavefunctions (which may contain configuration interaction) and one-electron functions $\theta_i(\underline{x}_k)$, together with N_b bound channel functions Φ_j, which are (N+1)-electron single configuration bound state functions constructed from the N-electron orbitals. The bound channels are introduced in order to satisfy orthogonality requirements (Burke and Seaton (1971)) but also allow short-range correlation effects to be accounted for. Their inclusion is of particular importance in the present context since they enable configuration mixing (and in the case of the final stae, autoionization) to be included even in a single free channel approximation. If the Coulomb-Born excitation amplitude were being calculated rather than the Coulomb-Born ionization amplitude, the final target state wavefunctions, Ψ_f, would also be of the form of Eq (2). In the case of ionization, however, the final state of the (N+1)-electron system is not a stationary state but is represented by

$$\Psi_f(\underline{x}_1 \ldots \underline{x}_{N+1}) = \sum_{\substack{LSM_LM_S \\ \ell_\gamma m_\gamma m_{s_\gamma}}} (-1)^{\ell_\gamma - L_\gamma - M_L + \frac{1}{2} - S_\gamma - M_S} \frac{2\pi}{\sqrt{\chi_\gamma}}\, i^{\ell_\gamma + 1}\, Y^*_{\ell_\gamma m_\gamma}(\hat{\chi}_\gamma)$$

$$\times \sqrt{(2L+1)(2S+1)} \begin{pmatrix} L_\gamma & \ell_\gamma & L \\ M_\gamma & m_\gamma & -M_L \end{pmatrix} \begin{pmatrix} S_\gamma & \frac{1}{2} & S \\ M_{S_\gamma} & m_{s_\gamma} & -M_S \end{pmatrix}$$

$$\times\ \Psi_\gamma\,(\alpha_\gamma LSM_LM_S\pi \,|\, \underline{x}_1 \ldots \underline{x}_{N+1}) \qquad (3)$$

where the wavefunctions Ψ_γ are of the form of Eq (2) and satisfy S-matrix boundary conditions. The sums over quantum numbers with subscripts γ implies a sum over all channels which are open at the energy considered for each L, S and π, which in turn implies a sum over all linearly independent solutions Ψ_γ for each LS. It is convenient to regard the functions Ψ_o and Ψ_γ as solution vectors of dimension N_c+N_b with components defined by the individual terms in Eq (2). The dimensionality of the solution vectors is then a function of L, S and π and depends on the states of the N-electron target which are used to generate the free channels. Since we are interested in cross-sections for a fixed initial electron energy and final state, it is only necessary to include in Eq (3) those linearly independent solutions which correspond to outgoing waves in channels for which the residual ion is in a state with energy E_γ. The other linearly independent solutions, corresponding to different states γ' with energy $E_{\gamma'}$ contribute to a different cross-section to that which is required and hence need not be considered. The omitted solutions could be used if desired to calculate cross-sections for processes in which the system is left in the state γ', corresponding to a different ejected electron energy.

Simplified expressions for the ionization cross-section in the various exchange approximations can be obtained by carrying out multipole expansions of the interelectronic potentials, coupling to total orbital and spin angular momenta, integrating over angular and spin variables and summing over the total spin angular momentum quantum numbers. The resulting expressions involve the reduced matrix elements $\langle S_oL_o \| T^\lambda \| S_oL\chi_\gamma \rangle$ where

$$T^{\lambda}_{\mu} = \frac{\sqrt{4\pi}}{\sqrt{2\lambda+1}} \int_0^{\infty} F_{\ell_0}(k_0, Z-N-1, r_{N+2}) \sum_{i=1}^{N+1} Y_{\lambda\mu}(\hat{r}_i) \frac{r^{\lambda}_{i<}}{r^{\lambda+1}_{i>}}$$
$$\times F_{\ell_1}(k, Z-N-1, r_{N+2})\, dr_{N+2} \tag{4}$$

is an irreducible tensor and the functions F are Coulomb or distorted wave radial wavefunctions. These reduced matrix elements are evaluated by a general computer program, COBION, which is similar in structure to the program PHOTUC which calculates photoionization cross-sections (Saraph (1980)). In practice the close-coupling wavefunctions are calculated at fixed values of χ_γ and k for a given k_o and hence the integral over the ejected electron energies can only be evaluated when the integrand has been evaluated at a sufficient number of energies. Because these calculations are expensive, even for the evaluation of a single point on the integration mesh which requires the inclusion of sufficient total angular momenta, L, S, and partial waves to bring the integrals to convergence, an efficient means of integrating must be devised. In this work we have chosen to use an N-point Gaussian quadrature to evaluate the integrand in all cases where the integrand is slowly and smoothly varying. This is equivalent to fitting the integrand to a (2N+1) order polynomial but only requires N values of the integrand. At low energies (<3 times the ionization energy) a three-point Gaussian was found to be sufficient while at the highest energies a six-point Gaussian was used. This ensures that the resulting cross-sections are in error by no more than 2% through the use of this numerical integration. This is not important since the input data to IMPACT are chosen so as to give a final accuracy of this amount.

When autoionization appears in the differential cross-section the Gaussian method is no longer appropriate. Furthermore, because of the expense of these calculations it is not practical to map out the structure in detail and hence it is useful to employ analytical techniques which enable the structure to be calculated with a minimum of effort and cost. The general technique for doing this has been described by Jakubowicz (1980) and is simply an extension of the application of quantum defect theory in the calculation of structures in photoionization cross-sections (Dubau and Wells (1973) and Pradhan and Seaton (1980)). Because, however, the results given here are obtained using only single channel wavefunctions, for which the more general theory is not required, we will now describe a simple method for calculating the autoionization effects, which in this case are due to the mixing of open and bound channels rather than the mixing of open and closed channels.

AUTOIONIZATION

In order to describe the behaviour of the ionization cross-section when autoionization is present it is necessary to consider the analytic behaviour of the reduced matrix elements

$$<\alpha_o L_o S_o \| T^\lambda \| L S_o \gamma(\chi_\gamma) > \tag{5}$$

as a function of the ejected electron energy. When multichannnel close-coupling wavefunctions are used this problem is computationally much more involved than the equivalent problem in photoionization because the operator T^λ is a function of the incident electron energy and is also a (slowly varying) function of the ejected electron energy. Fortunately, the variation due to the different ejected electron energies in the T^λ can be neglected over small energy ranges and the dominant energy variation results from the resonant behaviour of the continuum close-coupling wavefunctions. The analytic development must, however, be carried out separately for each incident energy and, because partial wave expansions are used, separately for each combination of partial waves contributing to a given resonant channel. Wavefunctions with values of $SL\pi$ which are not resonant can be treated as constant in the resonance region.

For single channel wavefunctions

$$\Psi = \Theta + \sum_j c_j \Phi_j \tag{6}$$

and any resonant behaviour results from the mixing of the free channel and the bound channels. Since, by definition, the bound channels are held constant, a considerable simplification is possible, the mixing manifesting itself in the variation of the coefficients c_j and the free channel Θ only. Since the matrix elements of Eq (5) become

$$<\alpha_o L_o S_o \| T^\lambda \| L S_o \gamma(\chi_\gamma)> = <\alpha_o L_o S_o \| T^\lambda \| \Theta> + \sum_j <\alpha_o L_o S_o \| T^\lambda \| \Phi_j> c_j \tag{7}$$

then, in order to obtain the analytic properties of the matrix element, it is only necessary to obtain the variation of the free channel matrix elements and the coefficients c_j (the bound channel matrix elements are assumed to be constant since the operators are slowly varying). It follows from quantum defect theory (Seaton (1978)) that, if the wavefunction is normalized such that

$$F \underset{r\to\infty}{\sim} \sin \xi + \cos \xi \; R \tag{8}$$

where R is the reactance matrix and is a single number for the single channel case, then the free channel matrix element, the coefficients c_j and R can be expanded in the form

$$\sum_{i=o}^{m} \alpha_i E^i + \sum_{j=1}^{n_p} \frac{\beta_j}{E-E_j} \qquad (9)$$

where E_j is the energy of the j-th of the n_p poles at which the resonances occur.

If it is further assumed that the resonant behaviour is dominated by the bound channel terms, since it is the bound channels which are causing the resonances, then a further simplification is possible because the resonant behaviour of all the matrix elements is determined by that of the coefficients c_j alone. This is equivalent to assuming that the coefficients β_j in an expansion of the free channel matrix element in the form of Eq (9) are much smaller than those for an expansion of the bound channel terms in Eq (7) and is valid for all the cases considered here. Thus, by fitting the coefficents c_j and R to expressions of the form of Eq (9), the energy variation of all the matrix elements in a given resonant channel is known at all incident energies and it is only necessary to calculate the matrix elements using wavefunctions normalized according to Eq (8) at a single point close to the resonance for each incident energy required. This technique will be discussed further when results are presented for lithium-like ions.

RESULTS FOR BERYLLIUM-LIKE IONS

The results of calculations of ionization cross-sections for the beryllium-like ions presented here are of the simplest kind because they do not include any effects due to autoionization. The states which were included as close-coupling wavefunctions were of the form ^1L, with L taking values from zero to eight, and consisted simply of a single free channel of configuration $1s^22sk\ell$ (ℓ=L) except in the case of the ^1S where additional bound channels of configuration $1s^22s^2$, $1s^22s3s$ and $1s^23s^2$ were included. The bound channel of configuration $1s^23s^2$ does in fact give rise to autoionization but its effect was found to be very small and was consequently neglected. The ground state wavefunction was of the same form as the ^1S continuum wavefunction. Partial waves up to ℓ=17 were included to describe the ionizing electron. Since autoionization is unimportant in this case, the Gaussian method was used to evaluate the complete integrals over ejected electron energies. It is expected that the numerical accuracy is limited only by the use of the Gaussian method and should be better than 2%, particularly at the lower energies. Results have been obtained for the ions C^{2+}, N^{3+}, O^{4+}, Ne^{6+} and Fe^{22+} in the DCBX, CBX, DCBOX, CBOX, DCB and CB approximations. The distorted wave results for C^{2+}, N^{3+} and O^{4+} are compared with experimental data in Figures 1, 2 and 3. The general form of the results is the same in all cases: CB is not a good approximation and, as discussed above, severly overestimates the cross-section. CBOX appears to be quite adequate and in

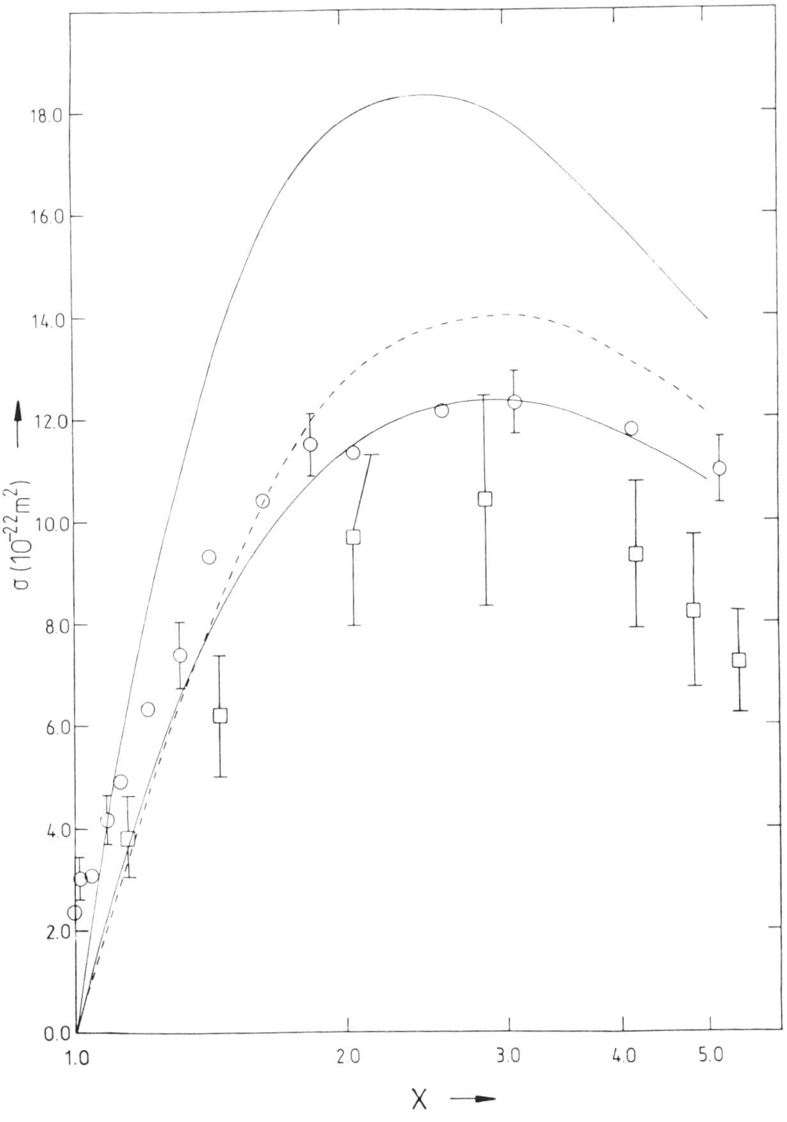

FIGURE 1

Ionization cross-section of C^{2+}. - - - DCBOX, —— (upper) DCB, —— (lower) DCBX. The ionization energy calculated by IMPACT was 3.373 Ryd. The circles are the crossed-beam data of Woodruff et al. (1978) and the squares trapped-ion data of Hamdan et al. (1978).

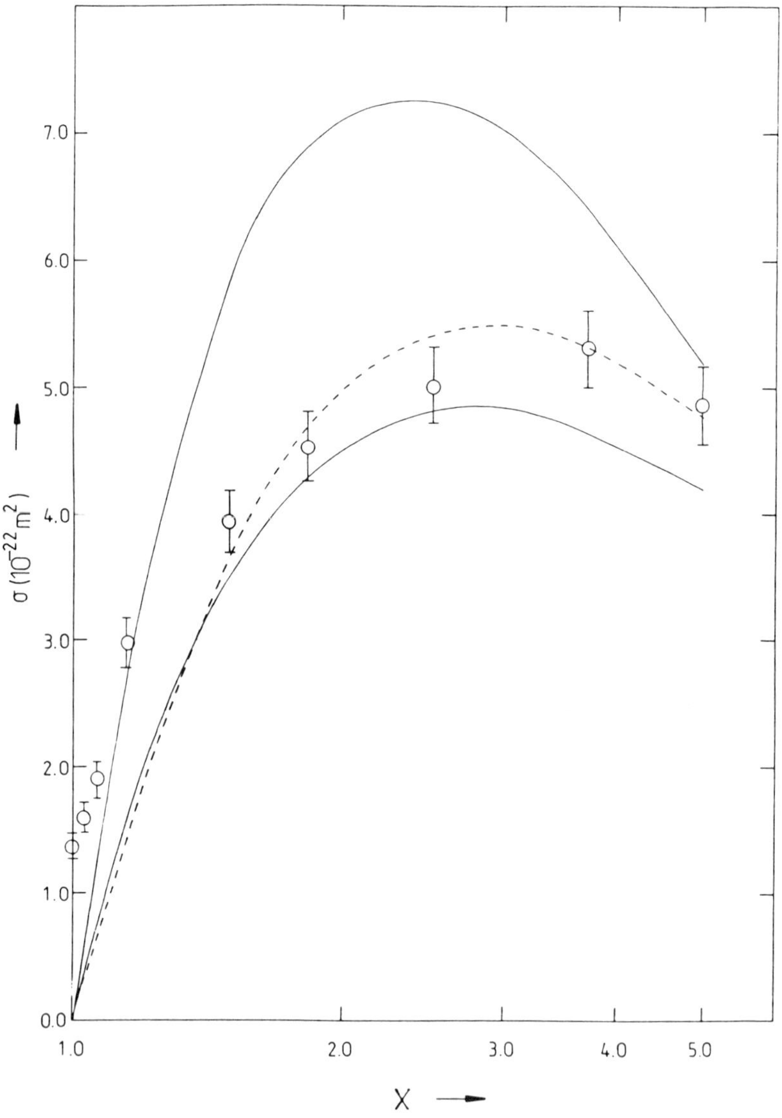

FIGURE 2

Ionization cross-section of N^{3+}. - - - DCBOX, —— (upper) DCB, —— (lower) DCBX. The ionization energy calculated by IMPACT was 5.519 Ryd. The experimental points are the crossed-beam data of Crandall, Phaneuf and Gregory (1979).

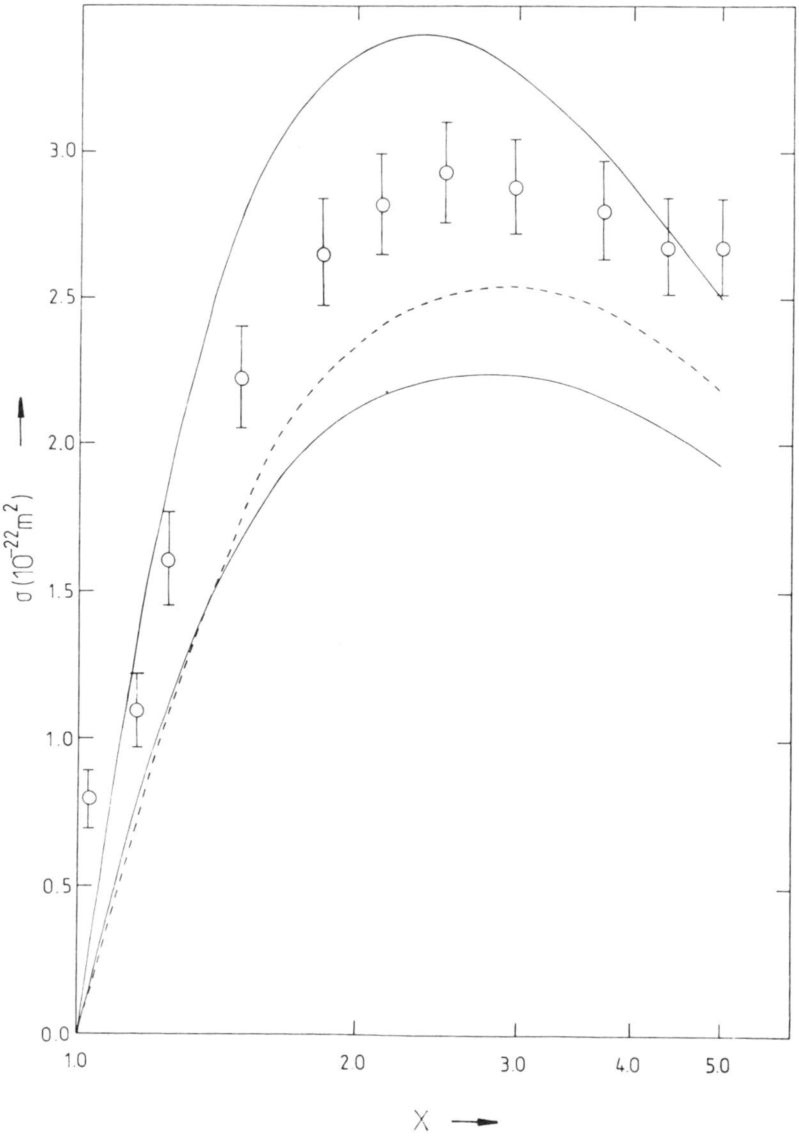

FIGURE 3

Ionization cross-section of O^{4+}. - - - DCBOX, —— (upper) DCB, —— (lower) DCBX. The ionization energy calculated by IMPACT was 8.167 Ryd. The experimental points are the crossed-beam data of Crandall, Phaneuf and Gregory (1979).

these cases is very similar to the results of the Lötz formula (Lötz (1967)). CBX, which is expected to be the best approximation in this group, yields results which are similar to those given by the infinite-Z method of Golden and Sampson (1977), indicating that the Golden and Sampson formula gives a reasonable approximation to the true Coulomb-Born exchange results for finite nuclear charge. The fact that the results of these two methods are virtually the same in the case of Fe^{22+} also indicates that the calculations are behaving correctly in the limit of large nuclear charges (it should be noted, however, that relativistic effects may be important for this ion and are not included here). Differences between the CBOX results obtained here and those of Moores (1978) on C^{2+} and N^{3+} can only be due to the different approximations which are made for the target states (Moores uses distorted waves to describe the target) and emphasise the importance of using accurate target states for complex ions even though the states used here are not as accurate as might be desired (see below). The effect of allowing for distortion and exchange in the wavefunctions describing the ionizing electron before and after the collision by using a Thomas-Fermi-Dirac-Amaldi statistical model potential (Eissner and Nussbaumer (1969)) is to decrease the cross-section from its udistorted value in all cases. The effect is most marked for the lower charged ions and at low energies where it is expected that the effect of the interelectronic potentials in the total Hamiltonian of the system is relatively more important. This is most pronounced in the exchange and interference terms since these include wavefunctions which describe the ionizing electron as moving more slowly than the ejected electron. In particular, the interference term is affected most of all because the reduction in magnitude of the exchange amplitude is magnified by the larger direct amplitude. As a result the DCBX and CBX results differ from each other by much more than either the DCBOX and CBOX or the DCB and CB results. These conclusions are also borne out by the results of Younger (1980a) on helium-like and lithium-like ions.

A measure of the quality of the close-coupling wavefunctions used can be obtained by comparing the value of the ionization energy calculated by IMPACT with the experimental value. In all cases the calculated value is lower than the observed value because the state of the three-electron ion ($1s^2 2s\ ^2S$) which is coupled to the extra electron to generate the wavefunction has an artificially high energy in the single state approximation. This means that configuration mixing, in this case predominantly from the $1s^2 2p^2\ ^1S$ state, which has been neglected in these calculations, is important for these systems and the $1s^2 2p\ ^2P$ configuration should be included to generate accurate wavefunctions. This has not been attempted for the present because of the complexity of treating the resonances resulting from the mixing of the channels when the higher channel is closed.

The comparison with experimental data is complicated by the fact that in all cases the absolute crossed-beam data (Woodruff et al.

(1978) for C^{2+} and Crandall, Phaneuf and Gregory (1979) for N^{3+} and O^{4+}) were obtained using beams contaminated with metastable components consisting of $1s^2 2s2p\ ^3P$ states which contribute to the measured cross-sections in unknown amounts; this accounts for the finite values of the measured cross-sections at threshold. In their experiments on N^{3+} and O^{4+} Crandall et al. estimated that the fraction of ions initially in the ^3P metastable state was about 50%. When a comparison is made between their results and theoretical results calculated for a 50-50 mixture of ground and metastable states, they obtain best agreement with the Lötz formula (within 10%). Away from threshold the metastable contributions should, however, be less important, although without rigorous measurements or calculations of the cross-sections it is not possible to say with certainty whether this is the case. With this caveat it appears that the results which have the most sound foundation in theory, DCBX, give an excellent representation of the crossed-beam data for C^{2+} above about twice the threshold for ionization of the ground state, where the metastable contribution should have ceased to be important. For N^{3+} and O^{4+} the agreement becomes progressively worse and it appears that DCBOX (or CBOX) and the method of Lötz give the best agreement. This is disturbing because the DCBX results are expected to improve with increasing charge of the ion and because the experimental cross-sections are not scaling as theory would predict (it should be noted that the Lötz formula depends on this scaling for its validity). It is not possible to resolve these discrepancies at present and further theoretical and experimental work is required on these ions. The most obvious shortcomings of the results presented here is the neglect of configuration interaction in the close-coupling wavefunction. On the other hand, although the trapped-ion measurement of Hamdan, Birkinshaw and Hasted (1978) is not absolute and must also be viewed with caution, it does not have any metastable contribution and may indicate that, for C^{2+} at least, the crossed-beam results are too high.

RESULTS FOR LITHIUM-LIKE IONS

The undistorted Coulomb wave results for C^{3+}, N^{4+} and O^{5+} are plotted together with the experimental data of Crandall et al. (1979) in Figures 4, 5 and 6 respectively. These results were obtained using close-coupling wavefunctions of the form ^2L with L taking values from zero to eight which were formed by coupling the configuration $1s^2$ to the added electron to give a single free channel of configuration $1s^2k\ell$ (ℓ=L) together with bound channels of configuration $1s^2 2s$, $1s2s^2$, $1s2p^2$, $2s2p^2$ in the ^2S state and $1s^2 2p$, $1s2s(^1S)2p$, $1s2s(^3S)2p$, $2s^2 2p$, $2p^3$ in the ^2P state. All of the bound channels except the $1s^2 2s$ and $1s^2 2p$ states give rise to autoionization but the states $2s2p^2$, $2s^2 2p$ and $2p^3$ do not contribute at the energies of interest here. The autoionization can be seen as large increases in the measured ionization cross-section in the CB and CBX cross-sections at about four times the ionization threshold. The ground state wavefunctions were of the same form as the ^2S continuum states. Below the autoionization threshold

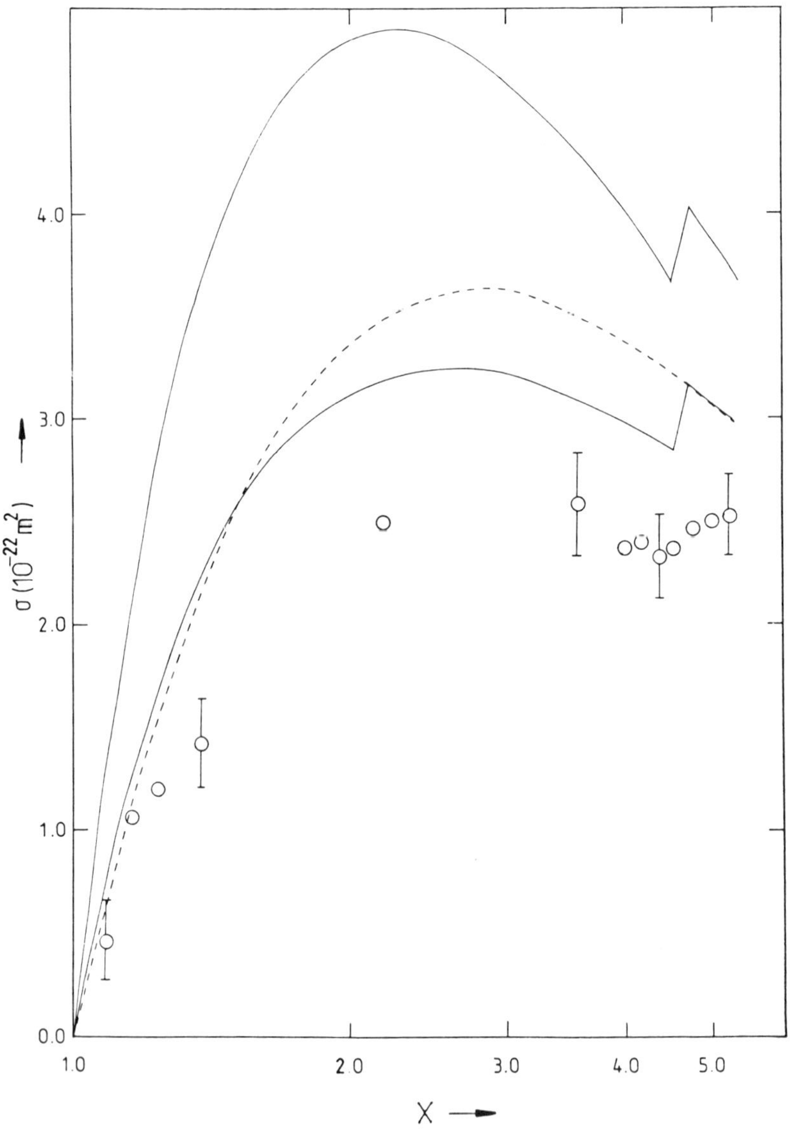

FIGURE 4

Ionization cross-section of C^{3+}. - - - CBOX, ——— (upper) CB, ——— (lower) CBX. The ionization energy calculated by IMPACT was 4.730 Ryd. The experimental points are the crossed-beam data of Crandall et al. (1979).

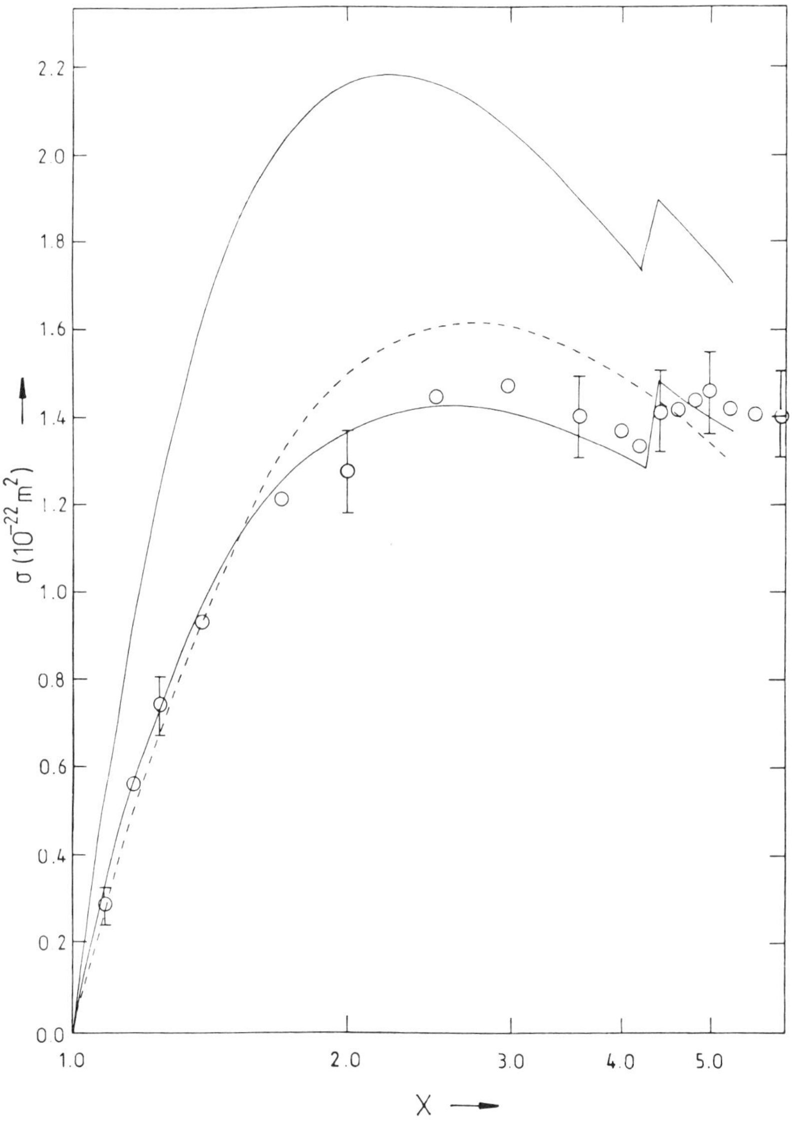

FIGURE 5

Ionization cross-section of N^{4+}. - - - CBOX, —— (upper) CB, —— (lower) CBX. The ionization energy calculated by IMPACT was 7.195 Ryd. The experimental points are the crossed-beam data of Crandall et al. (1979).

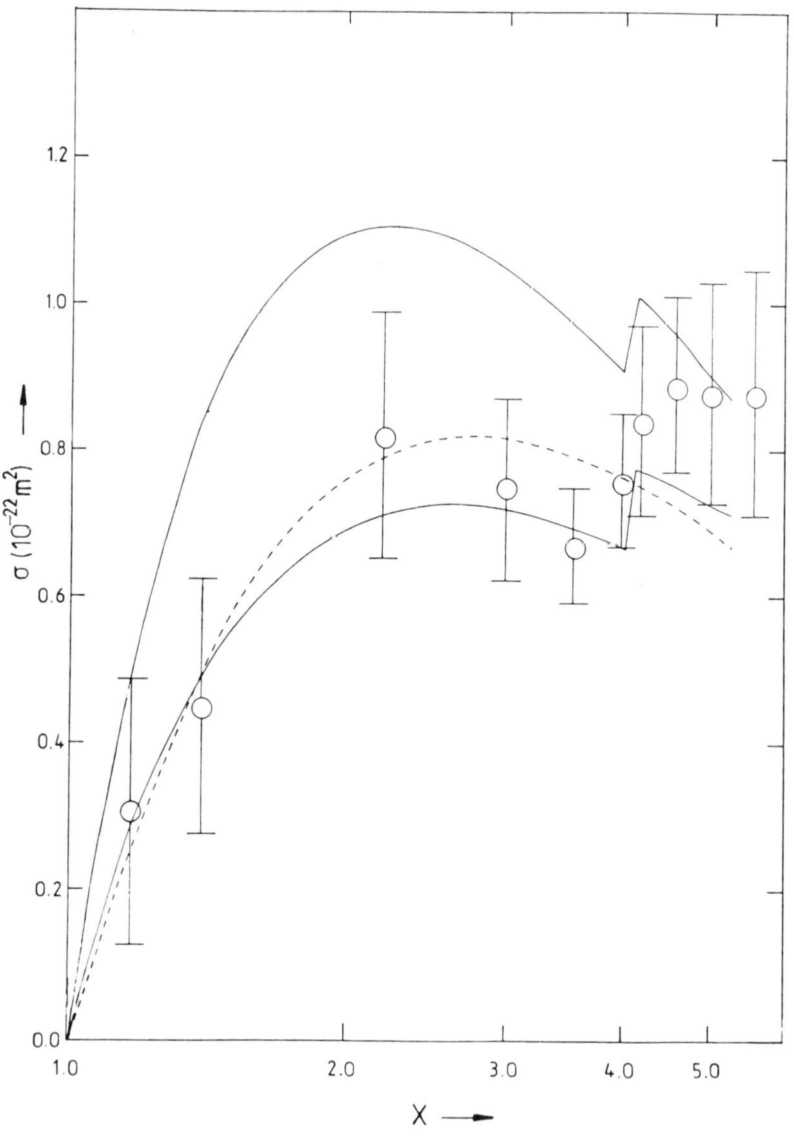

FIGURE 6

Ionization cross-section of O^{5+}. - - - CBOX, —— (upper) CB, —— (lower) CBX. The ionization energy calculated by IMPACT was 10.152 Ryd. The experimental points are the crossed-beam data of Crandall et al. (1979).

the differential cross-section is a slowly and smoothly varying function of the ejected electron energy and the integration over the ejected electron energies was performed using the Gaussian method. When the autoionization threshold is crossed sharp peaks appear in the differential cross-section as the flux increases through the mixing of the bound channels with the continuum. This is illustrated in Figure 7 where the calculated differential cross-section for Ne^{7+} at the incident energy of five times the ionization energy is plotted as a function of the ejected electron energy in the regions where the $1s2s(^1S)2p\ ^2P$, $1s2s(^3S)2p\ ^2P$ and $1s2s^2\ ^2S$ states give rise to autoionization. In the case of the $1s2s(^1S)2p\ ^2P$ state an increase of four orders of magnitude is found. The curves in Figure 7 were generated using the technique described above. The R matrices and c_j coefficients for the ^2S and ^2P channels were obtained at a series of energies about the resonances and fitted to expressions of the form of Eq (9). The reduced matrices were then calculated at a point close to the resonance of interest and hence, using the known variation of the R matrix and c_j coefficients, the structures could be generated. In the case of Figure 7, the $1s2s^2\ ^2S$ resonance was generated using data calculated at 48.5 Ryd while both the ^2P structures were calculated using data obtained at 49.5 Ryd. It can be seen that the method accurately reproduces the peaks in the differential cross-section both in height and shape. There are small discrepancies away from the resonances due to the neglect of variations in the contributions from the free channel terms, the non-resonant channels and the operators T^λ, all of which are not accounted for in the method. These discrepancies do not give rise to any significant errors when the integration over the differential cross-section is carried out, in this case using the trapezoidal rule. The contributions to the integrals away from the resonances are calculated using the Gaussian method. Because the resonances only appear in the exchange amplitude at the energies of interest here, the CBOX results do not contain any contribution from the autoionization and the continuity of the CBOX cross-section across the threshold enables the accuracy of the numerical integration to be checked. This also indicates that it is a good approximation to treat the direct amplitude as being constant in the resonance region. The general form of the results is the same in all cases except that the autoionization occurs at relatively lower energies and is increasingly important as the charge of the ion increases. Below the autoionization threshold the results are essentially similar to those for the beryllium-like ions where no autoionization is present. Again, CB is poor, CBOX is adequate and CBX is probably the best of the approximations. The effect of using distorted waves is less pronounced than in the case of the beryllium-like ions, partly because the charge on the target ion is higher in these cases and partly because there are fewer electrons in the ions. For O^{5+} and Ne^{6+} the effect of distortion is expected to be very small and hence distorted wave calculations were not carried out on these ions. The DCBX and CBX calculations of Younger (1980a) on O^{5+}, which are similar to our own, and on Mg^{8+} confirm this. Similarly, at the high energies corresponding to the

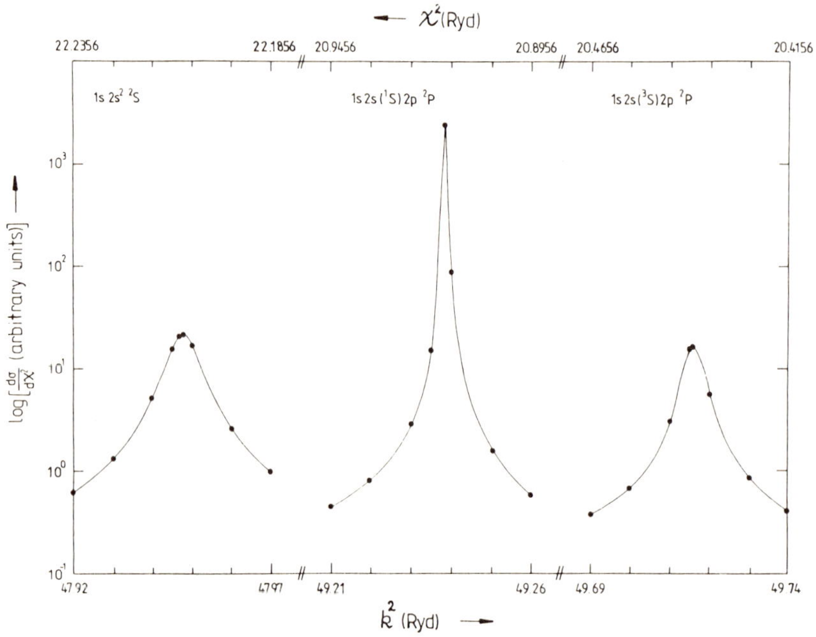

FIGURE 7

Calculated resonance structures in the differential cross-section for ionization of Ne^{7+} at an incident energy of five times the threshold energy. The points were calculated using the program COBION and the solid curves were obtained using quantum defect theory.

onset of autionization, distortion effects will be small and only Coulomb wave results were obtained. Above the autoionization threshold the dominant contributions to the cross-sections are from the 'knockout' process and the autoionzation from the $1s2s(^2S)2p\ ^2P$ resonance. The other resonances contribute to a much smaller degree. As each resonance is crossed the total cross-section should increase quasi-stepwise. No attempt has been made to map out this structure in these calculations and results were obtained immediately below the threshold of the $1s2s^2\ ^2S$ resonance and immediately above the threshold of the $1s2s(^3S)2p\ ^2P$ resonance (the $1s2p^2\ ^2S$ resonance lies still higher in energy but gives very little contribution to the cross-section). The positions of the resonances calculated by IMPACT are all in good agreement with the experimentally determined thresholds and the close-coupling results of Henry (1979) which do not, however, allow for mixing between the excited states and the continuum.

The comparison between the theoretical and experimental results is simpler in this case than for the beryllium-like ions since there is no contribution from metastables. The experiments show a marked increase in the cross-section above about four times the ionization energy due to the contribution of the autoionization, the relative importance of the effect increasing with the charge of the ion. Crandall et al (1979) have attributed the autoionization to the states $1s2s^2\ ^2S$, $1s2s(^1S)2p\ ^2P$, $1s2s(^3S)2p\ ^2P$ and $1s2s2p\ ^4P$. In fact, at higher energies states of configuration $1s2\ell n\ell'$ (ℓ=0,1,ℓ'>n) will also contribute and could be accounted for in an appropriate multichannel calculation with COBION as a manifestation of the mixing between open and closed channels. The metastable $1s2s2p\ ^4P$ state is of particular interest because it is forbidden in the Coulomb-Born approximation since the (N+1)-electron system must have the same total spin as the ground state after the collision. These states can, however, be excited if the ionization process is considered as an excitation of the three-electron ion by the ionizing electron, as in a conventional scattering calculation, with only one electron in the continuum in the final state. The corresponding excitation cross-section is then quite large (Henry (1979)) and, because the lifetime of the state is very long and it is weakly coupled to the continuum, it is possible that this state contributes significantly to the autoionization. The importance of this state should however decrease with increasing energy and Henry's results indicate that the dominant contribution is nevertheless from the $1s2s(^1S)2p\ ^2P$ state. With the exception of C^{3+}, the agreement between the CBX results and experiment is good, although the large error limits in the O^{5+} results preclude any definite conclusions from being made in this case. In the case of C^{3+} all the Coulomb-Born results, including those of Moores (1978), are consistently higher than experiment and it is not clear if the approximation is breaking down in the case of this low-charged ion since the experimental results in this case do not scale in the same way as those for the higher charged ions. The somewhat low value of the CBX cross-section in the autoionization region for O^{5+} is less worrying in view of the poorer quality of these measurements. One source of the discrepancy could be the omission of higher-lying autoionizing states, although the results of Sampson and Golden (1979) would indicate that the effect of these states is small. Alternatively, the form of CBX used may be overestimating the effect of interference in this case. The results for Ne^{7+} indicate that the increase in importance of the autoionization with increasing charge of the ion is not as great as predicted by Crandall et al (1979). The results of Sampson and Golden (1979) are in very good agreement with those obtained here despite the different physics of the two approaches. Sampson and Golden use scaled hydrogenic collision strengths calculated for a theoretical ion with infinite nuclear charge to obtain values of the appropriate excitation cross-sections which are then added to the ionization cross-sections as calculated by a similar method (Golden and Sampson (1977)). This method cannot account for the mixing between the excited state and continuum which is the source of the autoionization. Another important difference between this method and the one presented here is that, whereas in calculating excitation cross-sections to be added to an ionization cross-section a calculation of the scattering of an electron by an (N+1)-electron ion is carried out, when close-coupling wavefunctions are used to describe the (N+1)-electron system a calculation of the scattering by an N-electron ion is carried out. This means that in the limiting case of ionization of neutral lithium, the excitation calculations would predict a zero contribution from the autoionization at threshold while the method presented here would predict a finite contribution at threshold. An experimental test of this difference would be interesting.

One of the assumptions which is implicit in the form of the close-coupling wave functions used in these calculations is that all the autoionizing states only decay into the continuum and cannot decay by radiative emission. In order to allow correctly for such effects within the framework of the present formulation it would be necessary to include radiative terms in the Hamiltonian (Davies and Seaton (1969)), a procedure which would rapidly lead to the problem becoming numerically intractible. It is apparent however, from the work of Gabriel (1972) on the autoionization and radiative transition probabilities of the states considered here that the loss of flux from the ionization process by radiative decay is only important for ions with nuclear charge greater than about 1o. The good agreement between the experimental values predicted by IMPACT indicates that the single channel approximation is a reasonable one. This is because the configurations which can contribute to the ^2S ground state are widely separated in energy and the dominant configuration is the $1s^2 2s\ ^2S$, contributions from the nearest configurations being adequately accounted for through the bound channels. The same considerations apply to the continuum states and it can be expected that no significant errors are introduced into the calculations for the lithium-like ions through the use of only single channel wavefunctions. This is confirmed by results for O^{5+} at 1.5 and 2.5 times the ionization energy which were obtained using the more complete five-state-coupling wavefunctions. The five-state results however, do predict small changes in position and width of the resonances.

CONCLUSIONS

The principal conclusions which may be drawn from this work are the following: the CB approximation is poor and overestimates the cross sections; CBOX gives reasonable results provided the energy is not such that autoionization is important since it gives incorrect threshold energies. The CBX is the best of the Coulomb-Born approximations considered here. The infinite-Z method of Golden and Sampson gives a good approximation to the full CBX method of autoionization (when necessary) is included in for former, eben for C^{2+}. The effect of using distorted waves in place of Coulomb waves is to decrease the

CBX results at low energies but away from threshold are not of major importance in Li and Be-like ions for which high partial waves, which are affected little by distortion, provide the principal contribution to the cross-section. When the effects of metastable states are taken into account the agreement with experiment for Be-like ions is reasonably good although, unlike theory the experimental results do not scale along the isoelectronic sequence. For Li-like ions very good agreement is obtained with experiment for N^{4+} and O^{5+}; The C^{3+} results are larger than experiment (see Fig. 1) by an amount increasing from threshold to about 20% at the peak. The autoionization effects calculated here are at the same energy and of the same magnitude as predicted by experiment for C^{3+} and N^{4+} but the observed effect of O^{5+} is apparently underestimated by our calculations. In all cases, the agreement between the best DCBX calculation and the Lötz formula is better than about 25%. In view of the agreement with the crossed-beam experimental work, one is tempted to conclude that the results presented in this paper are correct to at least 20%, making a conservative estimate.

In future calculations it is intended to extend the work to the case of ions with outer p electrons and to allow for the effect of closed channel resonances in the final state as well as bound channels.

REFERENCES

Burgess A. and Rudge M.R.H., 1963, Proc. Roy. Soc. A 273, 372-86.

Burke P.G. and Seaton M.J., 1971, Meth. Comp. Phys. 10, 1-56.

Cheng C.C., Feldman U. and Doschek G.A., 1979, Ap. J. 233, 736-40.

Crandall D.H., Phaneuf R.A. and Gregory D.C., 1979, Oak Ridge Nat. Lab. Report No. ORNL/TM-7020.

Crees M.A., Seaton M.J. and Wilson P.M.H., 1978, C.P.C. 15, 23-83.

Davies P.C.W. and Seaton M.J., 1969, J. Phys. B Atom. Molec. Phys. 2, 757-65.

Dolder K.T., Harrison M.F.A. and Thonemann P.C., 1961, Proc. Roy. Soc. A 264, 367-78

Dubau J. and Wells J., 1973, J. Phys. B Atom. Molec. Phys. 6, 1452-60.

Eissner W. and Nussbaumer H., 1969, J. Phys. B Atom. Molec. Phys. 2, 1028-43.

Golden L.A. and Sampson D.H., 1977, J. Phys. B. Atom. Molec. Phys. 10, 2229-37.

Gabriel A.H., 1972, M.N.R.A.S. 160, 99-119.

Hamdan M., Birkinshaw K. and Hasted J.B., 1978, J. Phys. B. Atom. Molec. Phys. 11, 331-8.

Henry R.J.W., 1979, J. Phys. B. Atom. Molec. Phys. 12, L 309-13.

Jones M., 1974, C.P.C. 7, 353-67.

Jakubowicz H., 1980, Thesis, University of London

Jakobowicz H. and Moores D.L., 1980, Comments in Atom. Molec. Phys. Vol. 9 No. 2, 55-70.
Lötz W., 1967, Ap. J. Suppl. 14, 207-38; 1968, Z. Phys. 216, 241-7.
Moores D.L., 1978, J. Phys. B. Atom. Molec. Phys. 11 L, 403-5.
Moores D.L. and Nussbaumer H., 1970, J. Phys. B. Atom. Molec. Phys. 3, 161-72.
Peart B., Walton D.S. and Dolder K.T., 1969, J. Phys. B. Atom. Molec. Phys. 2, 1347-52.
Peterkop, 1977 "Theory of Ionization of Atoms by Electron Impact" Colorado Associated University Press.
Pradhan A.K. and Seaton M.J., 1980, to be published.
Rudge M.R.H., 1968, Rev. Mod. Phys. 40, 564-90.
Rudge M.R.H. and Seaton M.J. 165, Proc. Roy. Soc. A 283, 262-90.
Rudge M.R.H. and Schwartz S.B., 1966, Proc. Phys. Soc. 88, 563-78 and 579-85.
Sampson D.H. and Golden L.A., 1979, J. Phys. B. Atom. Molec. Phys. 12, L 785-91.
Saraph H.E., 1980, C.P.C., submitted for publication.
Seaton M.J., 1974, J. Phys. B. Atom. Molec. Phys. 7, 1817-40.
Seaton M.J., 1978, J. Phys. B. Atom. Molec. Phys. 11, 4067-93.
Woodruff P.R., Hublet M.C. and Harrison M.F.A., 1978, J. Phys. B. Atom. Molec. Phys. 11, L 305-8.
Younger S.M., 1980, a phys. Rev. A 22, 111-7,
1980 b Phys. Rev. A, in press.

ASPECTS OF ELECTRONIC CONFIGURATION INTERACTION IN MOLECULAR PHOTOIONIZATION

P. W. Langhoff

Department of Chemistry
Indiana University
Bloomington, Indiana 47405 USA

and

Max Planck Institut für Physik und Astrophysik
Institut für Astrophysik
Karl-Schwarzschild-Strasse 1
8046 Garching bei München
Federal Republic of Germany

INTRODUCTION

Experimental studies have been reported recently of partial-channel photoionization cross sections in diatomic and polyatomic molecules[1-3]. Many of the measured cross sections can be understood and clarified quantitatively on basis of theoretical studies in the so-called separated-channel static-exchange approximation[4]. In this approach, Hartree-Fock functions are employed for the electronic portions of target ground states, and corresponding continuum molecular eigenfunctions are constructed in the orbital approximation employing non-central static-exchange potentials. Consequently, the calculated partial-channel cross sections so obtained refer simply to the separate removal of electrons from the individual molecular orbitals of a molecule, and they do not take into account the configuration mixing possible among the various ionization channels. There are important cases, however, in which the effects of electronic configuration interaction do not simply provide small corrections to separated-channel static-exchange results, but, rather, dominate the spectral characteristics of partial-channel cross sections[5-7]. In these cases it is generally necessary to include initial and final-state configuration-mixing effects, as well as scattering-function coupling, in theoretical studies of molecular partial-channel photoionization cross sections.

In the present work, methods are described for performing molecular photoionization calculations in which configuration-mixing effects between scattering functions, as well as in the initial and final ionic states, are potentially important. The theoretical development, which incorporates aspects of time-dependent Hartree-Fock (TDHF) or random-phase approximation with exchange (RPAE) calculations of scattering functions[8-11], and Green's function and configuration-interaction calculations of ionic-state positions and corresponding spectroscopic amplitudes[12-14], is described in Section II. Illustrative calculations of valence-shell photoionization in molecular nitrogen are reported in Section III. It is seen that configuration-mixing between V_π and V_σ intravalence states in molecular nitrogen effects significantly the outer-valence $(3\sigma_g^{-1})X^2\Sigma_g^+$ and $(1\pi_u^{-1})A^2\Pi_u$ photoionization channels[6], and that final-ionic-state configuration mixing clarifies the spectrum of ionic channels observed in the inner-valence region, where the Koopmans approximation fails[7,15]. Concluding remarks are made in Section IV.

II. THEORETICAL DEVELOPMENT

In the present development, the vertical-electronic Born-Oppenheimer approximation[16] is employed in conjunction with rotational and vibrational closure in construction of partial-channel photoionization cross sections for the formation of specific parent molecular-ionic states. The resulting cross sections for incident photon energy $\varepsilon = h\nu$ can be written in the form[17]

$$\sigma_\alpha(h\nu) = (2/3)\varepsilon\,|\langle k\Phi^{(N)}_{\varepsilon-\varepsilon_\alpha}|\underset{\sim}{\mu}|\Phi_o^{(N)}\rangle|^2 \quad , \tag{1}$$

where $\Phi_o^{(N)}$ is the N-electron ground-state wave function, $\underset{\sim}{\mu}$ is the N-electron dipole moment operator, and $k\Phi^{(N)}_{\varepsilon-\varepsilon_\alpha}$ is an appropriate scattering function for the channel α at energy $\varepsilon-\varepsilon_\alpha$ above the threshold value ε_α (18). Because closure is employed, the partial-channel cross sections of Eq. (1) refer to averages over initial and sums over final rotational and vibrational states, and the corresponding wave functions and channel labels α refer to electronic degrees of freedom only. This facilitates comparisons with measured partial-channel cross sections[1-3], which generally do not achieve vibrational resolution. Moreover, non-Franck-Condon effects on cross sections summed over final vibrational and rotational states are generally small in light molecules, even in the case of time-delay in photoejection due to shape-resonance effects[19-21].

In accordance with the general theory of multichannel scattering[22-24], the electronic Born-Oppenheimer continuum states $k\Phi^{(N)}_{\varepsilon-\varepsilon_\alpha}$ of Eq. (1) can be determined from solutions of a set of coupled equations in which correct vertical-electronic wave functions $\Phi_\alpha^{(N-1)}$ define the appropriate asymptotic ionic channels of specific energy ε_α. In order to simplify solution of this formidable problem, it is necessary to identify the important physical features of the photoionization process, in which connection it is helpful to clarify the natures of the vertical-ionic states $\Phi_\alpha^{(N-1)}$. These are conveniently written in hole-particle expansions of the form[7]

$$\Phi_\alpha^{(N-1)} = \sum_{i=1}^{N} a_i^{(\alpha)}\Phi_i^{(N-1)} + \sum_{i=1}^{N}\sum_{j=1}^{N}\sum_{k>N}^{\infty} a_{ijk}^{(\alpha)}\Phi_{ijk}^{(N-1)} + \ldots, \quad (2)$$

where $\Phi_i^{(N-1)}$ are single-hole states formed by removing a single spin orbital from the ground-state Hartree-Fock wave function $\Phi_{HF}^{(N)}$, $\Phi_{ijk}^{(N-1)}$ are corresponding two-hole (i and j) one-particle (k) states, and higher-order hole-particle states are included as required. Neutral-molecule occupied and virtual canonical Fock spin orbitals are employed in the expansion of Eq. (2), and the energies ε_α and coefficients $a_i^{(\alpha)}$, $a_{ijk}^{(\alpha)}$,... are obtained from appropriate configuration-interaction (25,26) or Green's function[12-15] calculations.

Previously reported theoretical studies of the ionic states of light diatomic and polyatomic molecules[12-15], as well as corresponding measured photoelectron spectra[27-28] indicate there are only states of largely single-hole or Koopmans character in the so-called outer-valence regions ($\varepsilon_\alpha \simeq$ 10-20eV), whereas in the inner-valence region ($\varepsilon_\alpha \simeq$ 20-40eV) there are large numbers of mixed-configurational states. In the former case the ionic thresholds ε_α are determined by the ionization potentials of the occupied orbitals, although the effects of relaxation and reorganization play a small role also[12-15]. By contrast, the positions of the ionic states in the inner-valence regions are determined by strong configuration mixing between certain of the single-hole, two-hole one-particle, and possibly higher-order hole-particle states appearing in Eq. (2). Because of the clear distinction between outer-valence and inner-valence photoionization in light molecules, it is convenient to deal with each case separately, since the important physical features are different in the two cases.

(A) Outer-valence-shell molecular photoionization

Ionic states in the outer-valence region of light molecules are largely Koopmanslike. Consequently, the channel label $\alpha(\rightarrow i)$ in this case specifies the single-hole state produced $[\Phi_\alpha^{(N-1)} \rightarrow \Phi_i^{(N-1)}]$ in the ionization process, and corresponding scattering states $k\Phi_{\varepsilon-\varepsilon_i}^{(N)}$ can be written to good approximation in the forms

$$k\Phi_{\varepsilon-\varepsilon_i}^{(N)} = (\Phi_i^{(N-1)} \otimes k\phi_i), \quad (3)$$

where the $k\phi_i$ are scattering orbitals obtained from coupled-channel scattering equations[24] in the presence of Koopmans or single-hole-state potentials. Such calculations correspond to use of single-excitation configuration-interaction wave functions in which the ionic hole states involved are fixed by the neutral-molecule ground-state Hartree-Fock calculation. This approximation is improved upon with little additional effort by introducing initial-state configuration-interaction effects in the form of double excitations from the Hartree-Fock ground state. In this case the $k\Phi_i$ are obtained from time-dependent Hartree-Fock (TDHF) or random-phase-approximation-with-exchange (RPAE) calculations[8-11].

Introducing the scattering function of Eq. (3) into Eq. (1), and employing $\Phi_{HF}^{(N)}$ as the ground-state wave function gives

$$\sigma_i(h\nu) = (2/3)\varepsilon\,|\langle k\phi_i|\underset{\sim}{\mu}_i|\phi_i\rangle|^2 \tag{4}$$

for the partial-channel cross section corresponding to removal of electrons from the occupied spin orbital ϕ_i. The appropriate scattering orbital $k\phi_i$ of Eq. (4) is obtained from the expression

$$k\phi_i = k\phi_{i+} + k\phi_{i-}\ , \tag{5}$$

where the $k\phi_{i\pm}$ satisfy the RPAE equations[10,11],

$$(h - \varepsilon_i \mp \varepsilon)k\phi_{i\pm} + v_\pm\phi_i = 0\ , \tag{6}$$

with h, ε_i, and ϕ_i the Fock operator an ith canonical energy and spin orbital, respectively. In Eq. (6)

$$v_\pm = \sum_{j=1}^{N} \langle k\phi_{j\pm}|(1/r_{12})(1-P_{12})|\phi_j\rangle + \langle\phi_j|(1/r_{12})(1-P_{12})|k\phi_{j\pm}\rangle \tag{7}$$

are potentials coupling the various scattering channels[10]. The $k\phi_{i\pm}$ are identified as hole-particle/particle-hole contributions from the ith occupied spin orbital to the RPAE coupled mode of (discrete or continuum) excitation energy ε. When coupling between the different scattering orbital contributions ($i \neq j$) is small, the potential of Eq. (7) take the form

$$v_\pm \to \langle\phi_i|(1/r_{12})(1-P_{12})|k\phi_{i\pm}\rangle \tag{8}$$

in which case Eq. (6) reduces to the customary separated-channel static-exchange equation[4,29]. Of course, this latter approximation will be inadequate when there is strong coupling between the scattering functions $k\phi_{i\pm}$ and $k\phi_{j\pm}$ associated with different occupied orbitals ($i \neq j$), a situation of particular interest in the present study.

(B) Inner-valence-shell molecular photoionization

In the inner-valence region, the molecular ionic states of Eq. (2) include large two-hole one-particle and higher-order hole-particle contributions, as well as single-hole-state components. Because photoejection is sudden, even at incident photon energy near threshold[7], scattering orbitals are appropriately determined in the fields of initial-state single-hole potentials, with ionic-state relaxation taking place when outgoing electrons are far from the residual ions produced, and moving in a largely Coulombic field. Consequently, the appropriate N-electron scattering functions of Eq. (1) in this case can be written to good approximation in the forms

$$k\Phi^{(N)}_{\varepsilon-\varepsilon_\alpha} = (\Phi^{(N-1)}_{\alpha} \otimes k\phi_1,\ k\phi_2,\ \ldots k\phi_N)\ , \tag{9}$$

where the ionic states $\Phi^{(N-1)}_{\alpha}$ of Eqs. (2) in the inner-valence region are obtained from final-state calculations, and the scattering orbitals $k\phi_i$ to which they are coupled are determined from separate scattering calculations in the presence of initial-state single-hole potentials.

Introducing the function of Eq. (9) into Eq. (1), and employing the Hartree-Fock function $\Phi^{(N)}_{HF}$ for the ground state, gives

$$\sigma_\alpha(h\nu) = (2/3)\varepsilon \left| \sum_{i=1}^{N} a_i^{(\alpha)} \langle k\phi_i | \underset{\sim}{\mu}_i | \phi_i \rangle \right|^2 \tag{10}$$

for inner-valence-shell partial-channel photoionization cross sections. This expression corresponds to the so-called intensity-borrowing sudden-approximation[7,30], in which the correct ionic states α borrow intensity from dipole-allowed orbital transition moments in an amount specified by the single-hole amplitudes $a_i^{(\alpha)}$. When only one of the $a_i^{(\alpha)}$ contributes substantially to Eq. (2), as is often the case, Eq. (10) takes the form

$$\sigma_\alpha(h\nu) = |a_i^{(\alpha)}|^2 \sigma_i(\varepsilon-\varepsilon_\alpha) \tag{11}$$

where $|a_i^{(\alpha)}|^2$ is a so-called spectroscopic factor[30], and $_i(\varepsilon-\varepsilon_\alpha)$ is the ith orbital cross section referred to the threshold ε_α of the correct ionic state.

In the event initial-state correlation effects are important[31], the $a_i^{(\alpha)}$ of Eqs. (10) and (11) are replaced by the spectroscopic amplitudes

$$\chi_i^{(\alpha)} = \langle \Phi_\alpha^{(N-1)} | \hat{a}_i | \Phi_o^{(N)} \rangle \quad , \tag{12}$$

where $\hat{a}_i$ annihilates the ith spin orbital, and the sum in Eq. (10) is extended to include all orbitals that contribute to the ground-state function $\Phi_o^{(N)}$ (12). Appropriate initial- and final-state configuration-interaction calculations, or corresponding Green's function studies, are employed in determinations of the amplitudes of Eq. (12)[7].

ILLUSTRATIVE CALCULATIONS

Applications of the development outlined in Section II are reported here in the case of valence-shell photoionization in molecular nitrogen. It is known from photoelectron spectroscopy[27,28] and corresponding theoretical studies[12-15] that there are three outer-valence ionic states in nitrogen $[(3\sigma_g^{-1})X^2\Sigma_g^+(15.6eV)$, $(1\pi_u^{-1})A^2\Pi_u(16.7eV)$, $(2\sigma_u^{-1})B^2\Sigma_u^+(18.8eV)]$, corresponding largely $[|a_i^{(\alpha)}|^2 > 0.9]$ to ionization of the indicated orbitals and having the indicated adiabatic ionization potentials. By contrast, in the inner-valence region there are significantly more ionic states. These are conveniently grouped into bands given the spectroscopic designations $C^2\Sigma_u^+(23\text{-}27eV)$, $F^2\Sigma_g^+(27\text{-}31eV)$, $G^2\Sigma_g^+(31\text{-}35eV)$, and $(2\sigma_g^{-1})^2\Sigma_g^+(35\text{-}45eV)$[7]. Theoretical and experimental cross sections corresponding to production of the outer-and inner-valence shell ionic states or bands in N_2 are presented in the following two subsections.

<u>(A) Outer-valence-shell photoionization in N_2</u>

In Figure 1 are shown calculated cross sections for $3\sigma_g \to k\sigma_u$, $1\pi_u \to k\pi_g$, $2\sigma_u \to k\sigma_g$, and $2\sigma_g \to k\sigma_u$ photoionization in N_2, obtained from the development of Eqs. (1) to (7) and appropriate computational methodology[32-35]. Also shown in the figure are the results of separated-channel static-exchange calculations, in which coupling among all four scattering channels is neglected [Eq. (8)]. There are evidently significant differences between the $1\pi_u \to k\pi_g$ channel cross sections obtained from the two approximations, whereas the $3\sigma_g \to k\sigma_u$, $2\sigma_u \to k\sigma_g$ and $2\sigma_g \to k\sigma_u$ results are generally similar in both approximations. The very large feature at ∿19eV in the $1\pi_u \to k\pi_g$ static-exchange cross section can be attributed to the presence of a strong spurious $1\pi_u(\pi) \to 1\pi_g(\pi^*)$, or intravalence $N \to V_\pi$, contribution to the continuous spectrum. By contrast, configuration-mixing with 3 $_g$( ) 3 $_u$(*), or intravalence N V , excitation in

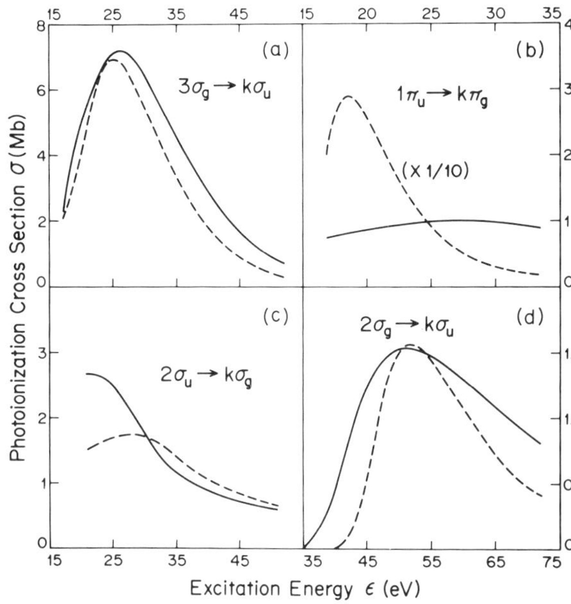

Fig. 1 Partial-channel cross sections in molecular nitrogen for (a) $3\sigma_g \rightarrow k\sigma_u$, (b) $1\pi_u \rightarrow k\pi_g$, (c) $2\sigma_u \rightarrow k\sigma_g$, and (d) $2\sigma_g \rightarrow k\sigma_u$ ionization; (——) RPAE and (----) separated-channel static-exchange results obtained from the present development, as discussed in the text. Note that the static-exchange $1\pi_u \rightarrow k\pi_g$ cross section (b) has been reduced by a factor of ten to aid its comparison with the much smaller corresponding RPAE result.

the RPAE calculations lowers the V_π contribution below the threshold, where it contributes to the resonance $X^1\Sigma_g^+ \rightarrow b'^1\Sigma_u^+$ transition at 14.4eV (36-38). Correspondingly, the intravalence V_σ contribution to the spectrum, which appears as a shape resonance ∿10eV above threshold in both $3\sigma_g \rightarrow k\sigma_u$ calculations, is shifted slightly to higher energy relative to the separated-channel static-exchange results by the effects of configuration mixing in the RPAE calculation.

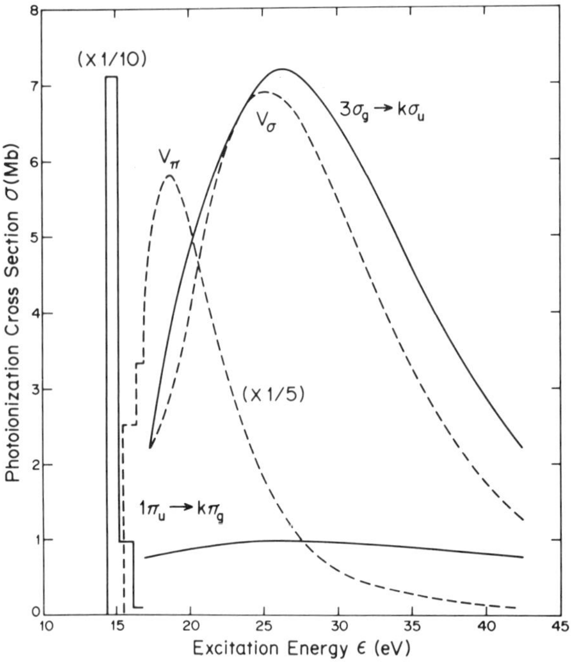

Fig. 2 Partial-channel cross sections in molecular nitrogen for $3\sigma_g \rightarrow k\sigma_u$ and $1\pi_u \rightarrow k\pi_g$ ionization; (——) RPAE and (----) separated-channel static exchange results obtained from the present development, as discussed in the text. The discrete portion of the RPAE $1\pi_u^-$ cross section is presented in the Stieltjes sense[34-35] in order to illustrate the presence of the strong $1\pi_u \rightarrow 1\pi_g$ contribution to the $X^1\Sigma_g^+ \rightarrow b'^1\Sigma_u^+$ resonance in the spectrum[36-38]. Note that this part of the RPAE $1\pi_u^-$ cross section is reduced by a factor of ten for ease of comparison, and the $1\pi_u \rightarrow k\pi_g$ static-exchange result is reduced by a factor of five. The symbols V_π and V_σ identify the approximate positions of the intervalence $N \rightarrow V_\pi$ and $N \rightarrow V_\sigma$ contributions to the separated-channel static-exchange $1\pi_u \rightarrow k\pi_g$ and $3\sigma_g \rightarrow k\sigma_u$ cross sections, respectively.

These points are summarized in Figure 2, where the repulsion of the intravalence $(\sigma \rightarrow \sigma^*)V_\sigma$ and $(\pi \rightarrow \pi^*)V_\pi$ contributions to the spectrum is evident in the RPAE cross sections when compared with the separated-channel results. The larger shift of the $N \rightarrow V_\pi$ excitation relative to that of the $N \rightarrow V_\sigma$ transition upon introduction of configuration mixing can be attributed to the greater spectral concentration of the former state, which corresponds to a discrete level spuriously mixed in the photoionization continuum. By contrast, the $N \rightarrow V_\sigma$ intravalence transition correctly corresponds to a shape resonance in the photoionization continuum, and consequently has a significantly lower spectral concentration than does the $N \rightarrow V_\pi$ excitation. Note in Figure 2 that the $1\pi_u \rightarrow k\pi_g$ static-exchange result has been reduced by a factor of five, and the discrete portion of the corresponding RPAE result, which is shown in the so-called Stieltjes sense[34,35], has been reduced by a factor of ten.

Although the RPAE and static-exchange $2\sigma_u \rightarrow k\sigma_g$ and $2\sigma_g \rightarrow k\sigma_u$ cross sections of Figure 1 are in general agreement, it is seen there are differences in detail. Specifically, the RPAE $2\sigma_u \rightarrow k\sigma_g$ cross section is larger at threshold than is the static-exchange result. Inspection of the corresponding experimental values reveals the presence of a noticable feature in the $2\sigma_u^{-1}$ cross section near threshold[39-42], suggesting that the RPAE result is in better accord with measurement than is the static-exchange result. The RPAE $2\sigma_g \rightarrow k\sigma_u$ result is seen to be noticeably broader than the corresponding static-exchange cross section, primarily as a consequence of (hole-particle) excitation mixing with the three other channels.

In order to further clarify the natures of the RPAE spectra of Figures 1 and 2, single-excitation or Tamm-Dancoff approximation (TDA) calculations, in which the first sum of (particle-hole) terms in Eq. (7) is neglected, are performed of $X^1\Sigma_g^+ \rightarrow k^1\Sigma_u^+$ photoionization in N_2. The resulting partial-channel cross sections are found to be in generally good agreement with the RPAE calculations, suggesting that ground-state correlation, which gives rise to the particle-hole terms in Eq. (7), has little effect on photoionization in N_2 at the equilibrium internuclear separation.

(B) Inner-valence-shell photoionization in N_2

Approximations to the inner-valence-shell ionic states of Eq. (2) in the case of N_2 are obtained from single-excitation (SE) and polarization (POL) configuration-interaction (CI) calculations employing the single-hole states as reference vectors[26]. A (10s, 6p,2d)/[5s,3p,2d] Gaussian basis set that includes individual s,p, and d diffuse functions is employed in both sets of calculations[32]. The SECI calculations correspond to use of neutral-molecule ground-state Fock orbitals in Eq. (2), and include contributions from all single-hole and all two-hole one-particle states. The POLCI calculations employ a $(3\sigma_g^{-1} 1\pi_g)^1\Pi_g$ Fock basis and include the limited number of three-hole two-particle terms in which at least one electron is in the valence $1\pi_g$ orbital, in addition to all the single-hole and two-hole one-particle terms. Because all single-hole and two-hole one-particle terms are employed in both developments, the two results differ primarily by the presence of valence-like three-hole two-particle terms in the POLCI calculations. Such terms, of course, are generally referred to as double excitations relative to the ionic single-hole states.

Ionic-state energies and spectroscopic factors obtained as indicated above are combined with appropriate corresponding dipole transition moments in accordance with Eq. (10) in calculations of molecular partial-channel photoionization cross sections in N_2. It is important to recognize that the relative signs of the spectroscopic amplitudes and of the corresponding dipole transition moments are generally important in evaluation of Eq. (10). In the case of photoionization in N_2 reported here, however, Eq. (11) is generally the appropriate one, since most of the ionic states determined are found to borrow intensity largely from only one hole state[7].

A comparison of the SECI and POLCI results in N_2 is made in Figure 3, where calculated and recently measured[43] photoelectron spectra at incident photon energy of $h\nu = 50.3$eV are presented. The calculated values are given by the expression

$$P_{h\nu}(\omega) = \sum_{\alpha} (\sigma_{\alpha}(h\nu)/\omega_{\alpha}/\overline{\pi})\exp[-(\omega-\varepsilon_{\alpha})^2/\omega_{\alpha}^2] \ , \qquad (13)$$

where the cross sections $\sigma_\alpha(h\nu)$ are obtained from Eq. (10), spectroscopic amplitudes, and orbital dipole transition moments, and the vibrational half widths $\omega_\alpha (\cong 1\text{-}2\text{eV})$ are estimated from previously reported calculations[15]. The Gaussian form given by Eq. (13) is known to be valid in the limit of strong vibrational excitation[44,45], in the present development it can be regarded simply as a convenient expedient for comparison with experiment. Although the two theoretical results of Figure 3 are on an absolute scale, the experimental spectrum refers to counts/second and, consequently, requires base-line and instrument-function corrections[43]. The experimental spectrum of Figure 3 evidently exhibits four distinct bands that are designated on basis of the present POLCI calculations as C $^2\Sigma_u^+$(23-27eV), F $^2\Sigma_g^+$(27-31eV), G $^2\Sigma_g^+$(31-35eV), and $(2\sigma_g^{-1})$ $^2\Sigma_g^+$(35-45eV), in general accord with previously suggested spectroscopic nomenclature[39-43]. The results of the POLCI calculations, which include nonnegligible three-hole two-particle contributions, are seen to satisfactorily reproduce the four bands, whereas the G band (31-35eV) is evidently missing from the SECI calculations, in which there are no significant three-hole two-particle contributions. Examination of the calculated POLCI spectrum indicates the C(23-27eV) and F(27-31eV) bands correspond to individual $^2\Sigma_u^+$(24.27eV) and $^2\Sigma_g^+$(28.51eV) states, respectively, whereas the G(31-35eV) band is comprised of three states of different symmetry [$^2\Sigma_g^+$(34.35eV), $^2\Sigma_u^+$(32.88eV), $^2\Pi_u$(32.77eV)]. Consequently, the G $^2\Sigma_g^+$ designation is made in this case on basis of the greater contribution to the band intensity from this state. Similarly, the $(2\sigma_g^{-1})$ $^2\Sigma_g^+$(35-45eV)

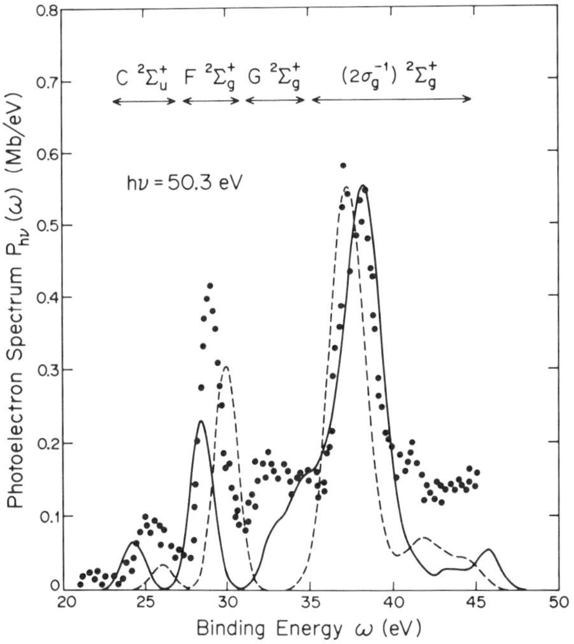

Fig. 3. Inner-valence-shell photoelectron spectrum in N_2 for $h\nu = 50.3$eV incident photons; (O) experimental values obtained employing a synchroton light source and an ∿0.2eV electron monochromator band pass[43]; (——) POLCI and (----) SECI calculations of Eq.(13) obtained employing appropriate energies and spectroscopic amplitudes[7], previously reported vibrational widths[15], and static-exchange orbital transition moments determined from the present development, as discussed in the test. Suggested spectroscopic assignments are based on the results of POLCI calculations; the C and F bands refer to individual electronic states of designated symmetry, whereas the G and the $2\sigma_g^{-1}$ bands are comprised of numbers of electronic states of various symmetry[7].

band intensity from this state. Similarly, the $(2\sigma_g^{-1})\ ^2\Sigma_g^+$(35-45eV) band is found to be comprised of large numbers of $^2\Sigma_g^+$, $^2\Sigma_u^+$, $^2\Pi_u$ ionic states converging on the ∿43eV N_2^{++} limit[46], although this band borrows most of its intensity from the $2\sigma_g^{-1}$ hole state. Both sets of calculations evidently reproduce the general features of the 50.3eV PES of Figure 3, although appropriate vibronic couplings must be introduced to clarify the details of the intensity distributions within the measured bands.

In Figure 4 are shown the $2\sigma_g \rightarrow k\ _u$ and $k\pi_u$ orbital cross sections from which the $(2\sigma_g^{-1})\ ^2\Sigma_g^+$ band borrows much of its intensity, and the calculated SECI and POLCI $(2\sigma_g^{-1})\ ^2\Sigma_g^+$(35-45eV) cross sections obtained from Eq. (10) and the appropriate amplitudes and transition moments. The $2\sigma_g \rightarrow k\sigma_u$ channel of Figure 4 evidently includes a $\sigma \rightarrow \sigma^*$ resonancelike feature centered ∿12eV above threshold, which, consequently, also appears in the $(2\sigma_g^{-1})\ ^2\Sigma_g^+$ band cross sections. Also shown in Figure 4 are experimental values obtained from suitably calibrated PES similar to those of Figure[39-43] 3, as well as estimates from dipole (e,e+ion) studies[47].

Evidently, the calculated $(2\sigma_g^{-1})\ ^2\Sigma_g^+$ cross sections of Figure 4 are in good mutual agreement and in quantitative accord with the measure results obtained from low-resolution[39] (e,2e) and higher-resolution synchrotron-radiation[43] photoelectron spectra. The estimates obtained from (e,e+ion) dissociative photoionization studies[47] are also seen to be in general accord with the calculated values. It is important to recognize that although the $2\sigma_g$ orbital cross section by itself is in some accord with measured values, the effects of intensity borrowing clearly spread the $2\sigma_g$ orbital cross section over a broad energy interval (Figure 3), and are largely responsible for the good agreement between theory and experiment shown in Figure 4. Moreover, the $3\sigma_g$, $2\sigma_u$, and $1\pi_u$ contributions to the cross section associated with ionic states in the 35 to 45eV interval are small but not negligible. Consequently, although the cross section of Figure 4 can be designed as largely of $(2\sigma_g^{-1})\ ^2\Sigma_g^+$ character[39-43], many mixed-configurational ionic states, in fact, contribute, and comparisons of the $2\sigma_g^{-1}$ orbital cross section itself with measured values are somewhat fallacious[43]. The quantitative results of Figure 4 verify completely the previously indicated importance of intensity-borrowing effects in the inner-valence-shell photoionization region in N_2 (15).

In Figure 5 are shown theoretical and experimental values for cross sections corresponding to the C $^2\Sigma_u^+$, F $^2\Sigma_g^+$, and G $^2\Sigma_g^+$ bands in N_2^+, obtained as in the case of the $(2\sigma_g^{-1})\ ^2\Sigma_g^+$ results of Figure 4. Evidently, the SECI results for the C $^2\Sigma_u^+$ band are smaller by a factor of two than both the POLCI calculations and the high-resolution PES measurements in the ∿45-65eV interval[43]. Estimates obtained from (e,e+ion) studies are seen to be in accord with the two latter results, although below ∿35eV the (e,e+ion) results

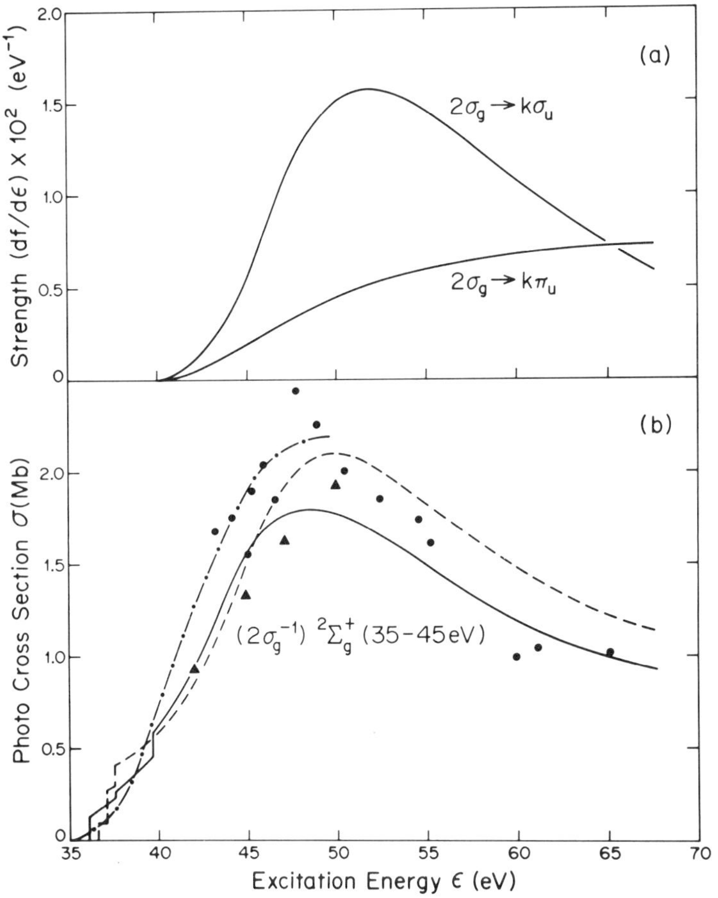

Fig. 4. Static-exchange orbital oscillator strengths for $2\sigma_g \rightarrow k\sigma_u$ and $k\pi_u$ ionization in N_2 obtained from the present development, as discussed in the text, and inner-valence-shell $(2\sigma_g^{-1})\ ^2\Sigma_g^+$(35-45eV) band partial-channel photoionization cross section in N_2; (0) synchroton-radiation measurements[43]; (Δ) dipole (e,2e) measurements[39]; (-·-·-) dipole (e,e+ion) measurements[47]; (——) POLCI and (----) SECI calculations obtained from Eq.(10) and appropriate spectroscopic amplitudes and static-exchange orbital transition moments.

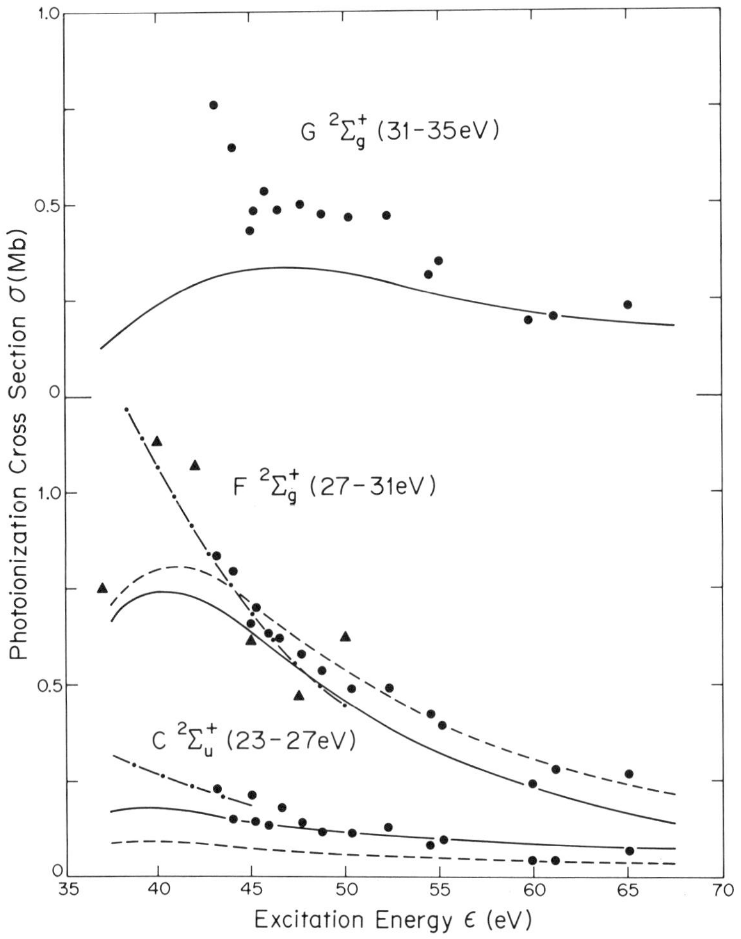

Fig. 5. Inner-valence-shell C $^2\Sigma_u^+$ -, F $^2\Sigma_g^+$ -, and G $^2\Sigma_g^+$ - band partial-channel photoionization cross sections in N_2; (0) synchroton-radiation measurements[43]; (Δ) dipole (e,2e) measurements[39]; (-·-·-·)diploe (e,e+ion) measurements[47]; (——) POLCI and (----) SECI calculations obtained from Eq.(10) and spectroscopic amplitudes and static exchange orbital transition moments. The C- and F-band cross sections refer to individual electronic states of designated symmetry, whereas the POLCI G-band is comprised of three distinct states of $^2\Sigma_g^+$(34.35eV), $^2\Sigma_u^+$(32.88eV) and $^2\Pi_u$(32.77eV) symmetry.

increase rapidly near threshold (not shown), and are not in agreement with the calculated values. The three experimental values for the F $^2\Sigma_g^+$ - band are seen to be in general agreement, and in good accord with both sets of calculations. Since this band borrows its intensity from the $3\sigma_g$ and $2\sigma_g$ cross sections, the presence of a resonancelike maximum in this channel is in accord with expectation. Finally, the POLCI results for the G-band are seen to be in general accord with the high-resolution PES measurement, although, as indicated above (Figure 3), the SECI calculations do not give rise to states in the G-band region. The results of Figure 5 suggest the intensity-borrowing sudden approximation can provide quantitatively reliable estimates of even the weak partial-channel cross sections in diatomic molecules that correspond to ionic states dominated by two-hole one-particle and three-hole two-particle configurations.

IV. CONCLUDING REMARKS

The present report illustrates the importance of configuration-mixing effects in molecular photoionization employing the nitrogen molecule as a specific example. In the outer-valence region in N_2 there is evidence of strong mixing between $N \rightarrow V_\pi$ and $N \rightarrow V_\sigma$ intravalence transitions, giving rise to coupling between $3\sigma_g \rightarrow k\sigma_u$ and $1\pi_u \rightarrow k\pi_g$ photoionization channels. Time-dependent Hartree-Fock calculations are seen to correctly include this mixing, and to give cross sections in good accord with experiment. By contrast, neglect of configuration mixing gives a $1\pi_u \rightarrow k\pi_g$ cross section that includes spurious $N \rightarrow V_\pi$ contributions in the photoionization continuum. Similar configuration-mixing effects, can arise in other molecules having two or more unoccupied valence orbitals, and so channel coupling in molecular photoionization is expected to be particularly widespread in these cases.

In the inner-valence region in N_2 final-state configuration-mixing is seen to give rise to a large number of ionic channels in photoionization. These are conveniently grouped into bands the intensities of which are accounted for satisfactorily on basis of the intensity-borrowing sudden approximation. In this approach, single-excitation and polarization configuration-interaction calculations are combined with orbital dipole transition moments in construction of photoelectron spectra and corresponding partial-channel cross sections. The results so obtained are seen to be in good agreement with recent synchrotron radiation and dipole (e,2e) measurements. Because dissociative photoionization in small molecules proceeds primarily through ionic states in the inner-valence region, further clarification and investigation is in order in other systems.

ACKNOWLEDGEMENTS

It is a pleasure to acknowledge the support of the National Science Foundation, the Petroleum Research Fund administered by the American Chemical Society, and the National Aeronautics and Space Administration through the auspices of the National Research Council. The kind hospitality of Jürgen Hinze and The Center for Interdisciplinary Research in Bielefeld, and of G.H.F. Diercksen and the Max Plank Institute for Physics and Astrophysics in Garching, is also gratefully aknowledged.

REFERENCES

1. J.A.R. Samson, Phys. Rept. 4, 303 (1976).
2. G.V. Marr, Daresbury Rept. DL/SRF/P417 (1978).
3. C.E. Brion and A. Hamnett, Advan. Chem. Phys. 45,2 1981.
4. P.W. Langhoff, N. Padial, G. Csanak, T.N. Rescigno, B.V. McKoy, J. de Chimie Physique 77, 589 (1980).
5. T.N. Rescigno, A. Gerwer, B.V. McKoy, and P.W. Langhoff, Chem. Phys. Lett. 66, 116 (1979).
6. Geoffry R.J. Williams and P.W. Langhoff, Chem. Phys. Letters 78, 21 (1981).
7. P.W. Langhoff, S.R. Langhoff, T.N. Rescigno, J. Shirmer, L.S. Cederbaum, W. Domke, and W. von Niessen, Chem. Phys. 58, 71 (1980).
8. P.A.M. Dirac, Proc. Cambridge Phil. Soc. 26, 376 (1930).
9. A.D. McLachlan and M.A. Ball, Rev. Mod. Phys. 36, 844 (1963).
10. P.W. Langhoff, S.T. Epstein, and M. Karplus, Rev. Mod. Phys. 44, 602 (1972).
11. Geoffry R.J. Williams and P.W. Langhoff, Chem. Phys. Letters 60, 201 (1978).
12. L.S. Cederbaum and W. Domcke, Advan. Chem. Phys. 36, 205 (1978).
13. J. Schirmer and L.S. Cederbaum, J. Phys. B11, 1889 (1978).
14. M.F. Herman, K.F. Freed, and D.L. Yeager, Advan. Chem. Phys. 48, 1 (1981)
15. J. Schirmer, L.S. Cederbaum, W. Domcke, and W. von Niessen, Chem. Phys. 26, 149 (1977).
16. I.N. Levine, Quantum Chemistry (Allyn and Bacon, Boston, 1970), Vol.II.
17. D.R. Bates, Mon. Not. Roy. Astron. Soc. 106, 432 (1946).
18. H.A. Bethe and E.E. Salpeter, Quantum Mechanics of One-and Two-Electron Atoms (Springer, Berlin, 1957).
19. R. Stockbauer, B.E. Cole, D.L. Ederer, J.B. West, A.C. Parr, and J.L. Dehmer, Phys. Rev. Letters 43, 757 (1979).
20. J.L. Dehmer, D. Dill, and S. Wallace, Phys. Rev. Letters 43, 1005 (1979).
21. G.Raseev, H. LeRouzo, and H. Lefebvre-Brion, J. Chem. Phys. 72, 5701 (1980).
22. M.L. Goldberger and K.M. Watson, Collision Theory (Wiley, NY, 1964).

23. K. Takayanagi and Y. Itikawa, Adv. At. Mol. Phys. 6, 105 (1970).
24. N.F. Lane, Rev. Mod. Phys. 52, 29 (1980).
25. I. Shavitt, "The Method of Configuration Interaction", in: Methods of Electronic Structure Theory, ed. H. F. Schaefer III (Plenum, NY, 1977), Vol.3, Chap.6.
26. J.P. Hay and T.H. Dunning, Jr., J. Chem. Phys. 64, 5077 (1976).
27. K. Siegbahn, C. Nordling, G. Johansson, J. Hedman, P.F. Heden, K. Hamrin, U. Gelius, T. Bergmark, L.O. Werme, R. Manne, and Y. Bauer, ESCA Applied to Free Molecules (North-Holland, Amsterdam, 1969).
28. D.W. Turner, C. Baker, A.D. Baker, and C.R. Brundle, Molecular Photoelectron Spectroscopy (Wiley, New York, 1970).
29. P.W. Langhoff, M. Karplus and R.P. Hurst, J. Chem. Phys. 44, 505 (1966).
30. T.A. Carlson, X-ray Photoelectron Spectroscopy (Academic, New York, 1978), Part III.
31. R.L. Martin and D.A. Shirley, Phys. Rev. A13, 1475 (1976).
32. T.H. Dunning and J.P. Hay, in Modern Theoretical Chemistry, H.F. Schaefer III, Editor (Plenum, NY, 1976), Vol.3, Chap.1.
33. H.F. Schaefer III, The Electronic Structure of Atoms and Molecules (Addison-Wesley, reading, MA, 1972).
34. P.W. Langhoff, "The Stieltjes-Tchebycheff Approach to Molecular Photoionization Studies," in Electron-Molecule and Photon-Molecule Collisions, T.N. Rescigno, B.V.Mc koy, and B. Schneider, Editors (Plenum, N.Y., 1979), pp. 183-224.
35. P.W. Langhoff, "Stieltjes-Tchebycheff Moment-Theory Approach to Photoeffect Studies in Hilbert Space", in Theory and Application of Moment Methods in Many-Fermion Systems, B.J. Dalton, S.M. Grimes, J.P. Vary, and S.A. Williams, Editors (Plenum, NY, 1980), pp. 191-212.
36. J. Geiger and B. Schroder, J. Chem. Phys. 50, 7 (1969).
37. R.S. Mulliken, Accounts Chem. Res. 9, 7 (1976).
38. R.S. Mulliken and W.C. Ermler, Diatomic Molecules (Academic, NY, 1977).
39. A. Hamnett, W. Stoll, and C.E. Brion, J. Electron Spectry. 8, 367 (1976).
40. J.A.R. Samson, G.N. Haddad, and J.L. Gardner, J. Phys. B10, 1749 (1977).
41. P.R. Woodruff and G.V. Marr, Proc. Roy. Soc. A358, 87 (1977).
42. E.W. Plummer, T. Gustafsson, W. Gudat, and D.E. Eastman, Phys. Rev. A15, 2339 (1977).
43. S. Krummacher, V. Schmidt, and F. Wuilleumier, J. Phys. B 13, 3993 (1980).
44. M.D. Frank-Kamenetskii and A. Lukashin, Soviet Phys. Usp. 18, 391 (1976).
45. L.S. Cederbaum and W. Domcke, J. Chem. Phys. 60, 2878 (1974); 64, 603, 612 (1976).
46. M. Cobb, T.F. Moran, R.F. Borkman, and R. Childs, J. Chem. Phys.72, 446 (1980).
47. G.R. Wight, M.J. van der Wiel, and C.E. Brion, J. Phys. B9, 675 (1976).

THE TIME-EVOLUTION OF PHOTO-IONIZATION BY MULTIPLE ABSORPTION OF PHOTONS

F.H.M. Faisal

Fakultät für Physik
Universität Bielefeld
D 48oo Bielefeld 1 - FRG

I shall be concerned with one relatively new aspect of the problem of photo-ionization which has just begun to be investigated. I shall be considering the time evolution of photo-ionization in presence of a coherent electromagnetic field. This aspect of photo-ionization may now be investigated in the laboratory with available strong monochromatic laser sources like Nd^{+++}-glass, Ruby or the dye lasers. These sources have generally sub (ionization)-threshold frequencies for most atoms. But they provide sufficiently high photon density to allow photo-ionization through multi-photon absorption processes. The topic is of course of interest in atomic physics, which is our main concern on this occasion. It is also of interest in other fields such as coherent electrodynamics or quantum statistical physics. For example, the problem of photon coherence as given by the correlation functions of increasing orders can be investigated (order by order) by measuring the current, generated by photo-ionization due to absorption of one, two or several photons, respectively. Or the relation between the irreversible behaviour of macroscopically large systems and the reversible microscopic equations (governing the motion of the microscopic constituents) can be investigated from yet another angle where even the elementary microscopic process of photo-ionization of individual atoms leads to irreversible decay (into the ionization continuum). Furthermore the nature of this evolution for earlier times can also be influenced in the laboratory (by adjusting, for example, the frequency or density of the incident photon field).

Obviously, the evolution of photo-ionization is a non-stationary process and as such cannot be described only in terms of steady state parameters, e.g. cross sections. The problem requires a direct attack for obtaining the probability of ionization as a function of the interaction time. One way of doing this, which is both rigorous and practical, is via the non-Hermitian Schrödinger theory in the time domain.[1] In this approach one converts the initial Hermitian Hamiltonian of the total system of atom plus radiation field into an equivalent non-Hermitian Hamiltonian in the appropriate sub-space. This can be done quite conveniently as is done in the well known formal theory of optical potentials. If the non-Hermitian Hamiltonian is generated by imposing the causal or retarded boundary condition one obtains a corresponding Schrödinger equation in time which automatically governs the evolution of the initially given system towards the future. If on the other hand one should use the advanced boundary condition, the related Schrödinger equation in time would describe the final value problems.

Theoretical formulation

Let the total Hermitian Hamiltonian of the interacting (atom and radiation) system be H. We divide the space of the state vectors into two parts P and Q[2,3] such that part P contains at least the initial state of the system and the part Q contains the whole or a part of the continuum states of the system. P contains, for example, the discrete initial ground state of the atom plus the discrete number state of the photon field. Q, for example, can contain the ionized state of the atom and the final state of the photon field. P and Q satisfy the usual projection properties ($P^2 = P$, $Q = 1-P$, $Q^2 = Q$, $PQ = 0$). The elimination of the Q-part leads as usually to the reduced Hamiltonian $\hat{H}$ in the P-part alone where

$$\hat{H} = H_{PP} + H_{PQ} \frac{P.V.}{E - H_{QQ}} H_{QP} - i\pi H_{PQ} \delta(E - H_{QQ}) H_{QP} \tag{1}$$

We can write the corresponding equation of motion for the state vector $|\Psi_P\rangle$ in the P-part as

$$i \frac{\partial}{\partial t} |\Psi_P\rangle = \hat{H} |\Psi_P\rangle \tag{2}$$

One notes that the sign of the imaginary part in $\hat{H}$ is non-positive since the delta-functional distribution is even. This is a direct consequence of the choice of the causal or retarted boundary condition and the equation of motion (1) governs all initial value problems. The corresponding equation for the final value problems is

$$i \frac{\partial}{\partial t} |\Psi_P\rangle = \hat{H}' |\Psi_P\rangle \tag{3}$$

where $\hat{H}'$ corresponds to the non-Hermitian Hamiltonian obtained with the advanced boundary condition, which changes the sign of the delta part in $\hat{H}$. Equations (2) and (3) are distinct and they describe the evolution of the system separately in different directions of time, (like the macroscopic Fourier (heat)-equation or what has been called the 'Anti-Fourier equation', respectively). They are an expression of the breaking of the symmetry of time reversibility in the microscopic equations themselves (c.f. Prigogine et al., 1973).[4] The non-Hermitian nature of $\hat{H}$ given by Eq. (1) depends critically on the existance of a continuum part of the spectrum of (the original Hamiltonian) H, which allows the delta functional distribution in (1) to be meaningfully defined as an integration operation. The total energy E must be degenerate with this continuum in order that the delta integration can provide a non-vanishing contribution which makes $\hat{H}$ to be non-Hermitian. Thus, the necessity of the existance of the continuum for the phenomenon of decay, as required by the theorem of Fock and Krylov (1947),[5] is explicitly incorporated in this formalism.

The solution of (2) is obtained in a very convenient form by diagonalizing the non-Hermitian Hamiltonian $\hat{H}$. Thus

$$
\begin{aligned}
|\psi_p\rangle &= e^{-i\hat{H}t}\,|\psi_p(0)\rangle \\
&= S\,D^{-\frac{1}{2}}\,e^{-i\lambda t}\,D^{-\frac{1}{2}}\,R^{T}\,|\psi_p(0)\rangle
\end{aligned}
\tag{4}
$$

$$
\begin{aligned}
\text{where}\quad & \hat{H}S = S\lambda \\
\text{and}\quad & \hat{H}^{T}R = R\lambda \\
\text{and}\quad & R^{T}S = D \text{ (diagonal)}
\end{aligned}
\tag{5}
$$

Note that generally for non-Hermitian Matrices one requires the non-unitary transformation S and R which diagonalise $\hat{H}$ and its simple transpose, $\hat{H}^T$, respectively. λ is the complex eigenvalue matrix. Properties and nature of the transformations used above have been discussed by us in greater detail elsewhere.[1] If (as it often happens) the resultant non-Hermitian matrix is symmetric then R can be replaced by S^T but the complex diagonal matrix $D = S^TS$ must be retained.

The transition amplitudes for excitation within the space of P can be readily obtained by projecting onto the unperturbed initial and final states $|i\rangle$ and $|f\rangle$, respectively. Thus

$$A^{P}_{f\leftarrow i}(t) = \sum_{s} S_{fs} \frac{e^{-i\lambda_s t}}{D_s} R_{is} \tag{6}$$

where $D_s = D_{ss}$ (diagonal).

If the final states lie in the Q-part, then it is necessary to project back on the Q-part, to obtain the transition amplitude

$$A^{Q}_{f\leftarrow i}(t) = \langle f|\psi(0)\rangle - i\, e^{-i\epsilon_f t}\Big[\sum_{sp} V_{fs} S_{sp} D^{-\frac{1}{2}} (\epsilon_f-\lambda_p)^{-1} D^{-\frac{1}{2}} \left(e^{i(\epsilon_f-\lambda_p)t} - 1\right) R_{ip}\Big] \tag{7}$$

Explicit construction of the non-Hermitian multiphoton Hamiltonian

To construct the non-Hermitian multiphoton Hamiltonian in the discrete P-space we first write the total Hermitian Hamiltonian of the 'radiation plus atom' system (see e.g. Faisal 1976)[6] as

$$H = H_A + H_R + H_{AR} \tag{8}$$

where H_A is the isolated atomic Hamiltonian which defines the atomic eigenstates $|j\rangle$ with energy ϵ_j by $H_A|j\rangle = \epsilon_j|j\rangle$. H_R is the radiation Hamiltonian which defines the photon number states $|n\rangle$ with eigenvalues $n\omega$ by

$$H_R|n\rangle = \omega a^{\dagger} a|n\rangle = n\omega|n\rangle$$

where ω is the field frequency. The interaction between the atom and the radiation in the dipole limit is

$$H_{AR} = \vec{D}\cdot\vec{\epsilon}\,(a^{+} + a)\left(\frac{2\pi\omega}{L^3}\right)^{1/2}$$

where $\vec{D}$ is the dipole operator of the system, $\vec{\epsilon}$ is the unit polarization vector and a^{+} and a are the photon creation or destruction operators, respectively. L^3 is the quantization volume.

Next, the Hamiltonian (9) is written out explicitly in its

matrix representation in the product (Schrödinger-Fock) space with

$$|j\rangle \otimes |n\rangle = |jn\rangle$$

and projected onto the P and Q parts separately. The Q part is then eliminated to obtain the reduced non-Hermitian Hamiltonian in the discrete P-space as (Schematic of an example, Fig. 1)

$$\hat{H} = |jn\rangle E_j^n \langle jn| + |jn\rangle \beta_{jj'}^{nn'} \langle j'n'|$$
$$+ |jn\rangle \left(\delta_{jj'}^{nn'} - \tfrac{1}{2} i \, \gamma_{jj'}^{nn'} \right) \langle j'n'|$$

where

$$E_j^n = \epsilon_j + n\omega$$

$$\beta_{jj'}^{nn'} = \alpha_{jj'} \left(\sqrt{n+1}\, \delta_{n',n+1} + \sqrt{n}\, \delta_{n',n-1} \right) \left(\frac{2\pi\omega}{L^3} \right)^{1/2}$$

and

$$\alpha_{jj'} = \langle j | \vec{D} \cdot \vec{\epsilon} \, | j' \rangle$$

is the transition dipole matrix elements

$$\delta_{jj'}^{nn'} = 2\pi \sum{}' \beta_{jp_1}^{nm_1} \langle p_1 m_1 | S | pm \rangle \, \rho(\epsilon_{K_p}^m) (E_j^n - \nu_{pm})^{-1}$$
$$\langle pm | S^\dagger | p_2 m_2 \rangle \, \beta_{p_2 j'}^{m_2 n'}$$

and

$$\gamma_{jj'}^{nn'} = 2\pi \sum{}' \beta_{jp_1}^{nm_1} \langle p_1 m_1 | S | pm \rangle \, \rho(\nu_{pm}) \langle pm | S^\dagger | p_2 m_2 \rangle \beta_{p_2 j'}^{m_2 n'}$$

$$\sum{}' \equiv \sum (p_1 p_2 p \; m_1 m_2 m)$$

$E-2\hbar\omega$ $-\frac{1}{2}i\Gamma$	$\beta+\delta$	$-\frac{1}{2}i\Gamma$	δ	δ
$\beta+\delta$	$E-\hbar\omega$ $-\frac{1}{2}i\Gamma$	$\beta+\delta$	δ	δ
$-\frac{1}{2}i\Gamma$	$\beta+\delta$	$E+\delta$	$\beta+\delta$	δ
δ	δ	$\beta+\delta$	$E+\hbar\omega$ $+\delta$	$\beta+\delta$
δ	δ	δ	$\beta+\delta$	$E+2\hbar\omega$ $+\delta$

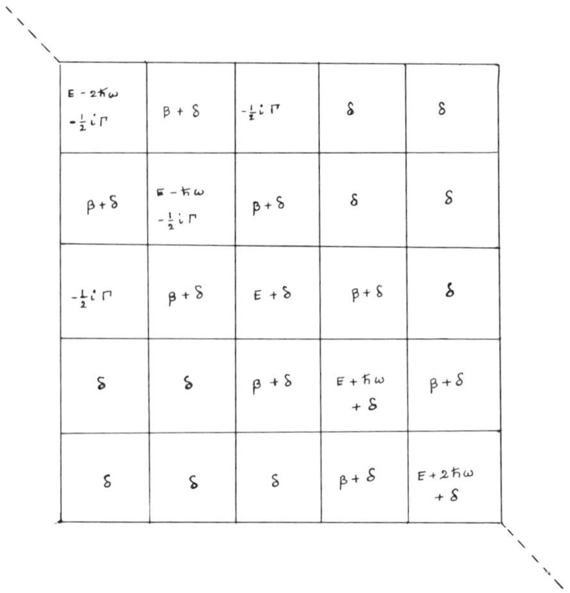

Fig. 1. A schematic view of the block structure of the non-Hermitian multiphoton matrix Hamiltonian. The 'photon-blocks' have the dimension J x J where J is the maximum dimension of the discrete space retained. The continuum space is folded into the complex sub-matrix Γ. The total dimension is $[(2N + 1)J]^2$, where N is the maximum number of photon-blocks retained. Note that the sub-matrices Γ appear both diagonally as well as non-diagonally. Depending on the threshold of ionization their imaginary part also vanish in the virtual domain of the full matrix. For the detailed mathematical structure, see the text.

The transformation matrix $\langle pm|S|p'm'\rangle$ and its adjoint and the corresponding eigenvalues ν_{pm} are obtained by prediagonalizing the submatrix

$$H_{\alpha Q} = |pm\rangle E_p^{in}\langle pm| + |pm\rangle \beta_{pp'}^{mm'}\langle p'm'|$$

where the matrix elements $\beta_{pp'}^{mm'}$ are evaluated for the atomic continuum states at energies ϵ_p and $\epsilon_{p'}$. The energies ϵ_p are chosen in terms of the discrete set of energy conserving points in the unperturbed continuum. This choice is a generalization of the well known Weisskopf-Wigner ansatz[7] in the theory of fluorescence and spontaneous emissions. In practice the whole calculation is performed with a finite matrix Hamiltonian built up from a maximum number, j_{max}, of atomic states and a maximum number, N, of photon exchange (real and virtual). Convergence with respect to both j_{max} and N are then looked for, for a fixed density and energy of the incident photons.

Once the non-Hermitian $\hat{H}$ is diagonalised and the convergence of the eigenvalues λ and the eigenvectors S with respect to the truncation size j_{max} and N are secured, the time evolution of the occupation probability of any of the discrete states are obtained from the squared modulus of (6). The evolution of the electron in the continuum is obtained from the amplitudes (7). Note that in the present problem the Hamiltonian $\hat{H}$ is symmetric and hence we can replace R by S^T and require to diagonalise $\hat{H}$ only once.

Results and Discussions

Before I discuss the results of application of the theory to the evolution of multiphoton ionization in the hydrogen atom, let me make a few remarks on the nature of convergence of the calculations.

In the present procedure, unlike in the usual perturbation theory, there are no divergencies with respect to resonances and the procedure converges in N roughly like

$$\left|\frac{eEa_0}{\hbar\omega}\right|^{|N|}$$

where eEa_0 is the measure of the atom-field interaction energy (E = electric field amplitude). The Nth order perturbation theory on the other hand converges like

$$\left|\frac{eEa_0}{\epsilon_{jj'} \pm n\hbar\omega}\right|^{|N|}$$

and hence readily breaks down near any intermediate resonance

$$\hbar\Delta_{jj'}^{(n)} \equiv \epsilon_{jj'} \pm n\hbar\omega = 0$$

The convergence with respect to inclusion of atomic states is more difficult to discuss generally. Nevertheless, inclusion of the Schrödinger-Fock states $|jn\rangle$ for which the detunings $\hbar\Delta_{jj'}^{(n)}$ are smaller than the field photon energy, $\hbar\omega$, or

$$0 \lesssim \hbar\Delta_{jj'}^{(n)} \lesssim \hbar\omega$$

is generally an excellent starting point for achieving numerical convergence in most cases of interest, i.e. when $\left|\frac{eEa_0}{\hbar\omega}\right| < 1$

For ultrahigh photon densities, $\left|\frac{eEa_0}{\hbar\omega}\right| > 1$,

very little is known, if at all, regarding the convergence properties of the theory (see, however, Ref. 8,9 and references therein, for discussions of convergence in certain asymptotic methods used at ultrahigh intensities).

Let me now discuss the results of explicit computation for ionization of a ground state hydrogen atom coupled to the continuum by a monochromatic radiation field. I would like to consider particularly (a) the toal ionization probability $P_{ion}(t)$ as a function of the interaction time; (b) the survival probability of the initial state as a function of time $P_{1s}(t)$ and (c) the evolution of the energy-spectrum of the ionized electrons.

In Fig. 2 (a) we present the toal resonance-ionization probability for an hydrogen atom. The frequency of the radiation is chosen to be $\omega = 0.375$ (a.u.) (corresponding to the 1s – 2p resonance frequency of the atom). The field amplitude E is taken to be 10^{-3} (a.u.) or 5.14×10^6 volt/cm. It would be seen that the ionization probability grows at first with time and then saturates towards the value of unity. The relevant scale of time is seen to be given by $1/\gamma = 261$ p sec.; γ is the width of the ground state due to ionization and corresponds roughly to the rate of ionization in the domain where the dependency of P_{ion} (t) is linear in t .

In Fig. 2 (b) the probability of finding the electron in the ground state is presented as a function of time. Starting from $P_{1s}(0) = 1$, it shows an over all irreversible decay with time. However, the instantaneous probability is highly oscillatory as can be seen from the (computer drawn) figure directly. We note that the corresponding ionization probability in Fig. 1 (a) shows no such oscillation. This is due to the fact that the probability distribution of the intermediate resonant 2p-state also oscillates rapidly but in nearly opposite phase to that of the ground state and effectively cancels the oscillating contribution of the ground state in the total probability. (The occupation probability of the non-resonant states although systematically incorporated in the computation are numerically very small compared to that of the resonant states).

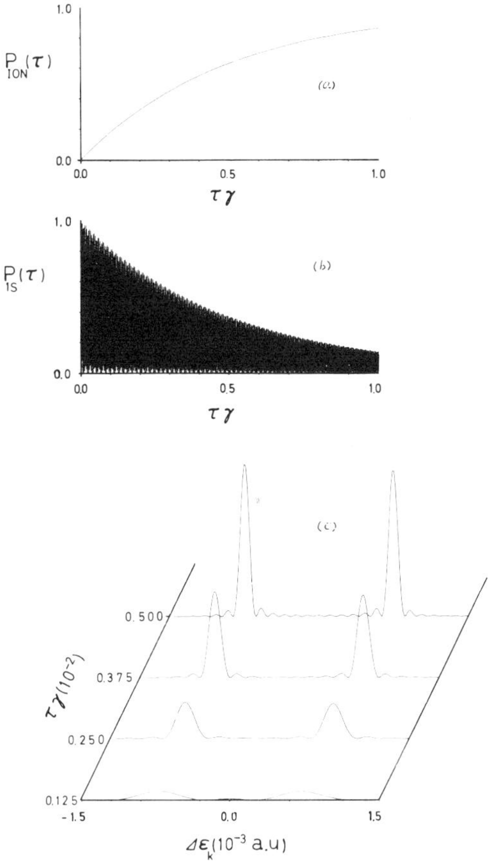

Fig. 2. Time dependence of resonant multiphoton ($n \geq 2$) ionization of the hydrogen atom. Shown are (a) the total ionization growth probability, (b) the survival probability of the initial ground state atom and (c) the evolution of the energy spectrum of the ionized electrons with increasing interaction time. The field photon energy $\hbar\omega$ = 0.375 a.u. corresponding to a single photon 1s-2p resonance. The parameter $1/\gamma$ = 261 psec, provides a convenient time scale of evolution of the system. Physically γ is the calculated width of the dressed state corresponding to the unperturbed initial state. The field strength corresponds to $E = 10^{-3}$ (a.u.) = 5.142×10^6 V/cm.

In Fig. 2(c) we present the energy-spectra of the ionized electron at different times. The first thing one notes about these spectra is their doublet structure. This is a consequence of the splitting of the intermediate resonant 2p-state under the influence of the dynamic stark effect induced by the coherent radiation field (an Autler-Townes effect). The doublet is seen to be placed symmetrically around the 'unperturbed energy' $\epsilon_k = 2\omega - E_{ion}$, where E_{ion} is the ionization threshold. One notes further that the doublet structure grows as well as narrows with increasing time. The height of the doublet is more than 15 times at $\gamma t = 0.125$ compared to that at $\gamma t = 0.5$ while the widths have shrunk by more than 3 times in the same time interval. This is qualitatively understood as the increase of probability of ionization with a diminishing 'rate' as time progresses.

Fig. 3(a) shows the total ionization probability for the two-photon ionization with an intermediate one photon near-resonance. The frequency of laser is chosen to be $\omega = 0.376$ (a.u.) corresponding to a detuning $\Delta = .001$ (a.u.) to the (near-resonant) 2p-state. The general growth of the ionization with time is similar to that in the resonant case (see Fig. 1(a)) but now to attain the same degree of ionization one requires longer than in the resonant case; this is reflected by the characteristic time $1/\gamma = 621$ p sec compared to 261 p sec in the resonant case.

In Fig. 3(b) we see the corresponding near-resonant probability of finding the electron in the ground state. This result differs from that of the resonant case (Fig. 2(a)) in that the rapid oscillations are confined mostly above the minimum value of zero, reflecting the general fact of incomplete Rabi precession for finite detuning from resonance. However, the envelope of the time evolution shows as before an irreversible decay with increasing time.

In Fig. 3(c) we show the evolution of the corresponding electron energy spectrum. The general structure of the spectrum is similar to that in the resonant case, Fig. 2(c), except that the Autler-Townes doublet is here placed asymmetrically around the electron energy $\epsilon_k = 2\omega - E_{ion} = 0.252$ (a.u.). Qualitatively, the widths of the electron energy spectrum indicates the rapidity with which the system ionizes. It would be seen that compared to the resonant case they are significantly narrower for the same (γt) values, reflecting relatively slower decay in the non-resonant situation. As in the resonant case, the ionization probability generally increases with a diminishing rate with increasing interaction time.

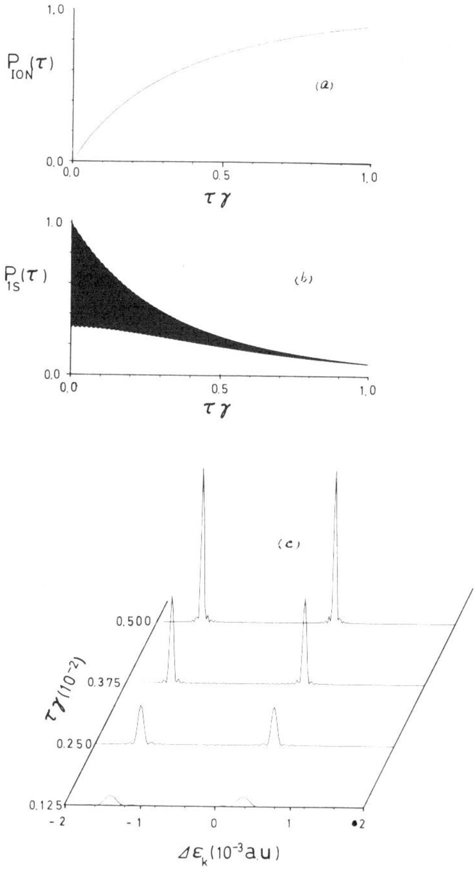

Fig. 3. Time dependence of non-resonant induced multiphoton decay of the hydrogen atom. Shown are, (a) the total ionization growth probability (b) the survival probability of the initial groundstate atom and (c) the evolution of the energy spectrum of the ionized electron. The photon energy, $\hbar\omega$ = 0.376 (a.u.), is .001 (a.u.) off-resonant with respect to the 1s-2p transition energy. The time scale of the evolution of the system is given by the parameter $1/\gamma$, where γ is the calculated width of the dressed state corresponding to the unperturbed initial 1s-state. In the present case $1/\gamma$ = 621 psec. The field strength E = 10^{-3} a.u. = 5.142 x 10^6 V/cm.

In Figs. (4) are presented the results for ionization by absorption of $n \geqslant 3$ photons. Fig. 4(a) shows the probability of resonant ionization. The frequency is $\omega = 0.1875$ (a.u.) so that there is a two-photon resonance with respect to the intermediate 2s-state. We note that the ionization probability in the early times passes through points of inflexion and then settles down to an uniform growth towards the saturation value of unity. In this circumstance the usual constant rate of transision - which requires a linear variation of probability with time - ceases to be a meaningful parameter of the process. Fig. 4(b) exhibits the decay of the ground state with increasing time. The decay here shows only a gentle modulation. This is in contrast to the rapid oscillations in the corresponding two-photon resonant case shown in Fig. 2(b). This difference is understood in terms of the coupling strengths involved between the ground and the resonant state and that between the resonant state and the ionization continuum. The two-photon resonant 1s-2p coupling is comparable to the coupling strength to the continuum from the 2s-level; both being of the order of E^2 (E is the field strength). Ionization in this case competes with the bound-bound Rabi-oscillation and effectively inhibits the latter. The one-photon resonant 1s-2p coupling is much stronger ($\propto E$) than the corresponding coupling to the 2p-continuum ($\propto E^2$) which allows many Rabi-oscillations to occur before ionization becomes effective. The severe inhibition of the Rabi-oscillation due to competition with the ionization decay is also revealed in the qualitative change in the doublet structures of the electron energy spectrum (Fig. 2(c) and 3(c)) into the singlet structure in Fig. 4(c). It shows indirectly the absence of the Autler-Townes splitting of the resonant state due to lack of sufficient Rabi oscillations. It would be noted however, that irrespective of the wide difference in the transient behaviour in the two cases, given long enough interaction time the ground states inevitably decay out into the continuum for any finite couling (no matter how small) - as can be seen from the probability envelopes in the two figures.

It is also interesting to investigate the dependence of ionization on simultaneous resonant coupling to two different degenerate states. Figs. (5) show the result of $n \geqslant 3$ photon ionization when the 1s state resonates simultaneously with the 3s and 3d states of the H-atom. An important consequence of such coupling is the prediction of beating in the transient regimes of the ionization probability (Fig. 5(a)). The decay probability of the ground state exhibits also the same beating effect (Fig. 5(b)). The origin of the beating effect lies in the somewhat different coupling strengths (both bound-bound and bound-free) associated with the 3s and 3d resonant states. This leads to slightly different "frequencies of ionization" along the two different resonant paths which interfere

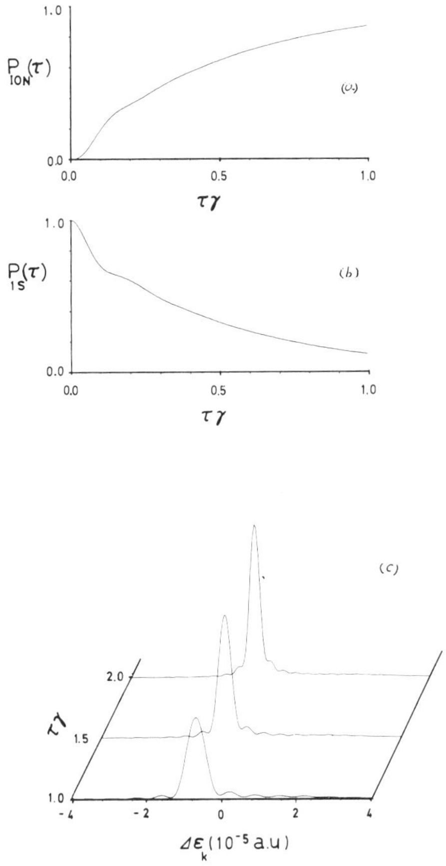

Fig. 4. Time dependence of resonant induced multiphoton ($n \geq 3$) decay of the hydrogen atom. Shown are, (a) the total ionization growth probability, (b) the survival probability of the initial (ground) state atom and (c) the evolution of the energy spectrum of the ionized electrons with increasing interaction time. The field photon energy, $\hbar\omega = 0.1875$ a.u. corresponds to a two-photon 1s-2s resonance. The time scale parameter $1/\gamma$ is given conveniently by the width of the dressed state corresponding to the unperturbed initial 1s-state; $1/\gamma = 24$ psec. The field strength $E = 10^{-3}$ a.u. $= 5.142 \times 10^6$ V/cm.

to generate the beat effect on the ionization probability and on the decay probability of the ground state. It should be noted that this effect is not due to the different ionization frequencies associated with the resonant and the 'non-resonant' paths, which are too different to lead to beating. (In fact the inclusion of more and more non-resonant paths in the computation*, for achieving quantitative convergence, showed no qualitative change of the beat structures which emerged from the beginning when no non-resonant path was allowed in the calculation.) The associated energy spectrum of the ionized electron as a function of the interaction time is shown in Fig. 5(c) which exhibits a lobsided doublet structure caused again by the competition between the comparable coupling strengths of the two resonant states to the continuum and to the initial bound state, respectively.

Summary

A relatively new aspect of photo-ionization, (generally) due to multiple absorption of photons, is the time evolution of the system decaying into the continuum. A systematic method of treating this problem is developed which converts the Hermitian Hamiltonian of the atom plus radiation system into an equivalent non-Hermitian Schrödinger problem in the time-domain. The practicability of the method is shown by detailed calculation for the evolution of a ground state hydrogen atom coupled to the ionization continuum by a strong radiation field of sub-(ionization) threshold photon energy. Interpretation of the results is facilitated due to the rigorous connection of the present method with the general theory of quantum decay. Our results reveal qualitatively different behaviour of the system depending on the photon energy and density and the atomic level structure. This fact provides a means of influencing the short-time evolution of the inevitable (long time) decay of the system. A number of interesting phenomena such as the splitting of the resonant and near-resonant discrete states of the atom, the associated fine structure of the energy spectrum of the ionized electrons, the existence of beat effects in the ion current or the competition between the excitation and the ionization channels in time, are demonstrated by accurate numerical calculations in hydrogen atoms.

---

*(in the present case the following atomic states and photon blocks have been included: 1s-2p-3p-4p-5p-6p-7p-8p-3s-3d-ϵs ϵp-ϵd-ϵf and $\Delta n = 0, \pm 1, \pm 2, \pm 3$)

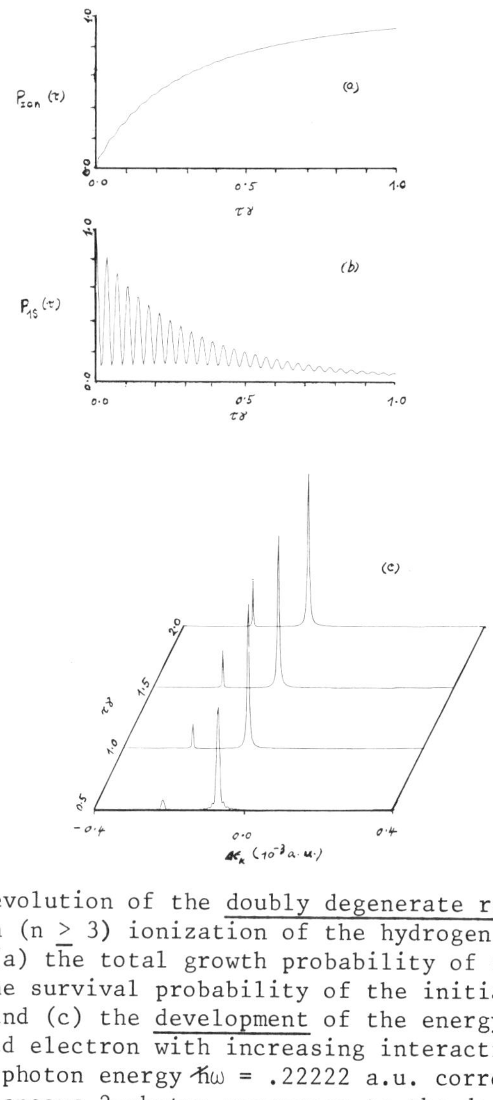

Fig. 5. Time evolution of the doubly degenerate resonant multi-photon ($n \geq 3$) ionization of the hydrogen atom. Shown are, (a) the total growth probability of ionization, (b) the survival probability of the initial ground-state atom and (c) the development of the energy spectrum of the ionized electron with increasing interaction time. The field photon energy $\hbar\omega$ = .22222 a.u. corresponds to a simultaneous 2-photon resonance to the degenerate 1s-3s and 1s-3d transitions. The time scale parameter $1/\gamma$ = 774 psec. is conveniently given by the width, γ, of the dressed state corresponding to the unperturbed initial state. The field strength, $E = 10^{-3}$ (a.u.) = 5.142×10^6 V/cm.

From the methodological point of view, the present approach treats the various partial waves in the continuum on equal footing and takes systematically into account, the propagation of the ionized electron in the continuum (by absorption of further photons after the ionization) through natural generalization of the Weisskopf-Wigner (energy-conserving) ansatz in the theory of spontaneous emission.

Acknowledgements

It is a pleasure to thank Prof. Jürgen Hinze for his invitation to present this talk at the workshop. The work presented here has been carried out jointly with Dr. J.V. Moloney. I would also like to thank Dr. E. Yurtsever for his many helps during the meeting.

References

1. F.H.M. Faisal and J.V. Moloney, "Time Dependent Theory of Non-Hermitian Schrödinger Equation: Application to Multiphoton Ionization Decay of Atoms", (to be published).
ESCOMP - European Study Conf. on Multiphoton Processes: Bénodet, France, June 18-22, Abstracts p.69 (1979).
ICOMP II - Int. Conf. on Multiphoton Processes: Budapest, Hungary, April 14-18, Abstracts p. B21-22 (1980).
2. C. Cohen-Tannoudji, Vol 2, in "Cargèse Lectures in Physics", Gordon and Breach, New York (1968).
3. H. Feshbach, Annals of Physics 5, 337 (1958).
4. I. Prigogine, "Time, Irreversibility and Structure" in "The Physicists Conception of Nature", ed. J. Mehra, Reidel Publishing Co., Boston, pp. 561-593 (1973).
5. V.A. Fock and S.N. Krylov, J. Exps. Theor. Fiz. 17, 93 (1947).
6. F.H.M. Faisal, J. Phys. B: Atom. Molec. Phys. 9, 3009 (1976).
7. V.F. Weisskopf and E.P. Wigner, Z. Physik 63, 54 (1930).
8. F.H.M. Faisal, J. Phys. B: Atom. Molec. Phys. 6, L 89 (1973).
Phys. Letts. 56A, 366 (1976).
9. H.S. Brandi and L. Davidovich, J. Phys. B: Atom. Molec. Phys. 12, L615 (1979).

THE LASER AS A TOOL TO SUPRESS THE BACKGROUND OF RESONANCES AND THRESHOLD EFFECTS IN ELECTRON SCATTERING

C.Jung* and H.S.Taylor

Department of Chemistry
University of Southern California
Los Angeles, California 90007, USA

ABSTRACT

It is demonstrated with the help of model calculations that a laser field can be used to supress the background in electron scattering cross sections and to pick out only those parts which are rapidly varying as λ function of the incoming electron energy. Therefore it becomes possible to measure the pure Breit-Wigner peaks and threshold effects independent of the fact that interference between the resonances and the background in radiationless electron scattering is strong. The results are interpreted within the low frequency limit for free-free transitions. The AC Stark shift of the resonance is also observed.

1. INTRODUCTION

Some of the most spectacular physics in electron-atom scattering is in the appearance of resonances and threshold effects. Unfortunately, in many cases these effects are small structures buried in a big background and their shape is usually modified by interference with the background.

Sometimes it is difficult to be sure if a measured structure is real or noise and sometimes it is hard to pull the interesting physical parameters out of the measured data. For a review of these effects and their experimental difficulties see[1]. If on the other

*Present address: Fachbereich Physik der Universität, 6750 Kaiserslautern, West Germany.

hand, an experimentalist were able, by some method, to supress all the background in the neighborhood of a resonance, then a pure Breit-Wigner peak would be observed and it would be straightforward to determine the position E_R and the width Γ of this resonance. The purpose of this paper is to explain a method of performing an electron scattering experiment in such a way that only those parts of the scattering amplitude lead to a signal, which vary rapidly as function of the incoming electron energy. Thereby we pick resonances and cusps out of the total scattering amplitude and get rid of all background influences.

The basic idea is to perform the electron scattering process within a strong laser field and to collect only those scattered electrons which have emitted or absorbed a certain number of photons. This method has been suggested first[2] in a treatment of free-free transitions in first order perturbation theory and low frequency limit. It will become clear in this paper that the idea works under much more general circumstances. Processes in which an electron scatters off an atom or molecule and emits or absorbs photons at the same time are known under the name free-free transitions. (For a review of these processes see[3]). In these processes three systems (electron, target, laser field) interact with each other and this makes a theoretical treatment so complicated that up to now only special cases and approximations have been investigated. If the photon energy is far away from all transitions energies of the target states involved in the process then the interaction between the laser and the target can to a high degree of approximation be neglected and the problem reduces to the description of electrons under the simultaneous influence of the target potential and the laser field. To check this the influence of the laser-target interaction on the processes considered here can be estimated by the following considerations: The external field induces an electric dipole moment in the target atom and this dipole field can be felt by the scattered electron and can thereby modify the electron-target interaction potential V. Next let us estimate the order of magnitude of this effect. In most of our calculations a photon energy of 0.005 a.u. is used which is close to the CO_2 laser photon energy of 117 meV. Because this photon energy is smaller by a factor 10 to 100 compared to the first excitation energy of the atomic ground states we use the static polarizability to calculate the induced dipole moment. (Thereby we overestimate the effect). For atoms the polarizability is between $2 \cdot 10^{-25}$ cm^3 (in e.s. units) for such a small closed shell atom like Helium and $400 \cdot 10^{-25}$ cm^3 for such a big alkali atom like Cesium. To see strong signals in free-free transitions we need laser power fluxes in the order of 10^8 Watt/cm^2 and for a CO_2 laser this corresponds to an electric field amplitude of 270 000 V/cm or 900 e.s.u.. Therefore the induced dipole moment is between $1.8 \cdot 10^{-22}$ e.s.u. and $3.6 \cdot 10^{-20}$ e.s.u. for the various atoms. A scattering electron at a distance of 10^{-8} cm, which is in the order of magnitude of the average distance of an electron in a resonance state from the atomic center, will feel a potential energy E_d of a magnitude between 10^{-3} eV and 10^{-1} eV. In electron scattering dipole effects are usually of importance only at very low energy where the electron-dipole interaction energy E_d is of the same order of magnitude as the kinetic energy E_i of the electron. In the case considered here $E_d/E_i \approx 0.01$ in the worst case. Therefore it seems to be justified to neglect the laser-atom interaction.

This neglect of the laser-target interaction does not of course apply at all to molecules which have vibrational transition energies close to the photon energy. In general, the laser frequency must be far out of resonance with the target system.

In addition the laser field strength considered here is small compared to the internal field in the atoms and therefore it is not necessary to worry about the ionization of the target.

Now even this simplified problem without laser-target interaction has not yet been solved in general. Various treatments have either trated the laser-field in perturbation expansion or have used the low frequency approximation. The method used here to supress the background can be interpreted theoretically in the low frequency limit. The relevant ideas of this approximation will be given later in section 3 (For some additional information about more general aspects of the low frequency approximation see[4-7]).

To fully appreciate the method it is perhaps better to forget about formal theories and instead to make some simple model calculations in the spirit of a numerical experiment. In section 2 we investigate in this way what may happen in an experiment since up to now the new method has not yet been realized in the laboratory. As such we construct a simple 1-dimensional model, which can be solved in terms of known functions. We do not use the low frequency approximation and we do not make a perturbation expansion of the electron-laser interaction. The approximations we must make are the neglect of the laser-target interaction, mentioned above, the dipole approximation and a cut off of the photon Fock space due to the inability to numerically handle more than a finite number of photon number states. In practice this is not a problem as all reported results, upon inclusion of further photon states in the model, do not significantly change.

In section 3 we present a theoretical interpretation of our numerical results. Section 4 contains conclusions and final remarks and some suggestions onto how to optimise possible experiments. In the appendix we explain our model in detail and show how we calculate all transition amplitudes in terms of known functions.

We use atomic units for all numbers throughout the whole rest of the paper.

2. RESULTS OF THE NUMERICAL EXPERIMENTS

We take a scattering problem in 1 dimension and represent the target by a 2-state system with an excitation energy $E_{12} = 0.6$. We assume that this target provides a 2-channel square well potential V to the incoming electron (see fig.1). In each figure below we have indicated the special choice of V used. We always start the process with the target in its lower state. If the laser is switched off and the incoming kinetic energy E_i of the electron is below 0.6 then the electron may either be transmitted or reflected without energy change. We denote this probability for transmission by T_{rl} and the probability for reflexion by R_{rl} (rl is an abbreviation for "radiationless"). If E_i is above 0.6 the electron may excite the target and leave it in its upper state. We denote the probability for excitation in radiationless transmission by S_{rl} and the probability for excitation radiationless reflexion by Q_{rl} · S_{rl} and Q_{rl} exist only for E_i 0.6.

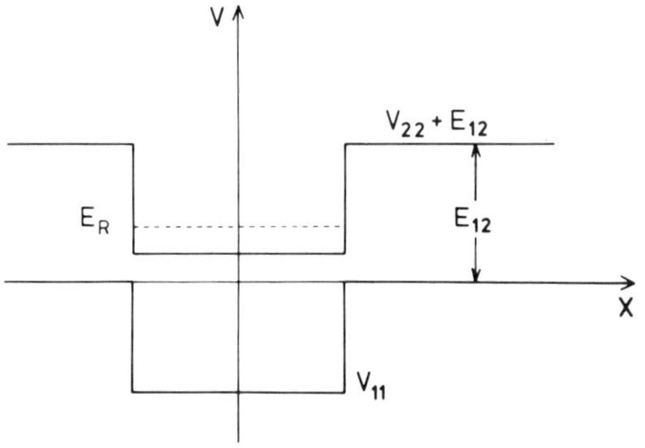

Fig. 1. Plot of the diagonal elements of the model potential for electron-target interaction used in all numerical calculations. The offdiagonal elements $V_{12} = V_{21}$ (not plotted in fig. 1) are of the same shape as V_{11} but of different depth. Bound states of the 1-channel potential $V_{22} + E_{12}$ become Feshbach type resonances of the full coupled 2-channel potential. Their width Γ is determined by the magnitude of V_{12}. For small V_{12} we find $\Gamma \propto (V_{12})^2$. See the two resonances in the upper lines of fig. 2 and 4 . A threshold effect occurs at the rim of the upper well i.e. at $E = E_{12}$.

If we switch on the laser field, then the electron can, in addition to the above, exchange energy with the field in integer multiples of the photon energy ω. We denote the probabilities for processes in which the electron gains (loses) the energy Nω from the field by Q_N, R_N, S_N, T_N respectively. $R_N(E_i)$ is the probability that an electron comes in with initial kinetic energy E_i, hits the target, is reflected leaving the target behind in its lower state and flies away with a final kinetic energy $E_f = E_i + N\omega$. Similarly, $Q_N(E_i)$ is the probability that the electron comes in with energy E_i, hits the target, is reflected leaving the target behind in its upper state and flies away with final energy $E_f = E_i \pm E_{12} + N\omega$. $T_N(E_i)$ and $S_N(E_i)$ are the probabilities for the corresponding processes in transmission. Of course, $Q_N(E_i)$ and $S_N(E_i)$ only exist for $E_f > 0$ i.e. $E_i + N\omega > E_{12}$. The quantities Q, R, S, T are the 1-dimensional analogs to the differential cross sections in 3 dimensions.

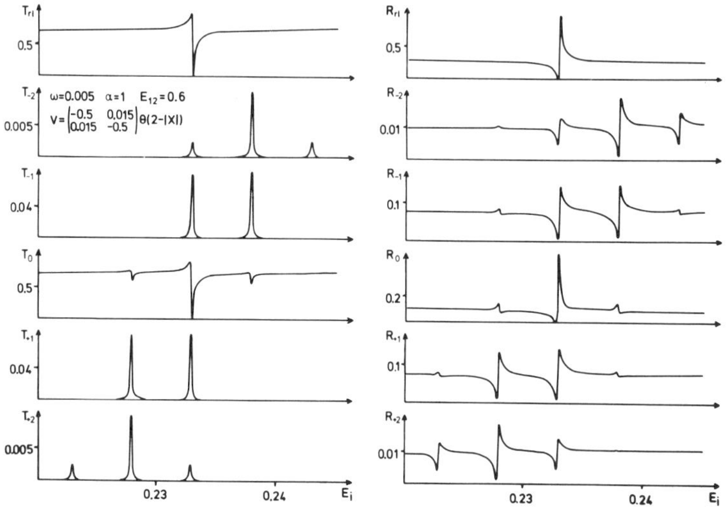

Fig. 2. Plots of resonance structures in radiationless scattering and in free-free transitions. For more expanations see main text.

In fig. 2 we have chosen a potential V which causes an elastic scattering resonance at an incoming energy of $E_r \approx 0.233$ with a width of $\Gamma \approx 0.0002$. This Γ is in the same order as the natural width of electron-rare gas atom resonances. The energy E_r is far below the threshold at $E_T = 0.6$ and therefore no excitation of the target is possible near this energy. In the top line of fig. 2 we show the quantities $T_{r1}(E_i)$ on the left and $R_{r1}(E_i)$ on the right as functions of E_i. Below we show the quantities $T_N(E_i)$ and $R_N(E_i)$

as functions of E_i for several values of N. The photon energy is always $\omega = 0.005$ which is close to the photon energy of a CO_2 laser. The power density of the field is so big that the quantity $\alpha = eA/mc + \omega$ has a magnitude of exactly 1 (e is the electron charge, m is the electron mass, c is the speed of light, A is the amplitude of the vector potential). This is still a moderate power density and therefore only one photon processes give a strong signal. Two photon transitions can just be seen. In fig. 3 we show the results for a calculation in which we have increased the laser power so that $\alpha = 2$ but left all other parameters unchanged.

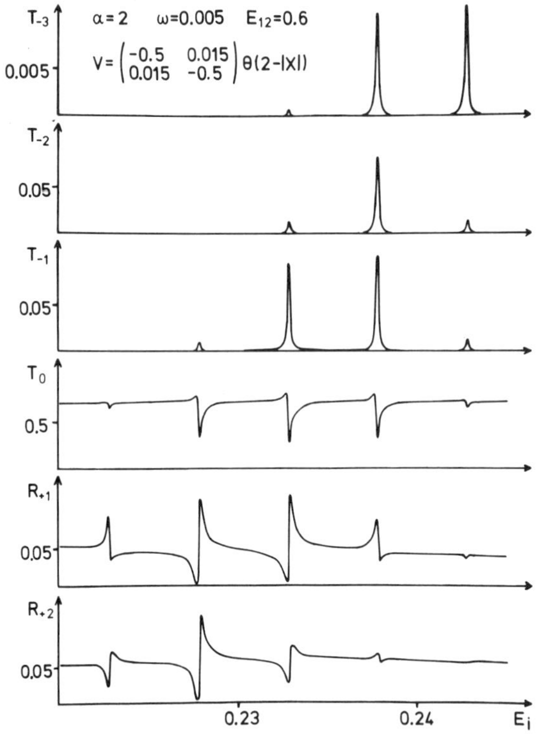

Fig. 3. Plots of resonance structures in free-free transitions. For more explanations see main text.

The important results of these figures are: The elastic resonance causes a whole series of resonances in each R_N and T_N but only a few ones are strong enough so that we can observe them immediately. The distance between two adjacent structures is exactly ω. The strength of the individual structures depends on the laser power. In general we see more resonances if we increase the laser power. The most striking observation is, that in T_N, $N \neq 0$ there is no background at all and the resonances appear as pure Breit-Wigner peaks. In all other cases (i.e. for T_0 and all R_N) the interference of the resonance with the background is either similar to the corresponding radiationless case only with the difference that the relative resonance effect is smaller or the resonance shape is reflected compared to the radiationless case. The best examples for this reflection are in R_{+1} in fig. 3. We see that it is easier to determine E_R and Γ out of T_{+1} or T_{-1} than out of T_{r1} or R_{r1}. In fig. 2 note that T_N and T_{-N} or R_N and R_{-N} look similar in shape but are shifted by $N\omega$.

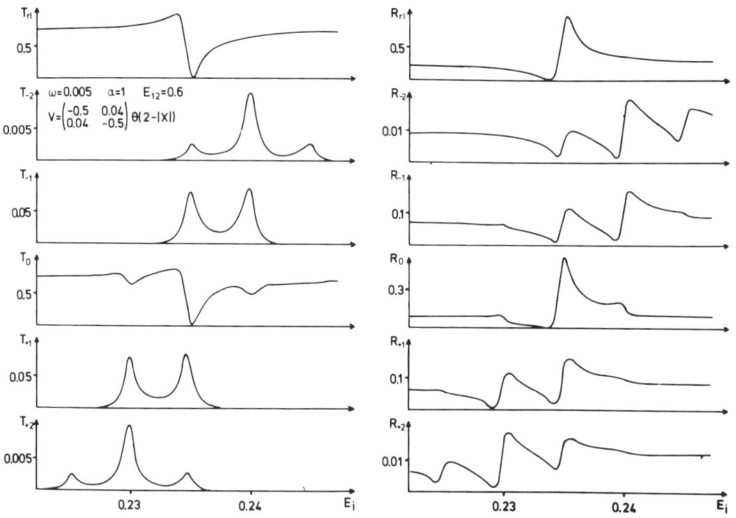

Fig. 4. Plots of resonance structures in radiationless scattering and in free-free transitions. For more explanations see main text.

In fig. 4 we have changed the offdiagonal elements of V so that the width Γ of the elastic resonance is now 0.0012 which is in the order of the resolution of common electron spectrometers. This change in V causes also a small shift in the position E_R of the resonance (see the top line in fig. 4). All other parameters are the same as in fig. 2. The width Γ is now so big that there are overlaps of the various resonance structures in R_N and T_N which influence the resonance shapes in general. For $T_N, N \neq 0$ the wings of two adjacent resonance peaks simply add without any visible interference. Again T_N and T_{-N} or R_N and R_{-N} look similar in shape but shifted by $N\omega$. In fig. 5 we show the resonance curves for $\alpha = 3$

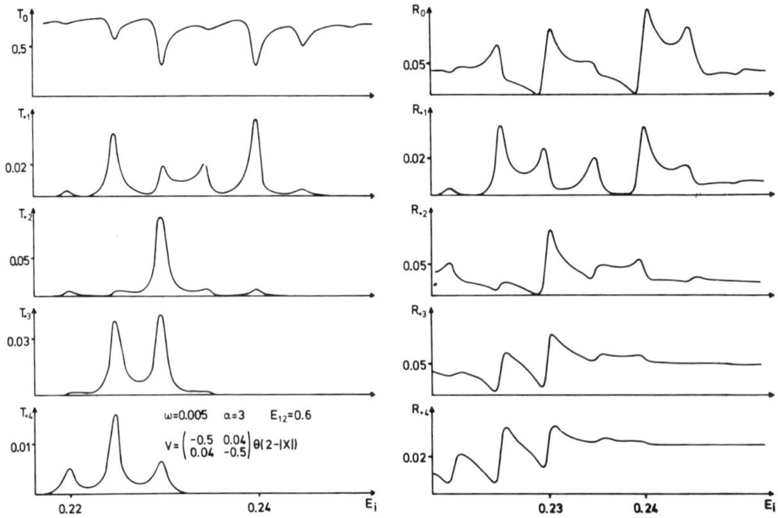

Fig. 5. Plots of resonance structures in free-free transitions. For more explanations see main text.

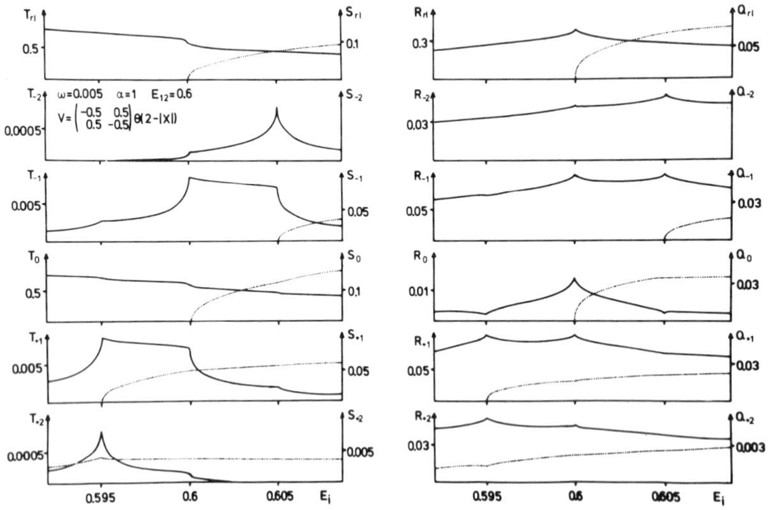

Fig. 6. Plots of threshold structures in radiationless scattering and in free-free transitions. For more explanations see main text.

and otherwise the same parameters as in fig. 4. We did not make calculations with another value of ω because the interesting quantity is the ratio between ω and Γ and an increase of ω would cause similar effects as a decrease of Γ.

The only threshold of our 2-channel model is at $E_T = 0.6$ and it is interesting to compare radiationless scattering and free-free transitions close to this energy. Fig. 6 shows radiationless scattering in the top line and free-free transitions below. In the drawings T and R are represented by solid lines and S and Q are represented by dotted lines. First we see that in free-free transitions the threshold for S_N and Q_N lies at $E_i = E_T - N\omega$. This is easy to understand because the energy which the electron gains/loses from the laser field goes into the energy balance for the excitation of the target by the electron impact and the new channel opens as soon as the final kinetic energy of the electron is above zero i.e. as soon as $E_i + N\omega - E_T > 0$ or $E_i > E_T - N\omega$. In T_{-1} we see a plateau between E_T and $E_T + \omega$ and in T_{+1} a plateau between $E_T - \omega$

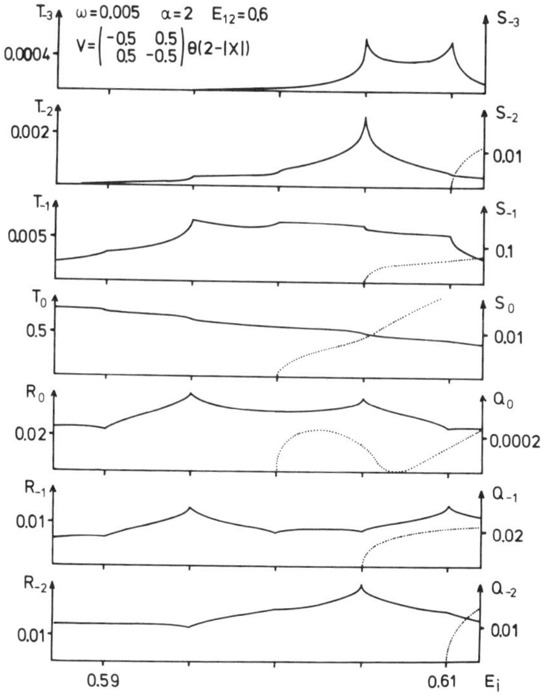

Fig. 7. Plots of threshold structures in free-free transitions. For more explanations see main text.

and E_T with a sharp drop on both sides. In addition there are smaller threshold effects shifted by ω away from the sharp drops. In T_{-2} and T_{+2} we see a sharp peak at $E_T + \omega$ or $E_T - \omega$ respectively and again smaller effects shifted away by ω. In the other cases (T_0 and all R_N) the threshold effects occur at various $E_i = E_T + n\omega$ and their shape is either of the same type as in the corresponding radiationless scattering or it is just turned upside down. In S_N and Q_N there are only very weak structures, too weak to be seen clearly in fig. 6.

In fig. 7 we see results for α = 2 and otherwise the same parameters as in fig. 6. Now the threshold effects are distributed over more energy values and for example in T_{-1} the main drop on both sides of the plateau is shifted outwards by ω on both sides compared to fig. 6. In the picture for Q_0 we see an "accidental" zero near $E_i = 0.607$ which will be explained below. In fig. 8 we see results for ω = 0.002 and otherwise the same parameters as in fig. 6. All curves look nearly the same as in fig. 6 only the energy scale

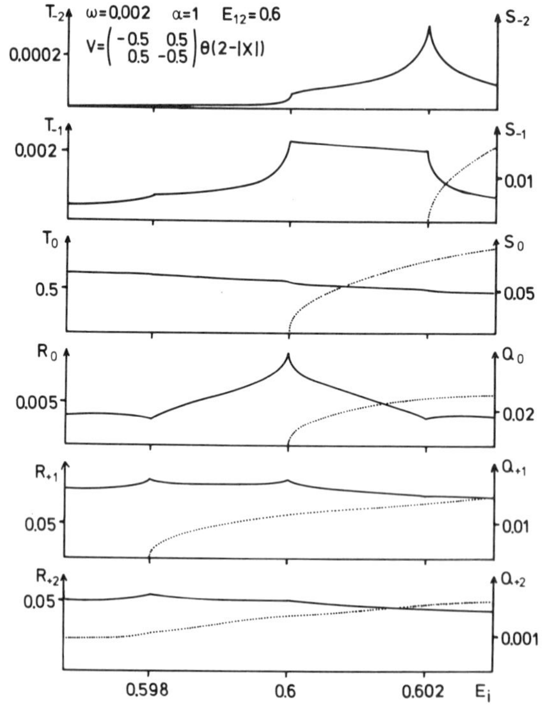

Fig. 8. Plots of threshold structures in free-free transitions. For more explanations see main text.

The most important results for the thresholds are in summary: A sudden drop (or increase) in the radiationless scattering as function of the incoming energy produces in the free-free probabilities T_{+1} and T_{-1} a plateau near the threshold energy and zero signal otherwise. At moderate laser power the length of the main structure is exactly ω. Thereby the presence or absence of a threshold makes a 100% change in the relative signal strength but the absolute signal strength is always very low. Depending on the values of ω and α the strong contributions to the threshold effects in T_N appear at different energies.

3. INTERPRETATION OF THE NUMERICAL RESULTS

A theoretical explanation of the results of section 2 can be given most easily within the low frequency approximation and therefore let us first explain its basic idea. In order to be more general we give all formulas for the 3-dimensional case and consider the numerical results from section 2 as the special cases of the scattering angles 0 and π. If the laser wavelength is very long compared to atomic distances, then we can separate the free-free transition into three steps. First a free electron moves within the laser field and can virtually emit and absorb photons. Let us assume that the laser field is single mode and in the pure number state $|N\rangle$ in absence of the electron. We denote the state of a free electron moving with momentum $\vec{p}$ by $|\vec{p}\rangle$. If we neglect photon depletion effects and use the dipole apppoximation, then the exact state of the electron in the field is given by (see appendix)

$$|\Phi\vec{p},N\rangle = \sum_n J_n(\vec{\alpha}\cdot\vec{p})\,|\vec{p}\,\rangle\,|N+n\,\rangle\,. \qquad (1)$$

J_n is the Bessel function of first kind and order n. $\vec{\alpha} = eA\vec{\varepsilon}/mc\overset{+}{h}\omega$ where ε is the polarization vector of the laser field. This $\vec{\alpha}$ is the 3-dimensional generalization of the α given above in section 2. The Bessel functions $J_n(\vec{\alpha}\vec{p})$ can be viewed as interaction coefficients for virtual n photon absorbtion/emission at absence of a target.

In dipole approximation there is no recoil of the electron during emission or absorbtion of a photon and therefore the electron momentum does not depend on how many photons the electron has absorbed or emitted i.e. $|\vec{p}\rangle$ in (1) is independent of n. The energy of the electron is different in its various states and the electron wave connected with the photon number state $|N+n\rangle$ is at energy $E = p^2/2m - n\omega$. Only the $n = 0$ term is on the energy shell and all other terms are off shell.

Then in the second step of the free-free transition this mixture of electron waves hits the target and is scattered. The main idea of the low frequency approximation is to neglect the laser-electron interaction during this second step and therefore the electron-target scattering is described by a radiationless scattering amplitude. But according to what was said above each term of the sum in (1) is scattered at its particular intermediate energy $E_i - n\omega$ and its scattering is therefore described by a scattering amplitude at this shifted energy.

In the third step the scattered electron waves interact again with the laser and evolve into their final states. Along these lines the following formula (3) has been derived first[8] and con firmed later[9] by another derivation. This formula gives the scattering amplitude $f_N(E_i,\vartheta)$ for an electron to come in with kinetic energy E_i, be scattered by an angle ϑ and to have a final kinetic energy E_f of

$$E_f = E_i + N\omega \tag{3}$$

$$(f_N(E_i,\vartheta) = \sum_k J_{N-k}(\vec{\alpha}\vec{p}_f) f_{rl}(E_i + k\omega,\vartheta) J_k(-\vec{\alpha}\vec{p}_i) \ . \tag{3}$$

f_{rl} is the corresponding amplitude for radiationless electron-target scattering. The three factors correspond to the three physical stages above. An analogous formula holds for scattering with excitation of the target if we take on the r.h.s. of (3) the corresponding scattering amplitude for radiationless excitation of the target and take the excitation energy into the energy balance (2).

Now let us decompose f_{rl} in (3) into a resonance part and a background part

$$f_{rl}(E_i + k\omega,\vartheta) = f^R_{rl}(E_i + k\omega,\vartheta) + f^{BG}_{rl}(E_i + k\omega,\vartheta) \ . \tag{4}$$

The Bessel functions decrease rapidly for increasing order and fixed argument as soon as the absolute value of the order becomes larger than the absolute value of the argument. Therefore only a few k give a strong contribution to the k sum in (3). Since f^{BG}_{rl} depends only weakly on its energy argument then we can neglect this dependence, take all f^{BG}_{rl} at the energy E_i, pull f^{BG}_{rl} out of the k sum and apply the addition theorem of the Bessel functions and find

$$f_N(E_i,\vartheta) = J_N(\vec{\alpha}(\vec{p}_f - \vec{p}_i)) f^{BG}_{rl}(E_i,\vartheta) + \sum_k J_{N-k}(\vec{\alpha}\vec{p}_f) f^R_{rl}(E_i + k\omega,\vartheta) J_k(-\vec{\alpha}\vec{p}_i) \ . \tag{5}$$

Our model calculation is independent of the low frequency approximation and so a comparison between the exact numerical results of sec. 2 and (5) can be viewed as test of (5).

The present state of the experimental verification of (5) is as follows[10]: $|f_N|^2$ has been measured as function of N for several values of $\vartheta, E_i, \vec{\varepsilon}$. All results are in qualitative agreement with (5). It has not been possible to check for quantitative agreement because the exact laser power distribution in space and time has not been known. For quantitative calculations its knowledge would be absolutely necessary since multiphoton transitions are a non-linear effects and depend essentially on the exact power distribution in space and time and not only on the average power (For these problems see[11]). In ref. 10 resonances did not play any role and therefore only the first term in (5) has been measured. From such a measurement we cannot, in contrast to the present work, learn anything new about electron-atom scattering because the result is essentially the product of the elastic background cross section and a Bessel function — both well known quantities.

In ref. 12, 13 $|f_{-1}|^2$ has been measured as function of E_i near the argon resonances at 11eV but unfortunately only in backward direction for the one scattering angle $\vartheta = 160°$. Since a weak laser has been used only the resonance structures at E_R and $E_R + \omega$ could be seen. Their shape and strength is in full agreement with (5).

For scattering without excitation of the target $|p_i| \approx |p_f|$ since $\omega \ll E_i$ and therefore in foreward direction we find $\alpha(\vec{p}_f - \vec{p}_i) \approx 0$. The background term in (5) disappears in foreward direction for $N \neq 0$ because $J_N(0) = \Delta_{N,0}$. This is the explanation why we did not see a background contribution for T_N, $N \neq 0$ in the model calculations. f^R_{rl} produces a resonance structure when its energy argument is at E_R and therefore we get a resonance structure in the k-sum in (5) every time when $E_i + k\omega = E_R$ for a k $\mathbf{Z}$. The magnitude of the resonance effect at $E_i = E_R - k\omega$ depends on the exact value of the two Bessel functions in the corresponding term in (5). If $\Gamma \ll \omega$ we get a series of well separated resonances with an energy spacing ω between adjacent peaks. This is exactly what we saw in fig. 2-5.

In backward direction $\vec{p}_f \approx -\vec{p}_i$ and in general $J_N(\alpha(\vec{p}_f - \vec{p}_i))$ is not near a zero and the background does not disappear. Depending on the relative signs of the various Bessel functions the relative phase between the resonance term and the background term is either the same as in the radiationless case or it is just shifted by π. Therefore the resonance shape is either the same as in the radiationless case or it is just reflected. The magnitude of the relative resonance effects depends again on the exact values of the corresponding Bessel functions and thereby on the parameters E_i, α. We saw that T_N and T_{-N} or R_N and R_{-N} look nearly the same in shape but are only shifted

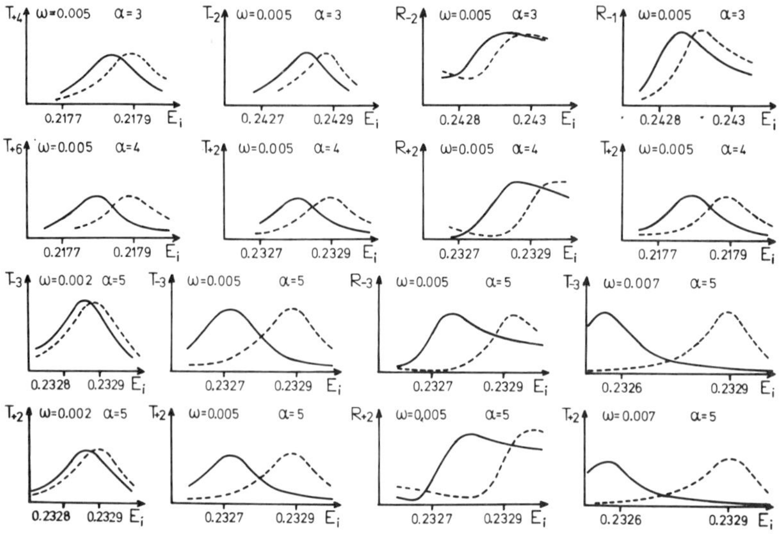

Fig. 9. Some examples for the AC Stark shift of resonances in free-free transitions. The solid lines are the results of numerical calculations. The broken lines are the results of eq. (3). For more explanations see main text.

by $N\omega$. This can be explained by the fact that $J_N = (-1)^{Nj} - N$.

One effect which is not contained in (5) is the AC Stark shift of the resonances. Let us look at fig. 9. In this figure we show resonance structures in T_N and R_N at various initial energies $E_i = E_R + k\omega$ and for various values of ω, α and N. The potential V is the same as in figs. 2 and 3. In each picture we show the low frequency approximation of (3) as broken line and the exact numerical result as solid line. We see a shift of the resonance structures which does not depend on E_i, N and ϑ but depends on ω and α. Besides this shift the low frequency limit gives the correct shape and height for all resonance peaks. The fact that the shift does not depend on the particular peak at which we look i.e. does not depend on E_i, N and ϑ is an indication that the resonance itself is shifted and not only its appearance in the free-free transitions. The Stark shift goes quadratically with ω and quadratically with α. This is in exact agreement with the results of ref. 9. In our examples in fig. 9 the shift $\Delta E_R \approx -0.25\omega^2\alpha^2$. According to ref. 9 the proportionality constant between ΔE_R and $\omega^2\alpha^2$ is a measure of the laser induced coupling between the resonance wave function φ_2 in channel 2 and the continuum wave function φ_2 in channel 1. Therefore a measurement of $\varphi E_R/\omega^2\alpha^2$ could give experimental information about the dipole matrix element $<\varphi_1|\vec{\varepsilon}\,\vec{p}|\varphi_2>$.

We could not see a change of Γ with increasing laser power. For this extra width caused by the Stark decay to become important much higher laser powers would be necessary.

Now let us try to explain the threshold effects in figs. 6-8 by making a quite crude model for a threshold effect in radiationless scattering and representing the amplitude for scattering without target excitation by

$$f_{rl}(E_i,\vartheta) = c(\vartheta) + \theta(E_i - E_T)\, d(\vartheta) \tag{6}$$

where c and d do not depend on E_i and θ is the unit step function. If we insert (6) into (5) we get

$$f_N(E_i,\vartheta) = c(\vartheta)\, J_N(\vec{\alpha}(\vec{p}_f - \vec{p}_i))$$

$$+ d(\vartheta) \sum_k J_{n-k}(\vec{\alpha}\vec{p}_f)\theta(E_i + k\omega - E_T)\, J_k(-\vec{\alpha}\vec{p}_i) \tag{7}$$

For $\vec{p}_f \approx \vec{p}_i$ and $N \neq 0$ the first term drops. The second term produces a step at each $E_i = E_T - k\omega$, k **Z.** The relative strength of the various steps depends on the exact values of the Bessel functions and thereby on the values of the parameters E_i and α.

If we set $\vartheta = 0$ in (7) and assume a moderate laser power for which only first order contributions are important, we obtain

$$f_{-1}(E_i,0) = \frac{1}{2}\alpha\,|p|\,d(0)\,\{\theta(E_i - E_T) - \theta(E_i - \omega - E_T)\}\ .$$

This is just a plateau between E_T and $E_T + \omega$ and zero otherwise.

An analogous calculation applies to T_{+1} and gives a plateau between $E_T - \omega$ and $E_T \cdot T_{rl}$ in fig. 6 does not have a sharp step at E_T but a rounded one and therefore also T_{+1} and T_{-1} have a rounded drop off on both sides of the plateau.

We see additional smaller threshold effects at other E_i which come from higher order contributions. For higher laser power as e.g. in fig. 7 higher order contributions may become more important than first order contributions and therefore the steps at other E_i become more important.

If we look at T_{-2} and take only second order contributions of (7) we get

$$f_{-2}(E_i,0) = d(0)\,\frac{1}{8}\,\alpha^2 p^2\ \{\theta(E - E_T)$$

$$- 2\theta(E_i - E_T - \omega) + \theta(E_i + E_T - 2\omega)\}$$

This is a positive plateau between E_T and $E_T + \omega$ and a negative plateau of the same absolute value between $E_T + \omega$ and $E_T + 2\omega$ and it is 0 otherwise. $T_{-2} = p_f/p_i \; |f_{-2}(E_i,0)|^2$ is then just a plateau between E_T and $E_T + 2\omega$. The step in T_{r1} in fig. 6 is rounded and therefore a wedge remains at $E_T + \omega$ in T_{-2} in fig. 6. This wedge is a left over between two rounded cut offs on both sides. For higher power density in fig. 7 again additional threshold effects at other E_i become important in T_{-2}. An analogous reasoning applies to T_{+2}.

For T_0 and R_N both terms in (7) contribute and the relative phase between both terms is either the same as in the radiationless scattering or it is shifted by π. Therefore the shape of the threshold structures is either the same as in the radiationless case or it is just turned upside down as e.g. in R_0 in fig. 6 at $E_T \pm \omega$.

For $E_i = 0.607$, $\alpha = 2$, $N = 0$, $\vartheta = \pi$ and excitation of the target the quantity $\vec{\alpha}(\vec{p}_f - \vec{p}_i)$ is just at the first zero of the Bessel function $J_0(\vec{\alpha}(\vec{p}_f - \vec{p}_i))$ and this explains the accidental zero of Q_0 in fig. 7.

DISCUSSIONS AND CONCLUSIONS

We have seen that the low frequency approxiamtion given in (3) explains all qualitative features of resonances and thresholds in free-free transitions except the AC Stark shift. The quantitative error of the low frequency approximation — corrected for AC Stark shift — has always been below a few percent[9]. But the dependence of this error on the various parameters will probably depend on the dimension of the space. Therefore it would not be of general interest to add here a detailed investigation of the numerical error of the low frequency approximation in our 1-dimensional model.

What advice can we give to an experimentalist, who wants to utilize free-free transitions in order to look for resonances and cusps in electron-atom scattering?

The most interesting property of free-free transitions shown in this paper is that only rapidly varying parts of the radiationless scattering causes any free-free signal in foreward direction. This fact may be used in an experiment to project out resonances and threshold effects. The price to pay for this removal of the background is a reduction of the number of electrons which contribute to the signal. As shown in figs. 2-5 the ratio between the resonance height in T_R, $N \neq 0$ and in T_{r1} is 0.1 in the most favorable cases.

In an experiment it is not possible to measure at an angle of exactly $\vartheta = 0$. By proper adjustment of the laser polarization vector $\vec{\varepsilon}$ it is possible to fulfil the relation

$$\vec{\varepsilon}(\vec{p}_f - \vec{p}_i) = 0 \tag{8}$$

for any choice of $\vec{p}_i$ and $\vec{p}_f$, just turn $\vec{\varepsilon}$ perpendicular to the momentum transfer $\vec{q} = \vec{p}_f - \vec{p}_i$. At the same time we want to see a strong resonance signal and according to (5) $\vec{\varepsilon}\vec{p}_f$ and $\vec{\varepsilon}\vec{p}_i$ should be big enough to give a big value for the Bessel functions. Therefore, $\vec{\varepsilon}$ should be as parallel to $\vec{p}_f$ and $\vec{p}_i$ as possible. All these conditions can be met best if $\vec{p}_f$ and $\vec{p}_i$ are as parallel to each other as the construction of the electron spectrometer allows and if $\vec{\varepsilon}$ is in the direction exactly in between $\vec{p}_f$ and $\vec{p}_i$.

As indicated by the figs. 2-5 resonance signals are biggest and clearest in T_{+2} or T_{-2} at $E_i = E_R - \omega$ or $E_i = E_R + \omega$ respectively if the laser power density is so big that $\vec{\alpha}\vec{p}_f$ and $\vec{\alpha}\vec{p}_i$ are in between 1.5 and 2.2 i.e. if $J_1(\vec{\alpha}\vec{p})$ is near its first maximum. In this case the reonance signal comes from the term $J_{\pm 1}(\vec{\alpha}\vec{p}_f)\, f_{r1}(E_i \pm \omega, \vartheta)\, J_{\pm 1}(\vec{\alpha}\vec{p}_i)$ in (5). This means: The electron comes in shifted by one photon energy away from the resonance energy. Then it absorbs/emits one photon and can go into the resonance state. The resonance decays and the outgoing electron absorbs/emits another photon on its way out. For this choice of the parameters the resonance signal in $T_{\pm 2}$ is about 10% of the resonance signal in the radiationless scattering, which is the best one can hope for. In addition, for these values of the parameters the other resonance structures in $T_{\pm 2}$ (i.e. the ones at E_R, $E_R \pm 2\omega$, $E_R \pm 3\omega$, etc.) are quite small and this decreases the sources of confusion if several elastic resonances occur close together.

For the investigation of thresholds choose a moderate power density of the laser so that only first order contributions give a strong signal to the free-free amplitudes and look at T_{+1} or T_{-1} for structures of the length ω. Two sharp drops in T_{+1} at $E_a - \omega$ and E_a and a smooth behavior in between indicate a threshold at E_a. A similar structure in T_{-1} between E_a and $E_a + \omega$ indicates the same threshold at E_a.

As indicated in figs. 6-8 the absolute signal in T_N, $N \neq 0$ is extremely weak but the relative threshold effect is 1 and can therefore be seen clearly. This is in sharp contrast to elastic scattering where the relative threshold effects are generally small. Usually these effects are observed in electron impact excitation, where their relative effect is bigger than in the elastic channel. Our idea is unusual in so far that it filters out threshold effects in a process which is elastic with respect to the electron-target interaction. The big disadvantage of our method is the extremely small absolute size of the signal and there is the possibility that in a real experiment machine noise will bury the weak threshold signal and make it impossible to utilize our idea in the laboratory. In any case for resonances or thresholds the photon energy should be chosen larger than the energy interval over which the radiationless scattering varies rapidly in order to avoid confusing overlaps of the structures in free-free transitions.

Can "accidental" zeros like the one in Q_0 shown in fig. 7 (i.e. those values of parameters for which $\vec{\alpha}(\vec{p}_f - \vec{p}_i)$ is at zero of some Bessel function) be used for anything? We don't think so, because their position depends strongly on the exact laser power and in an experiment the laser power varies in time and space and the electrons, collected in the detector, have experienced quite different laser powers during their scattering process. Therefore, such an accidental zero would be completely smeared out in any real experiment.

There is also the interesting possibility that a free-free experiment might be used to probe the laser field strength. In[11] it has been shown how the results of a measurement of $|f_N(E_i,\theta)|^2$ as function of N can be used to calculate the average power density of the laser field in which the experiment has been performed. It is probably possible to generalize the results of[11] and to work backwards to even more detailed information about the laser field.

APPENDIX

In this section we explain our model in detail and show how we calculate the amplitudes for all processes.

We work in a 1-dimensional space and choose x as space coordinate. We describe the electron-target interaction by a 2-channel square well potential of the form

$$\mathbf{V}(x) = \begin{pmatrix} V_{11} & V_{21} \\ V_{21} & V_{22} \end{pmatrix} \theta(R - x) . \tag{A1}$$

$V_{12} = V_{21} \varepsilon R$ so that $\mathbf{V}$ is a self adjoint operator. θ is the unit step function. The matrix of excitation energies is

$$\mathbf{B} = \begin{pmatrix} 0 & 0 \\ 0 & E_{12} \end{pmatrix} . \tag{A2}$$

We cut the x-axis into the three intervals

$$I_1 = (+R,\infty) ,$$
$$I_2 = (-R,+R) ,$$
$$I_3 = (-\infty,-R) .$$

The target is in its ground state initially and the electron comes in with momentum k_{in}. Then the matrix of the electron momentum in intervals I_1 and I_3 is

$$\mathbf{K} = \begin{pmatrix} k_o & 0 \\ 0 & h_o \end{pmatrix} = \begin{pmatrix} k_{in} & 0 \\ 0 & \sqrt{k_{in}^2 - 2mE_{12}} \end{pmatrix} . \qquad (A3')$$

In I_2 we set $\mathbb{M} = \mathbb{B} + \mathbb{V}$ and construct the orthogonal matrix $\mathbf{U}$ which diagonalizes $\mathbf{M}$ according to $\mathbb{U}^+\mathbb{M}\mathbb{U} = \mathbf{\Sigma}$ with

$$\mathbf{\Sigma} = \begin{pmatrix} \sigma_1 & 0 \\ 0 & \sigma_2 \end{pmatrix} . \qquad (A4)$$

$[k_{in}^2\mathbf{1} - 2m\mathbf{\Sigma}]^{\frac{1}{2}}$ is the diagonalized momentum matrix in I_2. We assume a single mode laser field and use the dipole approximation. The total Hamiltonian $\mathbf{H}$ for the motion of the electron under the simultaneous influence of the target and the laser field is

$$\mathbb{H} = \{\frac{p^2}{2m} + \hbar\omega a^+ a - \frac{ep}{mc}\beta(a^+ + a)\}\mathbb{1} + \mathbb{B} + \mathbb{V}(x) . \qquad (A5)$$

β is an abbreviation for $\beta = (2\pi c^2/\omega L^3)^{\frac{1}{2}}$ where L^3 is the quantization volume of the electromagnetic field. p is the electron momentum operator, a and a^+ are the annihilation and creation operators for a laser photon. We write $|N\rangle$ for an eigenstate of a^+a with eigenvalue N $\mathbb{N}$.

First let us look at eigenfunctions of $\mathbf{H}$ in the intervalls I_1 and I_3. There $\mathbf{H}$ is diagonal with respect to the target states. We neglect photon depletion effects in the sense that we set $\sqrt{N + n} = \sqrt{N}$ in the coupling strength between various photon number states, where N is the initial number of photons in the laser beam and N + n is the number of photons in intermediate or final states. Using the recursion formula of the Bessel functions

$$2nJ_n(y) = y\{J_{n-1}(y) + J_{n+1}(y)\} \qquad (A6)$$

we see that functions of the form

$$\varphi(k_o,x,N) = \exp(ik_o x) \sum_n J_n(\alpha k_o) \, |N+n\rangle \qquad (A7)$$

are eigenfunctions of the operator

$$\frac{p^2}{2m} + \hbar\omega a^+ a - \frac{ep}{mc}\beta(a^+ + a) \qquad (A8)$$

with eigenvalue

$$N\omega + k_o^2/2m \; . \tag{A9}$$

α is an abbreviation for $\alpha = e\sqrt{N}\, 2\beta/mc\, \hbar\omega$. The quantity $2\beta\sqrt{N}$ is the amplitude of the corresponding classical vector potential A of the laser field. Note that k_o is independent of n in (A7) because recoil effects of the electron are neglected in dipole approximation. In order to start from the most general eigenfunctions of **H** we must consider that the functions φ in (A7) are degenerated in two ways:

1. φ remains an eigenfunction of operator (A8) to the same energy (A9) if we reverse k_o i.e. if we replace k_o by $-k_o$.

2. We get an eigenfunction to the same energy if we replace N by N + L and simultaneously replace k_o by k_L where $k_L^2/2m + L\omega = k_o^2/2m$. Therefore the most general eigenfunction of **H** with eigenvalue E in the interval I_1 is the 2 component column vector

$$\Phi_1(E,x) = \sum_L \begin{pmatrix} r_{1,L}\varphi(k_L,x,L+N) + w_{1,L}\varphi(-k_L,x,L+N) \\ r_{2,L}\varphi(h_L,x,L+N) + w_{2,L}\varphi(-h_L,x,L+N) \end{pmatrix} \tag{A10}$$

where all k_L and h_L are given by

$$E = k_L^2/2m + (N+L)\omega = h_L^2/2m + (N+L)\omega + E_{12} \; . \tag{A11}$$

$r_{i,L}$ and $w_{i,L}$ are arbitrary complex constants to be fixed later by boundary conditions.

In a completely analogous way we find for the most general eigenfunction of **H** with eigenvalue E in interval I_3

$$\Phi_3(E,x) = \sum_L \begin{pmatrix} j_{1,L}\varphi(k_L,x,L+N) + t_{1,L}\varphi(-k_L,x,L+N) \\ j_{2,L}\varphi(h_L,x,L+N) + t_{2,L}\varphi(-h_L,x,L+N) \end{pmatrix} \; . \tag{A12}$$

$j_{i,L}$ and $t_{i,L}$ are again constants to be fixed later by boundary conditions.

In the interval I_2 **H** is not diagonal with respect to target states but $\mathbb{U}^+\mathbb{H}\mathbb{U} = \tilde{\mathbb{H}}$ is and therefore we first construct eigenfunctions to $\tilde{\mathbf{H}}$. In the same way as before we find for the most general eigenfunction of $\tilde{\mathbf{H}}$ with eigenvalue E

$$\tilde{\Phi}_2(E,x) = \sum_L \begin{pmatrix} a_{1,L}\varphi(\lambda_L,x,L+N) + b_{1,L}\varphi(-\lambda_L,x,L+N) \\ a_{2,L}\varphi(\mu_L,x,L+N) + b_{2,L}\varphi(-\mu_L,x,L+N) \end{pmatrix} \tag{A13}$$

where

$$\lambda_L^2/2m + (N+L)\omega + \sigma_1 = E = \mu_L^2/2m + (N+L)\omega + \sigma_2 \tag{A14}$$

with σ_1 and σ_2 given in (A4). $a_{i,L}$ and $b_{i,L}$ are again constants to be fixed later by boundary conditions.

Because of $\mathbb{H}\mathbb{U}\tilde{\Phi}_2 = \mathbb{U}\tilde{\mathbb{H}}\tilde{\Phi}_2 = E\mathbb{U}\Phi_2$ we see that $\Phi_2 = \mathbb{U}\tilde{\Phi}_2$ is an eigenfunction of $\mathbb{H}$ with eigenvalue E in interval I_2.

At $x = +\infty$ we choose the boundary condition that there is only one incoming wave with the target in state 1 and the laser and electron in state $\varphi(-k_{in},x,N)$ i.e.

$$w_{i,L} = \delta_{i,1}\delta_{L,0} \ . \tag{A15}$$

At $x = -\infty$ we choose the boundary condition that there is no incoming wave at all from the left i.e.

$$j_{i,L} = 0 \text{ for all } i \text{ and all } L. \tag{A16}$$

At $x = +R$ and $x = -R$ we require that the wavefunction Φ is continuous and has a continuous first derivate. This gives 4 equations for 2-component column functions or 8 equations for linear combinations of photon number states. These equations must be fulfilled for the coefficients of each photon state separately and therefore each photon state gives us 8 equations which connect the constants $a_{i,L}$, $b_{i,L}$, $t_{i,L}$, $r_{i,L}(i=1,2)$. Some of these equations are inhomogeneous.

Altogether, we get an infinite inhomogeneous system of linear equations to determine the free constants. In our model calculations we could only handle a finite number of equations with a finite number of unknowns and therefore we had to cut off the system of equations. Out of the coefficient matrix of the system of linear equations we have cut out the $(2l+1)8\times(2l+1)8$ matrix centered at the 8×8 block which comes from the L=0 terms in (A10, A12, A13) and the n=0 terms in (A7). Accordingly we have cut out of the inhomogenity vector a $(2l+1)8$ component piece centered at the block which comes from the n=0 terms in (A7). Then we have solved these $(2l+1)8$ coupled linear equations.

We have found rapid convergence of the results with increasing ℓ values as soon as $\ell>|\alpha p|$. For all calculations shown in figs. 2-9 it has been sufficient to choose a ℓ value between 5 and 10. The rapid convergence can be understood from the fact that Bessel functions decrease rapidly as soon as the absolute value of the order becomes larger than the absolute value of the argument. If we take a larger value of ℓ, then we include coefficients which contain Bessel functions of higher orders.

As a last step we calculate the quantities

$$R_L = \frac{h_L}{k_{in}}|r_{1,L}|^2 \quad , \quad Q_L = \frac{h_L}{k_{in}}|r_{2,L}|^2$$

$$T_L = \frac{h_L}{k_{in}}|t_{1,L}|^2 \quad , \quad S_L = \frac{h_L}{k_{in}}|t_{2,L}|^2$$

These are the quantities plotted in figs. 2-9 as function of the incoming electron energy E_i.

All these calculations can also be performed for the case that the target is in state 2 initially. The only change is to replace **K** of (A3) by

$$\mathbb{K} = \begin{pmatrix} k_o & 0 \\ 0 & h_o \end{pmatrix} = \begin{pmatrix} \sqrt{k_{in}^2 + 2mE_{12}} & 0 \\ 0 & k_{in} \end{pmatrix}$$

and to replace (A15) by

$$w_{i,L} = \delta_{i,2}\delta_{L,0}$$

ACKNOWLEDGEMENT

This work has been supported by NSF Grant CHE 76-15656 A02 and ONR Grant N000-14-77-C-0102.

We thank Prof. J. Hinze and the Zentrum für interdisziplinäre Forschung of the Universität Bielefeld for their hospitality while this work was being completed.

REFERENCES

1. G.Schulz, Rev. Mod. Phys. 45,378 (1973)
2. C.Jung and H.Krüger, Z. Physik A287,7 (1978)
3. M.Gavrila and M.Van der Wiel, Comments At. Mol. Phys. 8,1 (1978)
4. F.Low, Phys. Rev. 110,974 (1958)
5. L.Heller, Phys. Rev. 174,1580 (1968)
6. M.Mittleman, Phys. Rev. A21,79 (1980)
7. L.Rosenberg, Phys. Rev. A21,1939 (1980)
8. H.Krüger and C.Jung, Phys. Rev. A17,1706 (1978)
9. M.Mittleman, Phys. Rev. A20,1965 (1979)
10. A.Weingartshofer, E.Clarke, J.Holmes and C.Jung, Phys. Rev. A19,2371 (1979)
11. C.Jung, Phys. Rev. A21,408 (1980)
12. D.Andrick and L.Langhans, J. Phys. B11,2355 (1978)
13. L.Langhans, J. Phys. B11,2361 (1978)

INDEX